Communications in Computer and Information Science 2724

Rationale

The CCIS series is devoted to the publication of proceedings of computer science conferences. Its aim is to efficiently disseminate original research results in informatics in printed and electronic form. While the focus is on publication of peer-reviewed full papers presenting mature work, inclusion of reviewed short papers reporting on work in progress is welcome, too. Besides globally relevant meetings with internationally representative program committees guaranteeing a strict peer-reviewing and paper selection process, conferences run by societies or of high regional or national relevance are also considered for publication.

Topics

The topical scope of CCIS spans the entire spectrum of informatics ranging from foundational topics in the theory of computing to information and communications science and technology and a broad variety of interdisciplinary application fields.

Information for Volume Editors and Authors

Publication in CCIS is free of charge. No royalties are paid, however, we offer registered conference participants temporary free access to the online version of the conference proceedings on SpringerLink (http://link.springer.com) by means of an http referrer from the conference website and/or a number of complimentary printed copies, as specified in the official acceptance email of the event.

CCIS proceedings can be published in time for distribution at conferences or as post-proceedings, and delivered in the form of printed books and/or electronically as USBs and/or e-content licenses for accessing proceedings at SpringerLink. Furthermore, CCIS proceedings are included in the CCIS electronic book series hosted in the SpringerLink digital library at http://link.springer.com/bookseries/7899. Conferences publishing in CCIS are allowed to use our online conference service (Meteor) for managing the whole proceedings lifecycle (from submission and reviewing to preparing for publication) free of charge.

Publication process

The language of publication is exclusively English. Authors publishing in CCIS have to sign the Springer CCIS copyright transfer form, however, they are free to use their material published in CCIS for substantially changed, more elaborate subsequent publications elsewhere. For the preparation of the camera-ready papers/files, authors have to strictly adhere to the Springer CCIS Authors' Instructions and are strongly encouraged to use the CCIS LaTeX style files or templates.

Abstracting/Indexing

CCIS is abstracted/indexed in DBLP, Google Scholar, EI-Compendex, Mathematical Reviews, SCImago, Scopus. CCIS volumes are also submitted for the inclusion in ISI Proceedings.

How to start

To start the evaluation of your proposal for inclusion in the CCIS series, please send an e-mail to ccis@springer.com

Vaneet Aggarwal · David E. Bernal Neira ·
Denny Yu
Editors

Frontiers in Industrial Engineering

Purdue Quantum AI

Fifth Gavriel Salvendy International Symposium, PQAI 2025
West Lafayette, IN, USA, October 23–24, 2025
Proceedings

Springer

Editors
Vaneet Aggarwal
Purdue University
West Lafayette, IN, USA

David E. Bernal Neira
Purdue University
West Lafayette, IN, USA

Denny Yu
Purdue University
West Lafayette, IN, USA

ISSN 1865-0929 ISSN 1865-0937 (electronic)
Communications in Computer and Information Science
ISBN 978-981-95-7828-3 ISBN 978-981-95-7829-0 (eBook)
https://doi.org/10.1007/978-981-95-7829-0

This Springer imprint is published by the registered company Springer Nature Singapore Pte Ltd.
The registered company address is: 152 Beach Road, #21-01/04 Gateway East, Singapore 189721, Singapore

Preface

The Purdue Quantum Artificial Intelligence (PQAI) symposium was the 2025 iteration of the Gavriel Salvendy International Symposia on Frontiers in Industrial Engineering (GSIS-FIE) series. GSIS-FIE is a continuous, collaborative creation process with leaders in the field designed to provide dialogue and inspiration about the profession's evolution and impacts (https://engineering.purdue.edu/IE/Research/gsis). The 2025 symposium was held in person at Purdue University in West Lafayette, Indiana, from October 23 to 24, 2025 (https://engineering.purdue.edu/IE/PurdueQuantumAIWorkshop2025). It served as a forum to exchange contributions and visions for the rapidly evolving quantum artificial intelligence (AI) landscape with the world's leading minds at the intersection of **quantum computing, optimization, reinforcement learning, quantum machine learning, and algorithms**. An especially pioneering topic unique to this symposium was the dialogue on the intersection between quantum AI and human factors. It marks a significant gathering focused on the visionary theme of Quantum AI and inspiring discussions of challenges and opportunities at this rapidly evolving frontier.

The 2025 international program brought together scientists, engineers, and domain experts from academic and business organizations across the world. It attracted international participants across many disciplines to present and discuss current research activities and results. Contributions span the three core topics, including:

- Quantum Machine Learning and Models
- Quantum Optimization and Linear Systems
- Quantum Cognition and Human-Centered Computing

A total of 22 papers were submitted for peer review through OpenReview.net, an online platform that facilitates scientific peer review. Authors were given the option to anonymize their submissions, enabling either single- or double-blind review. Of these submissions, fourteen papers were selected for presentation at the event, four for poster presentations, and four were not accepted. A separate call was issued for late-breaking posters, resulting in thirteen submissions, of which twelve were accepted (not included in these proceedings). All submitted papers were evaluated by at least two reviewers, with each Technical Program Committee (TPC) member assigned a maximum of four papers. Late-breaking posters were reviewed by at least one Program Chair. Apart from the TPC members listed below, no additional external reviewers participated in the review process.

Upon acceptance, authors were given the option to submit a full paper, short paper, or extended abstract. We received 4 full papers, 7 short papers, and 2 extended abstracts. In addition, 7 researchers were invited to contribute invited papers, which are also included in these proceedings. In total, the proceedings contain 10 full and 8 short papers, and 2 extended abstracts. All full and short paper submissions were peer-reviewed by at least two TPC members, while invited papers were not subject to peer review. The views and content presented in the papers are solely the responsibility of the authors.

While past conferences on quantum AI reflect the exploratory nature of quantum computing and AI in the current state of quantum hardware and experimental platforms, the current proceedings highlight the cutting-edge advances of the field as well as the inspiring applications on humans and the challenges to realizing those impacts.

The core algorithmic advancements are showcased in the first two sections: Part I, Quantum Machine Learning and Models, explores frontiers in quantum neural networks, variational quantum algorithms (VQE), and quantum reinforcement learning (QRL), including new techniques for fine-tuning large language models. This is followed by Part II, Quantum Optimization and Linear Systems, which details new methods for solving complex problems like QAOA and QUBO, improvements to quantum linear systems algorithms (QLSAs), and also includes a comprehensive survey on building robust hybrid quantum-classical computing ecosystems. The promising human impacts of quantum AI and quantum computing are then showcased in Part III, Quantum Cognition and Human-Centered Computing. This unique section features topics centered on quantum theories advancing the modeling of human cognition (QToMNet), the development of Explainable Quantum Neural Networks (Q-XAI), and practical applications in domains like financial fraud detection. The volume concludes with Part IV, Backmatter, which provides technical and engineering insights, covering leveraging classical and quantum computing for process systems engineering applications and benchmarking hardware-accelerated Hamiltonian simulation using GPUs.

We extend our deepest gratitude to the invited guests, paper contributors, and symposium attendees for their inspiring discussions and meaningful engagement around the rapidly evolving frontier of quantum AI. We also express our sincere appreciation to the GSIS-FIE committee and the program committee for their dedicated support. Special thanks are due to Purdue Conferences (Erica D. Wilson and team) for their assistance in organizing the symposium and to the Purdue IE staff (including Abigail Rhee, Julia Sibley, and David Montgomery) for their support with website management. The valuable insights and suggestions provided by all contributors were instrumental in enhancing the clarity and overall consistency of this volume.

Finally, we thank our generous sponsors, Gavriel Salvendy, Purdue University Edwardson School of Industrial Engineering, and D-Wave Quantum Inc.

October 2025

Vaneet Aggarwal
David E. Bernal Neira
Denny Yu

Organization

Gavriel Salvendy International Symposia on Frontiers in Industrial Engineering Committee

Shimon Y. Nof (Chair)	Purdue University, USA
Vincent G. Duffy	Purdue University, USA
Yuehwern Yih	Purdue University, USA

Program Chairs

Vaneet Aggarwal	Purdue University, USA
David Bernal Neira	Purdue University, USA
Denny Yu	Purdue University, USA

Technical Program Committee

Fan Chen	Indiana University, USA
Sebastian Feld	Delft University of Technology, Netherlands
Rebekah Herrman	University of Tennessee Knoxville, USA
Mohsen Heidari	Indiana University, USA
Luigi Iapichino	Leibniz Supercomputing Centre, Germany
Pooyan Jamshidi	University of South Carolina, USA
Mohammad Ali Javidian	Appalachian State University, USA
Debanjan Konar	Samsung, India
Dongyang Li	Purdue University, USA
Dheeraj Peddireddy	Purdue University, USA
Shaswot Shresthamali	Kyushu University, Japan
Teague Tomesh	Infleqtion, USA
Takao Tomono	Keio University, Japan
Steffen Udluft	Siemens Technology, Germany
Xin Wang	Baidu, China
Junpeng Zhan	Alfred University, USA
Ruizhe Zhang	Purdue University, USA
Jianjun Zhao	Kyushu University, Japan
Shuchen Zhu	Duke University, USA
Yijie Zhu	Lancaster University, UK

Sponsoring Institutions

- Purdue University
- Edwardson School of Industrial Engineering, Purdue University
- D-Wave Quantum Inc.

Leveraging Classical and Quantum Computing for Process Systems Engineering Applications

Yirang Park and David E. Bernal Neira

Davidson School of Chemical Engineering, Purdue University, 480 Stadium Mall Drive, West Lafayette, IN, 47907 USA
dbernaln@purdue.edu

Abstract. Mixed-integer nonlinear programming (MINLP) is a modeling paradigm that combines discrete and continuous variables to model and solve a wide range of optimization problems. Its flexibility is especially useful for many real-world decision problems in engineering, operations, and finance, as these problems often involve discrete decisions and nonlinear system behaviors. Despite the ease of modeling, MINLP problems are challenging to solve as monolithic problems due to the combinatorial complexity of discrete variables and nonlinearities; however, they can be made manageable by adopting a decomposition strategy. Additionally, recent advances in computational hardware offer opportunities for addressing different parts of the problem more efficiently. Discrete subproblems can potentially benefit from quantum Ising solvers, while simulators and nonlinear solvers offer powerful tools for handling nonlinearities. To fully exploit these emerging computational capabilities, we propose an integrated approach that decomposes MINLP problems into discrete and continuous components and solves each subproblem using the most suitable computational method. In this work, two case studies are presented: an illustrative example involving the selection of an ionic liquid and its process design, and a more complex problem of drug substance manufacturing process optimization. The discrete subproblem in each case is formulated as an integer programming problem and solved using a commercial classical optimization solver, Gurobi. For comparative analysis, the same problem is reformulated as a quadratic unconstrained binary optimization and solved with simulated annealing, quantum annealing (QA), and entropy quantum computing (EQC). For the quantum methods, two different computing systems are used: D-Wave's specialized quantum annealer for QA, and Quantum Computing Incorporated (QCI)'s Dirac-1 quantum computer for EQC. The continuous subproblem is solved using Gurobi and a simulator-based optimization approach, respectively. In both examples, in terms of computational efficiency, Gurobi achieved the shortest runtime, whereas EQC took the longest, followed by QA and SA, in reaching feasible and optimal solutions. The heuristic methods demonstrated advantages in solution diversity compared to Gurobi's global search approach, identifying all or most of the feasible solutions in a single run and better capturing a broad solution space in a single run, while Gurobi provides global optimality guarantee and speed. This comparative analysis highlights the distinct strengths of each method and underscores the potential of this heterogeneous computing approach, which enables the use of different methods to address practical optimization problems.

Keywords: process design · optimization · heterogeneous computing · quantum computing · quantum annealing

Acknowledgments. This material is based upon work supported by the Center for Quantum Technologies under the Industry-University Cooperative Research Center Program at the US National Science Foundation under Grant No. 2224960.

Benchmarking Hamiltonian Simulation Using Graphical Processing Units

Anurag Ramesh[1], W. Michael Brown[2], Thomas Lubinski[3,4], and David E. Bernal Neira[1]

[1] Davidson School of Chemical Engineering, Purdue University, IN, USA
{rames102,dbernaln}@purdue.edu
[2] NVIDIA Corporation, CA, USA
michbrown@nvidia.com
[3] QED-C Technical Advisory Committee - Standards, VA, USA
tlubinski@quantumcircuits.com
[4] Quantum Circuits Inc., CT, USA

Abstract. Simulating quantum systems is a foundational application in quantum computing, particularly in fields such as computational chemistry. We present our use of a scalable framework, the Quantum Economic Development Consortium (QED-C) Application-Oriented Benchmark Suite (QED-C), to evaluate the performance of quantum algorithms across various hardware platforms. A key focus is leveraging NVIDIA CUDA-Q, a powerful GPU-accelerated platform for quantum-classical hybrid programming, to benchmark Hamiltonian simulation, Quantum Fourier Transform (QFT), and Phase Estimation (PE).

We simulate a range of physical systems within HamLib [3], including the transverse field Ising, Heisenberg, and Fermi-Hubbard models, as well as molecules such as H_2 using Suzuki-Trotter evolution. Simulations were executed on NVIDIA GPUs, including the A100, H100, GH200, and GB200 systems, at Purdue University [1] and Lawrence Berkeley National Laboratory (LBNL) [2], as well as in collaboration with NVIDIA. CUDA-Q's SpinOperator formalism enabled emulation of circuits for up to 38 qubits on the LBNL cluster, with performance up to $3\times$ faster than real quantum hardware. Strong scaling behavior is observed up to 32 GPUs, with execution times for some simulations reduced by more than 90%. For example, execution times for simulating a 33-qubit TFIM dropped from 19s (1 GPU) to 2s (32 GPUs).

Despite these gains, we observe classical HPC-like diminishing returns beyond 8 GPUs, due to inter-GPU communication bottlenecks. This impact is mitigated in the latest GB200 clusters that support extending the high-bandwidth NVLink GPU interconnect across multiple nodes. CUDA-Q proves especially effective for sampling-heavy workloads, offering near-linear scaling and improved parallel efficiency for PE and QFT. Our findings demonstrate that GPU-accelerated quantum Hamiltonian simulation with CUDA-Q provides a robust and high-throughput alternative to noisy intermediate-scale quantum (NISQ) devices, paving the way for future kernel-level optimizations and distributed quantum computing strategies.

Keywords. Hamiltonian simulation · CUDA-Q · GPU acceleration · Trotterization · multi-GPU scaling

References

1. McCartney, G., Hacker, T., Yang, B.: Empowering faculty: a campus cyberinfrastructure strategy for research communities. Educause Review (2014)
2. NERSC: Perlmutter (2022). https://www.nersc.gov/systems/perlmutter/
3. Sawaya, N., et al.: HamLib: a library of hamiltonians for benchmarking quantum algorithms and hardware. In: 2023 IEEE International Conference on Quantum Computing and Engineering (QCE), vol. 2, pp. 389–390. IEEE (2023)

Contents

Quantum Machine Learning and Models

Quantum Artificial Intelligence: Reinforcement Learning, Memory Models, and Self-programming Architectures

Samuel Yen-Chi Chen(✉)

Wells Fargo, New York, NY, USA
ycchen1989@ieee.org

Abstract. Quantum Artificial Intelligence (QAI) has emerged as a frontier discipline at the intersection of quantum computing and machine learning, with the potential to reshape the foundations of intelligent systems. This article provides a keynote perspective on several pioneering directions that collectively define a coherent research agenda for scalable and adaptive QAI. We begin with Quantum Reinforcement Learning (QRL), which introduced parameterized quantum circuits into sequential decision-making tasks and has since expanded into evolutionary, recurrent, and distributed variants. We then turn to Quantum Long Short-Term Memory (QLSTM) models, which extend quantum learning into temporal and sequential domains, enabling quantum-enhanced sequence modeling and federated learning applications. Building on these foundations, we present the concept of the Quantum Fast Weight Programmer (QFWP), a meta-learning framework that dynamically generates parameters and measurements for variational quantum circuits, paving the way for recursive and self-programming agents. Complementary to these advances, we highlight Differentiable Quantum Architecture Search (DiffQAS), which automates circuit design and contributes to the scalability of quantum neural networks. We conclude with a vision of QAI as a transformative force for the next technological epoch, outlining open challenges such as barren plateau mitigation, hybrid quantum-classical integration, and the development of distributed quantum architectures. Together, these contributions establish a foundation for QAI not only as a research field but as a paradigm for intelligent, self-improving quantum systems.

Keywords: Quantum Machine Learning · Quantum Neural Network · Quantum Artificial Intelligence

1 Introduction

Quantum computing (QC) promises substantial advantages over classical computation in solving problems that are believed to be intractable on conventional machines [54]. These potential breakthroughs range from quantum simulation of

V. Aggarwal et al. (Eds.): PQAI 2025, CCIS 2724, pp. 3–21, 2026.
https://doi.org/10.1007/978-981-95-7829-0_1

physical systems to optimization and cryptography. In parallel, artificial intelligence (AI) has undergone a revolution, producing systems capable of superhuman performance in strategic games such as Go [63] and state-of-the-art natural language understanding and generation [68]. A natural question thus arises: *can the capabilities of quantum computing and artificial intelligence be combined to create a new paradigm of intelligent systems?* This question motivates the development of quantum machine learning (QML) algorithms and, more broadly, the emerging discipline of quantum artificial intelligence (QAI).

Early progress in QML has largely centered on variational quantum circuits (VQCs) and quantum neural networks (QNNs), which provide a flexible framework for hybrid quantum–classical learning on noisy intermediate-scale quantum (NISQ) devices. Building upon this foundation, our research has advanced QAI along several pioneering directions. These include **quantum reinforcement learning (QRL)**, which integrates quantum circuits into sequential decision-making [16,19,28]; **quantum sequential models**, such as the quantum long short-term memory (QLSTM) architecture for time-series data [30]; and **quantum architecture search (QAS)**, which explores automated design of variational quantum circuits [22]. More recently, we have introduced the concepts of the **quantum fast weight programmer (QFWP)** and **learnable measurements**, which extend programmability from circuit parameters to measurement operators [17,25,26], opening the path toward self-programming quantum agents.

This article provides an overview of these contributions. We begin with a brief background on the foundations of QAI, followed by a detailed discussion of our pioneering contributions. We then turn to distributed and federated quantum learning. Finally, we conclude with a perspective on the key challenges and promising research directions that may define the next decade of QAI.

2 Background and Foundations of Quantum AI

2.1 Variational Quantum Circuits as the Foundation

Variational quantum circuits (VQCs), also known as parameterized quantum circuits (PQCs) have become the cornerstone of quantum machine learning on near-term devices. By parameterizing unitary transformations with tunable angles, VQCs provide a flexible family of models that can be trained using classical optimization, including gradient-based and gradient-free methods, while leveraging quantum superposition and entanglement. Compared to traditional, non-variational quantum algorithms, which often require large-scale fault-tolerant hardware, VQCs offer a pragmatic route to deploy quantum learning methods on noisy intermediate-scale quantum (NISQ) processors [1]. This hybrid quantum–classical paradigm forms the basis of many QAI architectures discussed in this paper, ranging from quantum reinforcement learning to memory-based recurrent models. A typical quantum neural network (QNN), illustrated in Fig. 1, can be understood as three stages: (i) a data encoding circuit that maps classical inputs to quantum states, (ii) a variational circuit with trainable parameters, and (iii) a measurement stage where observables define how quantum information is projected into classical outputs. This three-part structure–encoding, parametrization, and measurement–forms the foundation of most variational quantum learning models including reinforcement learning, memory models, and self-programming architectures.

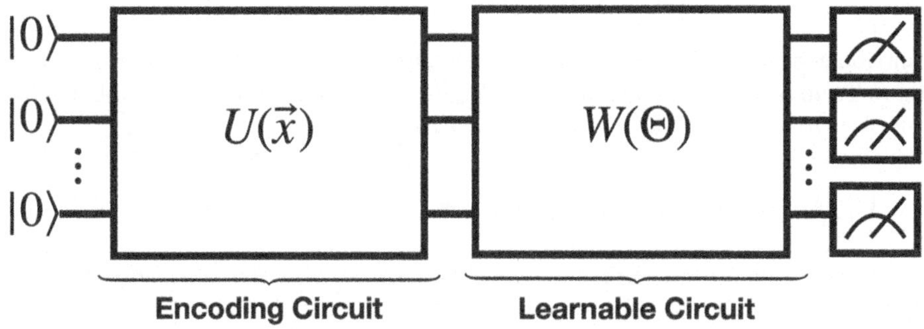

Fig. 1. Schematic of a quantum neural network (QNN). Given a feature vector $\boldsymbol{x}$, the encoding circuit prepares $U(\boldsymbol{x})|0\rangle^{\otimes n}$. The variational circuit applies M layers of unitary blocks, $W(\Theta) = \prod_{j=M}^{1} W_j(\boldsymbol{\theta}_j)$, producing $|\Psi\rangle = W(\Theta)U(\boldsymbol{x})|0\rangle^{\otimes n}$. Measurements of observables $\hat{B}_k$ at k-th qubit yield expectation values $\langle \hat{B}_k \rangle = \langle \Psi | \hat{B}_k | \Psi \rangle$, defining the quantum function $\boldsymbol{f}(\boldsymbol{x}; \Theta) = (\langle \hat{B}_1 \rangle, \ldots, \langle \hat{B}_K \rangle)$.

2.2 Classical Paradigms Informing Quantum AI

Several milestones in classical AI provide direct inspiration for their quantum counterparts. Reinforcement learning [67] has established itself as a framework for sequential decision making, where agents optimize long-term rewards in uncertain environments. Memory models such as recurrent neural networks and long short-term memory (LSTM) networks [36] introduced the ability to capture temporal dependencies across multiple scales, which remains essential for natural language and time-series analysis. Automated machine learning and neural architecture search [33,50] demonstrated that learning systems can optimize not only their parameters but also their structural design. Finally, federated learning [42] emerged as a paradigm for privacy-preserving and distributed training, reflecting the growing importance of scalability and collaboration in AI. These paradigms collectively set the stage for exploring how their quantum analogues can redefine intelligence in the quantum era.

2.3 The Interplay

The intersection of quantum computing and these core AI paradigms motivates the development of Quantum Artificial Intelligence. By combining variational quantum circuits with reinforcement learning, sequence modeling, architecture search, and federated learning frameworks, QAI extends the reach of machine intelligence into the quantum domain. Unlike classical AI, QAI must address challenges unique to quantum systems, including barren plateaus, decoherence, and limited qubit connectivity. At the same time, it offers unprecedented opportunities to harness quantum parallelism, non-classical correlations, and novel forms of programmability. These dual forces–constraints and opportunities–define the trajectory of QAI. In the following section, we outline our pioneering contributions along five axes: Quantum Reinforcement Learning, Quantum Long Short-Term Memory, Quantum Fast Weight Programming, Differentiable Quantum Architecture Search, and Learnable Measurements.

3 Pioneering Contributions

Quantum Artificial Intelligence has evolved rapidly over the last five years, but several core paradigms can be traced to our early contributions. In this section, I outline five pillars of this research program–Quantum Reinforcement Learning (QRL), Quantum Long Short-Term Memory (QLSTM), Quantum Fast Weight Programmer (QFWP), Differentiable Quantum Architecture Search (DiffQAS) and Learnable Quantum Measurements. Each represents not only a technical milestone but also a conceptual leap toward scalable and adaptive quantum learning systems.

3.1 Quantum Reinforcement Learning (QRL)

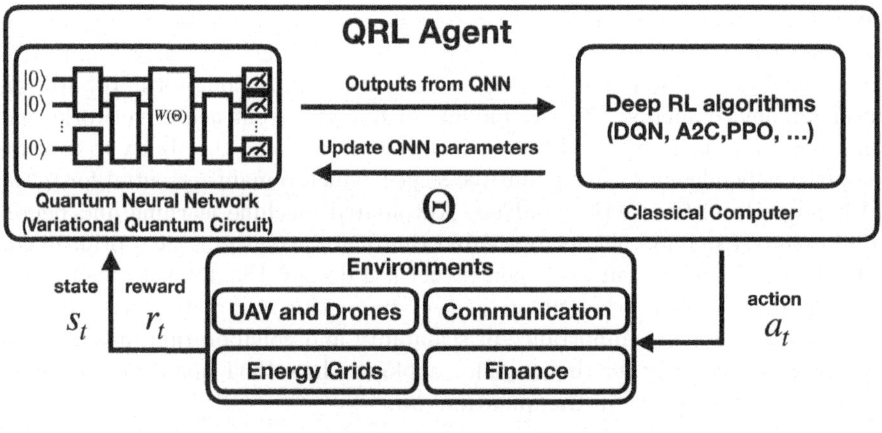

Fig. 2. Quantum reinforcement learning.

Reinforcement learning (RL) offers a natural testbed for exploring the capabilities of variational quantum circuits, as agents must resolve the trade-off between exploration and exploitation while operating in environments that demand long-horizon decision making. As depicted in Fig. 2, extending RL into the quantum regime gives rise to the notion of a *hybrid quantum-classical agent*, where quantum subroutines are embedded within the learning loop. In this setting, classical and quantum components interact synergistically, enabling potential quantum advantages in policy representation, value estimation, and optimization across a range of application domains. Our 2020 work on Variational Quantum Reinforcement Learning [28] was the first demonstration that parameterized quantum circuits could be integrated into deep reinforcement learning pipelines such as the deep Q-learning. This study established a foundation for QRL as a distinct research direction, showing how quantum circuits can approximate value functions and policy networks.

Subsequent developments expanded this framework. In [20], we introduced evolutionary optimization strategies for training variational quantum agents, overcoming local minima and issues that gradient-based methods struggling to learn, by leveraging population-based QNN parameters optimization. Later, in [12,15], we extended QRL to recurrent architectures, allowing agents to capture temporal dependencies beyond single-step policies. Recurrent QRL provides the ability to operate effectively in partially observable environments, where decision-making must rely on information aggregated across multiple past time steps rather than the current state alone. To further exploit parallel computing resources such as multi-core CPUs and quantum processing units (QPUs), we introduced asynchronous and distributed training frameworks for QRL [10,11]. In this setting, multiple replicas of the agent interact independently with their own instances of the environment, while periodically sharing gradient information with a centralized server. The asynchronous design improves overall efficiency, as individual models are not forced to wait for others to complete their episodes. In addition, incorporating prioritized experience replay across quantum agents enhances training stability, paving the way toward scalable multi-agent and distributed quantum learning systems.

These advances collectively position QRL as a cornerstone of Quantum AI: an experimental testbed for algorithmic innovation, a model of hybrid quantum-classical cooperation, and a potential route to practical applications ranging from finance [51] to space systems [38].

3.2 Quantum Long Short-Term Memory (QLSTM)

Sequential data processing is fundamental to natural language understanding, time-series forecasting, and signal analysis. Classical long short-term memory (LSTM) networks [36] have played a central role in these domains, yet their quantum counterparts remained largely unexplored until our introduction of the Quantum Long Short-Term Memory (QLSTM) model [30]. The QLSTM incorporates quantum-enhanced gating mechanisms, demonstrating that variational quantum circuits can effectively capture temporal dependencies across multiple time steps. As illustrated in Fig. 3, the conventional LSTM architecture is augmented by replacing classical deep neural network components with variational

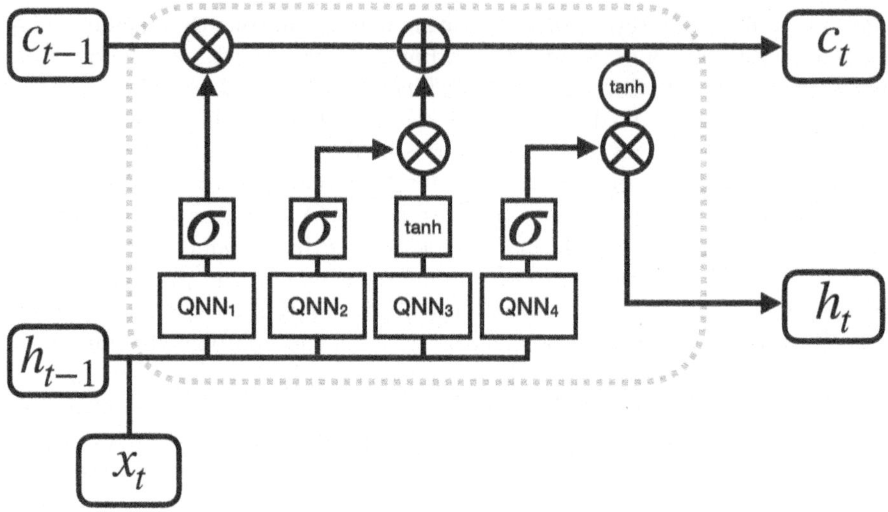

Fig. 3. Quantum long short-term memory.

quantum circuits (VQCs) or quantum neural networks (QNNs). The QLSTM can be formally described as:

$$f_t = \sigma\left(\mathrm{QNN}_1(v_t)\right) \tag{1a}$$

$$i_t = \sigma\left(\mathrm{QNN}_2(v_t)\right) \tag{1b}$$

$$\tilde{C}_t = \tanh\left(\mathrm{QNN}_3(v_t)\right) \tag{1c}$$

$$c_t = f_t * c_{t-1} + i_t * \tilde{C}_t \tag{1d}$$

$$o_t = \sigma\left(\mathrm{QNN}_4(v_t)\right) \tag{1e}$$

$$h_t = o_t * \tanh\left(c_t\right) \tag{1f}$$

where $v_t = [h_{t-1} x_t]$ represents the concatenation of input x_t at time-step t and the hidden state h_{t-1} from the previous time-step $t-1$. Previous work has demonstrated that QLSTM can achieve superior performance compared to classical LSTM models on a variety of time-series prediction tasks, particularly in settings where both models are constrained to a comparable number of trainable parameters [30].

Building on this foundation, we proposed distributed and federated extensions of QLSTM [4,9,37,47]. These advanced designs enable cross-node collaboration without compromising data privacy, and allow QLSTM models to process higher-dimensional inputs by partitioning data across multiple quantum chips. Collectively, these works represent the first systematic exploration of how sequential quantum models can scale in decentralized and heterogeneous environments. Applications of QLSTM span diverse domains where capturing subtle temporal correlations is essential, including gravitational wave signal detection [47], urban communication forecasting [6], and financial time-series modeling [7].

Beyond forecasting tasks, QLSTM has also shown promise in natural language processing [32,41,65] and in reinforcement learning under partial observability [12]. Taken together, these studies establish QLSTM as the second major pillar of Quantum AI–introducing memory and temporal reasoning capabilities into quantum learning systems.

3.3 Quantum Fast Weight Programmer (QFWP)

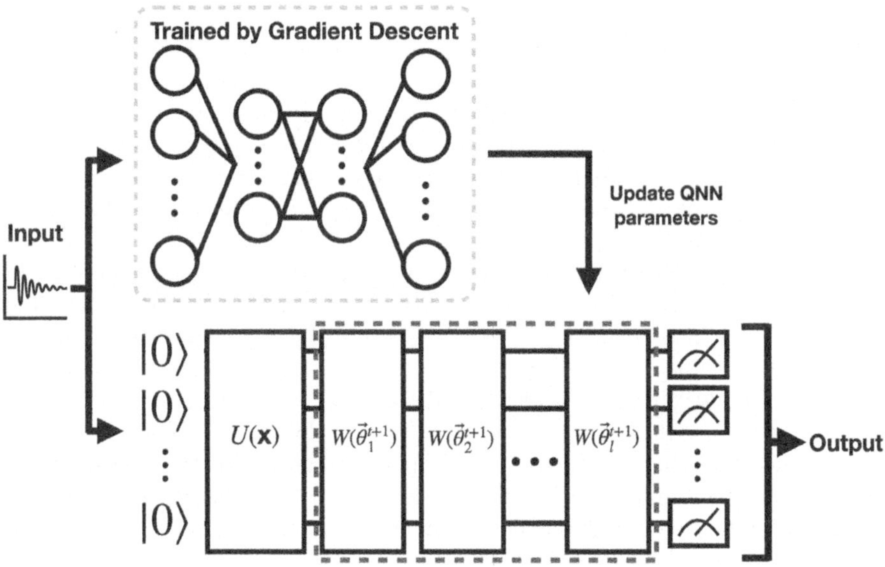

Fig. 4. Quantum fast weight programmer.

A central challenge in variational quantum learning lies in parameter adaptation. Conventional training requires gradual updates over many iterations, which limits responsiveness when environments or tasks change. To address this, we introduced the Quantum Fast Weight Programmer (QFWP) [17]. The QFWP is a meta-learning framework in which a secondary neural network acts as a programmer, generating variational parameters or the updates for these parameters for a quantum circuit in a single forward pass, as illustrated in Fig. 4. In QFWP, we adopt a hybrid quantum–classical architecture that integrates the complementary strengths of both domains. Classical neural networks serve as the *slow* networks, responsible for generating update signals that modulate the parameters of the *fast* networks, implemented as variational quantum circuits (VQCs). Given an input vector $\boldsymbol{x}$, the classical network produces outputs that adjust—rather than completely overwrite—the parameters of the quantum circuit. This design enables the QFWP to provide rapid, input-dependent adaptation of quantum models while preserving stability across training. At time step $t+1$, the parameters of the variational quantum circuit (VQC) are updated according to

$$\theta_i^{t+1} = f(\theta_i^t, \Delta_i),$$

where f denotes a combination rule that integrates the previous parameter value θ_i^t with the update signal Δ_i produced by the classical network. In the original formulation of QFWP [17], an *additive* rule was adopted,

$$\theta_i^{t+1} = \theta_i^t + \Delta_i,$$

so that information from prior time steps is preserved within the circuit parameters and can influence the behavior of the VQC when processing a new input $\boldsymbol{x}$. Numerical simulations on reinforcement learning and sequence modeling tasks [17] demonstrated that circuits equipped with QFWP adapt more flexibly and efficiently than conventional variational models. It was further observed that, even without recurrent connections, QFWP can attain performance comparable to that of QLSTM.

Think one-step further, the classical NN used to dynamically modify the QNN can be further enhanced, generated or compressed by another QNN. The Quantum-Train framework [49], in which a n-qubit QNN is used to generate the parameters of classical deep NN with up to 2^n trainable parameters, can be combined with the QFWP framework. The idea is that, another standalone QNN is trained to generate the parameters of the classical"slow programmer" [48]. With this enhancement, most of the trainable parameters are now in the quantum regime. Numerical simulation demonstrated that such enhancement can provide further model compression effects (smaller number of trainable parameters) without sacrificing models' performance.

More recently, we have begun to explore extensions of this idea beyond parameter generation. In particular, the QFWP can also serve as a mechanism for programming measurement operators, making the readout stage adaptive rather than static (see Sect. 3.5). Looking further ahead, one can envision recursive or self-referential architectures where programmers themselves evolve based on not only input signals but also prior outputs from the model itself. While these directions remain open, they highlight the potential of QFWP as a foundation for self-programming quantum systems.

3.4 Differentiable Quantum Architecture Search (DiffQAS)

Designing effective quantum circuits to perform designated tasks is notoriously difficult, as ansatz choices strongly influence performance. To address this, we pioneered Quantum Architecture Search (QAS) via reinforcement learning [39,70] to synthesize assigned quantum states. Beyond synthesizing quantum states, we generalize such methods into designing circuits for QML tasks [31]. In addition to classical RL, we can further generalize this idea into *quantum* RL-based QAS, as demonstrated in the work [13]. Notably, this QRL-based QAS framework can synthesis designated quantum states with small QNN architecture. While this is a simple proof-of-concept experiment, we see as a step towards the "self-improving" Quantum AI. Concurrently, we developed population-based optimization techniques to discover high-performing quantum circuit architectures, guided by information-theoretic criteria such as effective dimension [18]. These evolutionary strategies are naturally amenable to parallelization, making them well-suited for deployment on large-scale high-performance computing (HPC) infrastructures and, in the future, across multiple quantum processing units (QPUs). While RL and evolutionary optimization have demonstrated notable success in designing quantum circuit architectures for targeted tasks, several challenges remain. On the one hand, RL-based frameworks provide theoretical generality and policy-driven exploration, but in practice they often encounter difficulties such as high sample complexity, sensitivity to optimizer settings, and the inherent exploration–exploitation dilemma. On the other hand, population-based methods such as evolutionary optimization excel in their ease of parallelization, making them particularly well-suited for deployment on HPC clusters, where large batches of candidate architectures can be evaluated simultaneously. Nevertheless, both RL-based and evolutionary approaches are constrained by their reliance on discrete architecture search spaces, which rapidly become intractable due to the combinatorial growth of candidate configurations as the search space expands. Building on our early work and preliminary studies from

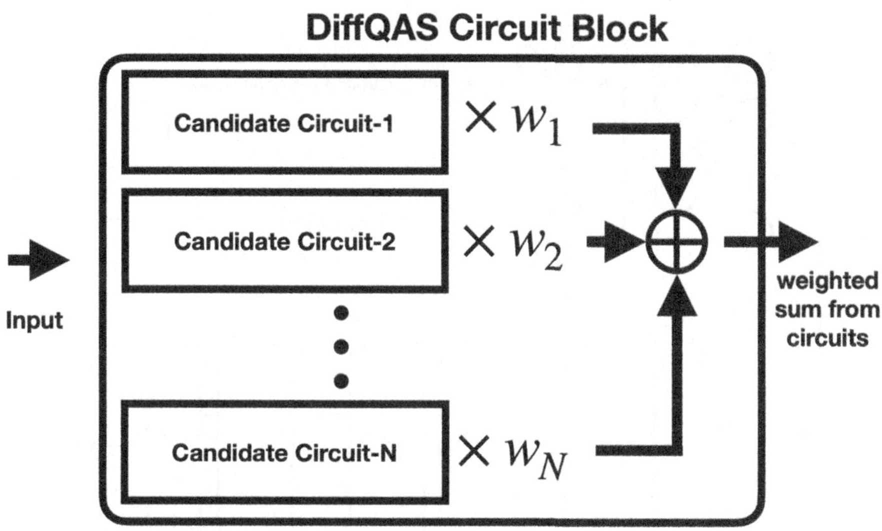

Fig. 5. Differentiable Quantum Architecture Search (DiffQAS). Within the DiffQAS framework, each candidate circuit $\mathcal{C}_j$ is associated with a trainable weight w_j. The overall output is expressed as an ensemble function, $f_{\mathcal{C}} = \sum_{j=1}^{N} w_j f_{\mathcal{C}_j}$, where $f_{\mathcal{C}_j}$ denotes the output of the j-th candidate circuit. Gradient-based optimization is then employed to update both the architecture weights $\{w_j\}$ and the conventional variational parameters of the quantum neural network in a unified manner.

other groups [66,71], we introduced Differentiable Quantum Architecture Search (DiffQAS) and applied it to a range of QML tasks, including distributed reinforcement learning [14], time-series modeling [24], and quantum-enhanced neural network parameter generation [23]. As illustrated in Fig. 5, DiffQAS relaxes the discrete circuit architecture space into a continuous domain by assigning trainable weights or probability distributions over candidate substructures. This relaxation enables joint, gradient-based optimization of both the architecture and its variational parameters within a unified, end-to-end differentiable framework. Numerical simulation results suggest that gradient-based exploration of circuit topologies can automatically find high-performance quantum circuit architectures for the designated tasks without substantial manual trial-and-error.

DiffQAS brings the automation revolution of AutoML into quantum computing: rather than relying on expert-designed ansätze, circuits can now be discovered, optimized, and deployed automatically. Combined with QRL and QLSTM, DiffQAS contributes to a coherent pipeline for self-designing quantum agents—a paradigm shift in how we build scalable quantum AI systems.

3.5 Learnable Quantum Measurements

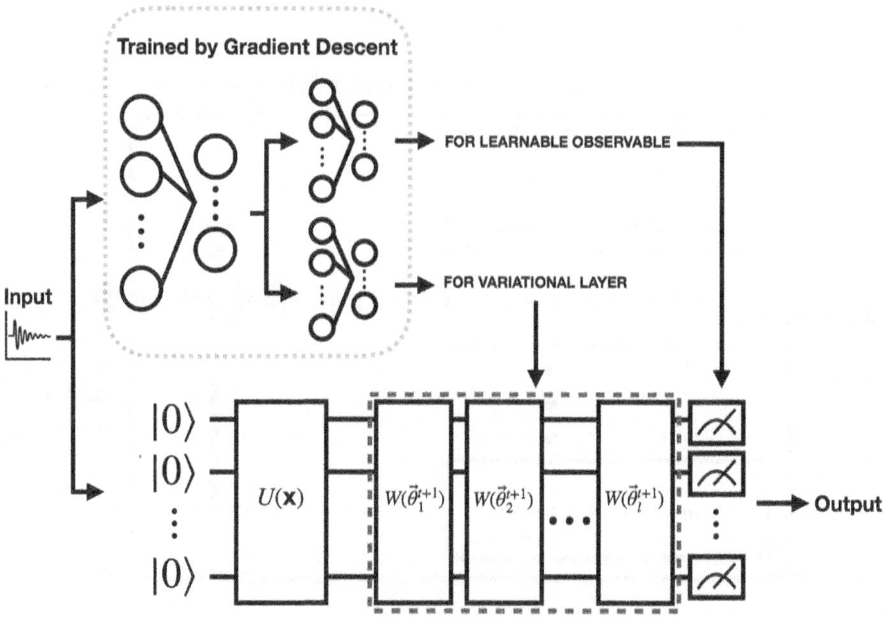

Fig. 6. Learning to program quantum measurements.

An additional dimension of programmability in Quantum AI lies in the measurement process itself. In conventional settings, measurements are typically specified as fixed observables chosen a priori, such as Pauli operators or simple projective bases. While effective for controlled experiments, this rigidity limits the adaptability of quantum models to more complex learning tasks. For instance, when Pauli observables are employed, their spectral range is restricted to $\lambda = \pm 1$. Consequently, the output of a variational quantum circuit (VQC) is confined to

$$\langle \psi | H | \psi \rangle \in [-1, 1],$$

independent of the circuit parameterization. By the Rayleigh quotient, we know that for any normalized state $|\psi\rangle$,

$$\lambda_{\min} \leq \langle \psi | H | \psi \rangle \leq \lambda_{\max}.$$

This observation suggests that expanding the spectral range of the measurement operator can enhance the expressivity of VQCs–an important property for a variety of quantum machine learning tasks.

We proposed the concept of learnable measurements, where measurement operators are optimized as part of the training loop. This can be realized through two complementary approaches. The first leverages differentiable optimization, directly updating the measurement operators using gradient-based methods, akin to DiffQAS but applied at the readout stage. The core idea is to parameterize the Hermitian matrix itself, and treat these "quantum measurement" parameters and the ordinary VQC rotation parameters in the same footing. Such direct optimization, although effective and simple, may not be the most capable approach. The measurement is fixed after the training and will not be adjusted according to the new input data, just like the ordinary feed-forward neural network. To resolve this rigidity, a second approach employs the Quantum Fast Weight Programmer (QFWP), illustrated in Fig. 6, to dynamically generate measurement operators conditioned on data, effectively allowing circuits to "program their own measurements" on the fly.

By integrating learnable measurements into QAI models, we elevate the measurement stage from a passive readout to an active, trainable component. This results in end-to-end differentiable pipelines where state preparation, evolution, and measurement are all subject to learning and adaptation. Our recent studies [25,26,43,44] highlight how this paradigm leads to improved performance across classification and reinforcement learning tasks.

4 Distributed and Federated Quantum Learning

As quantum hardware progresses, scalability will depend not only on single-device advances but also on distributed and federated paradigms that coordinate multiple quantum and classical resources. Much as classical machine learning has shifted toward federated and privacy-preserving settings, quantum learning must also evolve to operate across heterogeneous nodes connected through quantum and classical channels.

Our initial contribution in this direction was the proposal of Federated Quantum Machine Learning (FedQML) [29], which introduced the concept of collaboratively training quantum models without directly exchanging raw data (illustrated in Fig. 7). This framework extends the principles of classical federated

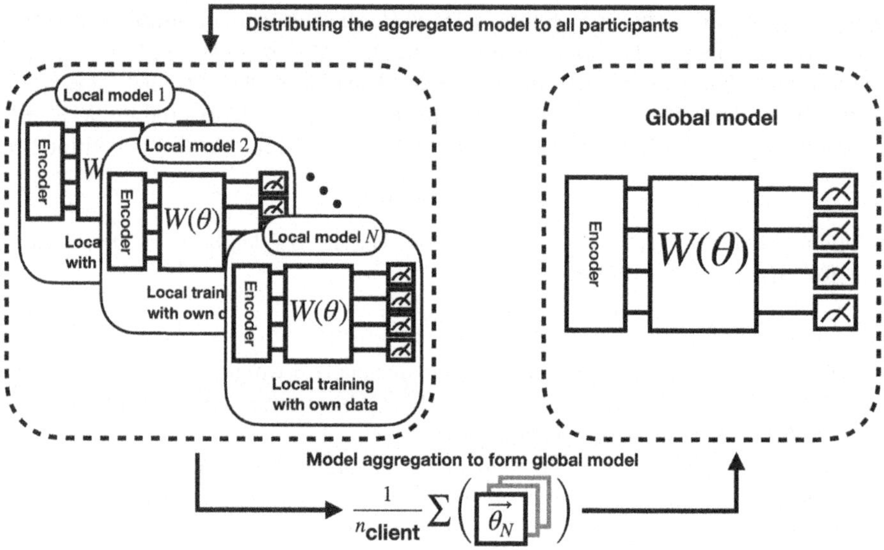

Fig. 7. Federated Quantum Machine Learning (FedQML).

learning to quantum neural networks, addressing issues of data privacy and locality while enabling joint optimization across distributed datasets.

This foundation has since expanded in several directions. In [3], we formalized the theoretical underpinnings of federated learning across both classical and quantum networks, while [62] and [69] incorporated differential privacy and teacher ensemble aggregation into federated QML pipelines. Extensions to sequence modeling, including Federated QLSTM [4,37,47], further demonstrated the feasibility of privacy-preserving training for temporal tasks such as gravitational wave detection and human activity recognition.

Beyond federated approaches, we also explored a range of distributed training mechanisms. These include parameter sharing via Quantum-Train architectures [8,46], distributed replay buffers for QRL [11], the use of multiple quantum chips to process higher-dimensional inputs [9], and asynchronous coordination and training of recurrent quantum agents [15]. Collectively, these methods demonstrate how distributed computation can accelerate training, enhance robustness, and enable scalability on modular quantum hardware.

5 Challenges and Open Problems

While variational quantum algorithms have demonstrated remarkable progress across a wide range of application scenarios in recent years, several fundamental challenges remain open.

5.1 Hardware Limitations

One major bottleneck is the gap between algorithmic development and the capabilities of current quantum hardware. Coherence times still constrain the feasible depth of quantum circuits, making it difficult to implement complex models in practice. Another limitation stems from device topology: restricted qubit connectivity complicates the design of QML circuits. Striking the right balance between entanglement (i.e., two-qubit operations) and robustness under hardware-level transpilation remains an active area of study. Furthermore, the limited number of available qubits poses a barrier for large-scale AI applications, where input dimensions can easily reach thousands or tens of thousands.

5.2 Hybrid Encoding and Model Partitioning

To address the dimensionality bottleneck, hybrid approaches have been explored in which classical models, such as deep neural networks [29,52,61] or tensor networks [20,21,60], compress the input into a latent representation suitable for quantum processing. However, such designs raise an important question: *does the classical component dominate the workload?* Understanding and quantifying the division of labor between classical and quantum modules remains a pressing research problem. An alternative strategy is to distribute the workload across multiple quantum chips, with each chip processing only a subset of the input features [9,56]. Yet, this multi-chip paradigm introduces new challenges, including the efficient creation of entanglement across devices and the aggregation of results without significant performance degradation.

5.3 Trainability and Barren Plateaus

Another critical obstacle is the vanishing gradient phenomenon, widely known as the *barren plateau* [40,53], which hampers the training of large-scale QNNs. Recent mitigation strategies fall into three main categories: (i) specialized parameter initialization methods [35,57,58], (ii) architectural designs and local cost functions [2,59], and (iii) advanced training paradigms such as layer-wise learning [64]. Despite progress, scaling QNN training to deep, high-dimensional circuits remains a formidable challenge.

5.4 Systems Integration and Data Interfaces

From a systems perspective, significant advancements are also required in hardware topology and control to enable the faithful execution of highly entangled QML circuits. Equally important are efficient interfaces between classical HPC systems and quantum processors, since most QAI applications originate from classical data sources. One outstanding challenge is how to efficiently load and encode large-scale classical data into quantum states, and subsequently extract useful information from the quantum device once computation has been carried out.

Together, these challenges highlight the need for co-design across algorithms, architectures, and hardware systems to unlock the full potential of quantum artificial intelligence.

6 Future Directions and Vision

Looking ahead, we believe that Quantum Artificial Intelligence (QAI) is still in its early stage, with enormous potential for both theoretical breakthroughs and real-world applications. Building upon the foundations of reinforcement learning, memory models, and self-programming architectures, we outline several promising future directions.

6.1 Scalable Quantum AI Architectures

One of the most pressing goals is the development of scalable QAI architectures capable of handling real-world AI tasks. Advances such as Differentiable Quantum Architecture Search (DiffQAS), Quantum Fast Weight Programmers (QFWP), and learnable measurements provide a pathway toward automated, self-programming quantum models. In the long term, multi-chip and distributed quantum neural networks may support higher-dimensional data, extending beyond the limitations of single devices.

6.2 Quantum Reinforcement Learning and Agents

Quantum Reinforcement Learning (QRL) has the potential to become a quantum-native paradigm for intelligence. Future research may focus on multi-agent quantum systems, recurrent quantum models with enhanced temporal or sequential reasoning, and recursive self-improving architectures. These developments open the possibility of designing autonomous QAI agents that adapt, learn, and even reprogram themselves.

6.3 Federated and Distributed Quantum Learning

Federated Quantum Machine Learning (FedQML) and distributed frameworks illustrate how quantum learning can be extended across networks of quantum devices. Key directions include privacy-preserving training with differential privacy, coordination across heterogeneous quantum nodes, and integration with large-scale high-performance computing infrastructures. Such distributed paradigms lay the foundation for a future "Quantum Internet of Intelligence."

6.4 Integration with Classical AI

Rather than positioning quantum and classical AI as competing paradigms, we foresee a deeper integration. Classical models may assist in feature compression, input encoding or quantum circuit adjustments, while quantum models contribute enhanced expressivity, optimization capabilities, or policy learning. A critical open question is how to rigorously quantify the contribution of quantum components within hybrid architectures.

6.5 Applications with Societal Impact

QAI is expected to play an increasingly important role in domains where data complexity and decision-making requirements are beyond classical reach. Potential applications include financial risk modeling, energy grid control and optimization, dynamic spectrum allocation in 6G communication networks, scientific discovery in physics [27], chemistry [34] and climate science [45], and biomedical signal analysis such as EEG [5] and fMRI [55]. These areas exemplify how QAI can impact critical infrastructures and society at large.

6.6 Long-Term Vision

Ultimately, we envision QAI converging toward recursive, self-programming agents that possess the ability to improve their own architectures and learning strategies. Coupled with distributed quantum networks, such systems could evolve into a large-scale ecosystem of autonomous quantum learners. This vision highlights a paradigm shift from QAI as a tool to QAI as a generative framework for understanding and creating intelligence itself.

In summary, the future of QAI depends on co-design across algorithms, architectures, and hardware, as well as strong integration with classical systems. By pursuing these directions, we may not only overcome the current barriers but also unlock new forms of machine intelligence rooted in the principles of quantum mechanics.

7 Conclusion

In this article, we present several recent advancements in quantum AI, especially the quantum reinforcement learning for sequential decision making tasks, quantum-enhanced memory models such as QLSTM and QFWP for time-series or sequential learning and differentiable quantum architecture search for automatically finding the high-performance circuit for designated tasks. Yet several challenges including the capability to handle wide and deep circuits, the vanishing gradient problem and the topology-aware design of QML circuits. We envision that the integration of concepts such as meta-learning or learning to learn, recursive learning, continual learning and external neural programmer architecture may help us gradually solving these challenges and pave a way toward large-scale quantum intelligence.

Disclosure of Interests. The views expressed in this article are those of the authors and do not represent the views of Wells Fargo. This article is for informational purposes only. Nothing contained in this article should be construed as investment advice. Wells Fargo makes no express or implied warranties and expressly disclaims all legal, tax, and accounting implications related to this article.

References

1. Bharti, K., et al.: Noisy intermediate-scale quantum algorithms. Rev. Mod. Phys. **94**(1), 015004 (2022)
2. Cerezo, M., Sone, A., Volkoff, T., Cincio, L., Coles, P.J.: Cost function dependent barren plateaus in shallow parametrized quantum circuits. Nat. Commun. **12**(1), 1791 (2021)
3. Chehimi, M., Chen, S.Y.C., Saad, W., Towsley, D., Debbah, M.: Foundations of quantum federated learning over classical and quantum networks. IEEE Netw. **38**(1), 124–130 (2023)
4. Chehimi, M., Chen, S.Y.C., Saad, W., Yoo, S.: Federated quantum long short-term memory (fedqlstm). Quant. Mach. Intell. **6**(2), 43 (2024)
5. Chen, C.S., Chen, S.Y.C., Tsai, A.H.W., Wei, C.S.: Qeegnet: quantum machine learning for enhanced electroencephalography encoding. In: 2024 IEEE Workshop on Signal Processing Systems (SiPS), pp. 153–158. IEEE (2024)
6. Chen, C.S., Chen, S.Y.C., Tsai, Y.C.: Benchmarking quantum and classical sequential models for urban telecommunication forecasting. arXiv preprint arXiv:2508.04488 (2025)
7. Chen, J.H., Huang, Y.C., Tsai, Y.C., Chen, S.Y.C.: Quantum-enhanced forecasting for deep reinforcement learning in algorithmic trading. arXiv preprint arXiv:2509.09176 (2025)
8. Chen, K.C., Chen, S.Y.C., Liu, C.Y., Leung, K.K.: Quantum-train-based distributed multi-agent reinforcement learning. In: 2025 IEEE Symposium for Multidisciplinary Computational Intelligence Incubators (MCII Companion), pp. 1–5. IEEE (2025)
9. Chen, K.C., Chen, S.Y.C., Liu, C.Y., Leung, K.K.: Toward large-scale distributed quantum long short-term memory with modular quantum computers. In: 2025 International Wireless Communications and Mobile Computing (IWCMC), pp. 337–342. IEEE (2025)
10. Chen, S.Y.C.: Asynchronous training of quantum reinforcement learning. Procedia Comput. Sci. **222**, 321–330 (2023)
11. Chen, S.Y.C.: Quantum deep q-learning with distributed prioritized experience replay. In: 2023 IEEE International Conference on Quantum Computing and Engineering (QCE), vol. 2, pp. 31–35. IEEE (2023)
12. Chen, S.Y.C.: Quantum deep recurrent reinforcement learning. In: ICASSP 2023-2023 IEEE International Conference on Acoustics, Speech and Signal Processing (ICASSP), pp. 1–5. IEEE (2023)
13. Chen, S.Y.C.: Quantum reinforcement learning for quantum architecture search. In: Proceedings of the 2023 International Workshop on Quantum Classical Cooperative, pp. 17–20 (2023)
14. Chen, S.Y.C.: Differentiable quantum architecture search in asynchronous quantum reinforcement learning. In: 2024 IEEE International Conference on Quantum Computing and Engineering (QCE), vol. 1, pp. 1516–1524. IEEE (2024)
15. Chen, S.Y.C.: Efficient quantum recurrent reinforcement learning via quantum reservoir computing. In: ICASSP 2024-2024 IEEE International Conference on Acoustics, Speech and Signal Processing (ICASSP), pp. 13186–13190. IEEE (2024)
16. Chen, S.Y.C.: An introduction to quantum reinforcement learning (qrl). In: 2024 15th International Conference on Information and Communication Technology Convergence (ICTC), pp. 1139–1144. IEEE (2024)

17. Chen, S.Y.C.: Learning to program variational quantum circuits with fast weights. In: 2024 International Joint Conference on Neural Networks (IJCNN), pp. 1–9. IEEE (2024)
18. Chen, S.Y.C.: Evolutionary optimization for designing variational quantum circuits with high model capacity. In: 2025 IEEE Symposium for Multidisciplinary Computational Intelligence Incubators (MCII Companion), pp. 1–5. IEEE (2025)
19. Chen, S.Y.C.: 1 - quantum reinforcement learning: Concepts and applications. In: Cheng, L., Saurabh, N., Mao, Y. (eds.) Quantum Computational AI, pp. 3–23. Morgan Kaufmann (2026). https://doi.org/10.1016/B978-0-44-330259-6.00009-8. https://www.sciencedirect.com/science/article/pii/B9780443302596000098
20. Chen, S.Y.C., Huang, C.M., Hsing, C.W., Goan, H.S., Kao, Y.J.: Variational quantum reinforcement learning via evolutionary optimization. Mach. Learn. Sci. Technol. **3**(1), 015025 (2022)
21. Chen, S.Y.C., Huang, C.M., Hsing, C.W., Kao, Y.J.: An end-to-end trainable hybrid classical-quantum classifier. Mach. Learn. Sci. Technol. **2**(4), 045021 (2021)
22. Chen, S.Y.C., Liang, Z.: Introduction to quantum machine learning and quantum architecture search. In: 2025 IEEE International Symposium on Circuits and Systems (ISCAS), pp. 1–6. IEEE (2025)
23. Chen, S.Y.C., et al.: Differentiable quantum architecture search in quantum-enhanced neural network parameter generation. arXiv preprint arXiv:2505.09653 (2025)
24. Chen, S.Y.C., Tiwari, P.: Quantum long short-term memory with differentiable architecture search. arXiv preprint arXiv:2508.14955 (2025)
25. Chen, S.Y.C., Tseng, H.H., Lin, H.Y., Yoo, S.: Learning to measure quantum neural networks. In: 2025 IEEE International Conference on Acoustics, Speech, and Signal Processing Workshops (ICASSPW), pp. 1–5. IEEE (2025)
26. Chen, S.Y.C., Tseng, H.H., Lin, H.Y., Yoo, S.: Learning to program quantum measurements for machine learning. arXiv preprint arXiv:2505.13525 (2025)
27. Chen, S.Y.C., Wei, T.C., Zhang, C., Yu, H., Yoo, S.: Quantum convolutional neural networks for high energy physics data analysis. Phys. Rev. Res. **4**(1), 013231 (2022)
28. Chen, S.Y.C., Yang, C.H.H., Qi, J., Chen, P.Y., Ma, X., Goan, H.S.: Variational quantum circuits for deep reinforcement learning. IEEE Access **8**, 141007–141024 (2020)
29. Chen, S.Y.C., Yoo, S.: Federated quantum machine learning. Entropy **23**(4), 460 (2021)
30. Chen, S.Y.C., Yoo, S., Fang, Y.L.L.: Quantum long short-term memory. In: ICASSP 2022-2022 IEEE International Conference on Acoustics, Speech and Signal Processing (ICASSP), pp. 8622–8626. IEEE (2022)
31. Dai, X., Wei, T.C., Yoo, S., Chen, S.Y.C.: Quantum machine learning architecture search via deep reinforcement learning. In: 2024 IEEE International Conference on Quantum Computing and Engineering (QCE), vol. 1, pp. 1525–1534. IEEE (2024)
32. Di Sipio, R., Huang, J.H., Chen, S.Y.C., Mangini, S., Worring, M.: The dawn of quantum natural language processing. In: ICASSP 2022-2022 IEEE International Conference on Acoustics, Speech and Signal Processing (ICASSP), pp. 8612–8616. IEEE (2022)
33. Elsken, T., Metzen, J.H., Hutter, F.: Neural architecture search: a survey. J. Mach. Learn. Res. **20**(55), 1–21 (2019)
34. Fedorov, D.A., Peng, B., Govind, N., Alexeev, Y.: VQE method: a short survey and recent developments. Mater. Theory **6**(1), 2 (2022)

35. Grant, E., Wossnig, L., Ostaszewski, M., Benedetti, M.: An initialization strategy for addressing barren plateaus in parametrized quantum circuits. Quantum **3**, 214 (2019)
36. Hochreiter, S., Schmidhuber, J.: Long short-term memory. Neural Comput. **9**(8), 1735–1780 (1997)
37. Hsu, Y.C., et al.: Federated quantum kernel-based long short-term memory for human activity recognition. arXiv preprint arXiv:2508.06078 (2025)
38. Kim, G.S., Chen, S.Y.C., Park, S., Kim, J.: Quantum reinforcement learning for coordinated satellite systems. In: ICASSP 2025-2025 IEEE International Conference on Acoustics, Speech and Signal Processing (ICASSP), pp. 1–5. IEEE (2025)
39. Kuo, E.J., Fang, Y.L.L., Chen, S.Y.C.: Quantum architecture search via deep reinforcement learning. arXiv preprint arXiv:2104.07715 (2021)
40. Larocca, M., et al.: Barren plateaus in variational quantum computing. Nat. Rev. Phys. 1–16 (2025)
41. Li, S.S., et al.: PQLM-multilingual decentralized portable quantum language model. In: ICASSP 2023-2023 IEEE International Conference on Acoustics, Speech and Signal Processing (ICASSP), pp. 1–5. IEEE (2023)
42. Li, T., Sahu, A.K., Talwalkar, A., Smith, V.: Federated learning: challenges, methods, and future directions. IEEE Signal Process. Mag. **37**(3), 50–60 (2020)
43. Lin, H.Y., Chen, S.Y.C., Tseng, H.H., Yoo, S.: Quantum reinforcement learning by adaptive non-local observables. arXiv preprint arXiv:2507.19629 (2025)
44. Lin, H.Y., Tseng, H.H., Chen, S.Y.C., Yoo, S.: Adaptive non-local observable on quantum neural networks. arXiv preprint arXiv:2504.13414 (2025)
45. Liu, C.Y., et al.: Quantum-enhanced parameter-efficient learning for typhoon trajectory forecasting. arXiv preprint arXiv:2505.09395 (2025)
46. Liu, C.Y., Chen, S.Y.C.: Federated quantum-train with batched parameter generation. In: 2024 15th International Conference on Information and Communication Technology Convergence (ICTC), pp. 1133–1138. IEEE (2024)
47. Liu, C.Y., Chen, S.Y.C., Chen, K.C., Huang, W.J., Chang, Y.J.: Federated quantum-train long short-term memory for gravitational wave signal. arXiv preprint arXiv:2503.16049 (2025)
48. Liu, C.Y., Chen, S.Y.C., Chen, K.C., Huang, W.J., Chang, Y.J.: Programming variational quantum circuits with quantum-train agent. In: 2025 International Conference on Quantum Communications, Networking, and Computing (QCNC), pp. 544–548. IEEE (2025)
49. Liu, C.Y., Kuo, E.J., Abraham Lin, C.H., Gemsun Young, J., Chang, Y.J., Hsieh, M.H., Goan, H.S.: Quantum-train: rethinking hybrid quantum-classical machine learning in the model compression perspective. Quant. Mach. Intell. **7**(2), 80 (2025)
50. Liu, H., Simonyan, K., Yang, Y.: Darts: differentiable architecture search. arXiv preprint arXiv:1806.09055 (2018)
51. Liu, Y.K., Pan, Y.H., Lu, P.F., Tsai, Y.C., Chen, S.Y.C.: Quantum-enhanced reinforcement learning with LSTM forecasting signals for optimizing fintech trading decisions. arXiv preprint arXiv:2507.12835 (2025)
52. Mari, A., Bromley, T.R., Izaac, J., Schuld, M., Killoran, N.: Transfer learning in hybrid classical-quantum neural networks. Quantum **4**, 340 (2020)
53. McClean, J.R., Boixo, S., Smelyanskiy, V.N., Babbush, R., Neven, H.: Barren plateaus in quantum neural network training landscapes. Nat. Commun. **9**(1), 4812 (2018)
54. Nielsen, M.A., Chuang, I.L.: Quantum Computation and Quantum Information. Cambridge university press, Cambridge (2010)

55. Park, J.J., et al.: Resting-state fmri analysis using quantum time-series transformer. arXiv preprint arXiv:2509.00711 (2025)
56. Park, J.J., Tseng, H.H., Yoo, S., Chen, S.Y.C., Cha, J.: It's-a-me, quantum mario: scalable quantum reinforcement learning with multi-chip ensembles. arXiv preprint arXiv:2509.00713 (2025)
57. Peng, Y., Li, X., Zhang, Z., Chen, S.Y.C., Liang, Z., Wang, Y.: Breaking through barren plateaus: reinforcement learning initializations for deep variational quantum circuits. arXiv preprint arXiv:2508.18514 (2025)
58. Peng, Y., Li, X., Zhang, Z., Chen, S.Y.C., Liang, Z., Wang, Y.: Can classical initialization help variational quantum circuits escape the barren plateau? arXiv preprint arXiv:2508.18497 (2025)
59. Pesah, A., Cerezo, M., Wang, S., Volkoff, T., Sornborger, A.T., Coles, P.J.: Absence of barren plateaus in quantum convolutional neural networks. Phys. Rev. X **11**(4), 041011 (2021)
60. Qi, J., Yang, C.H., Chen, P.Y.: QTN-VQC: an end-to-end learning framework for quantum neural networks. Phys. Scr. **99**(1), 015111 (2023)
61. Qi, J., Yang, C.H., Chen, S.Y.C., Chen, P.Y., Zenil, H., Tegner, J.: Leveraging pre-trained neural networks to enhance machine learning with variational quantum circuits. arXiv preprint arXiv:2411.08552 (2024)
62. Rofougaran, R., Yoo, S., Tseng, H.H., Chen, S.Y.C.: Federated quantum machine learning with differential privacy. In: ICASSP 2024-2024 IEEE International Conference on Acoustics, Speech and Signal Processing (ICASSP), pp. 9811–9815. IEEE (2024)
63. Silver, D., et al.: Mastering the game of go with deep neural networks and tree search. Nature **529**(7587), 484–489 (2016)
64. Skolik, A., McClean, J.R., Mohseni, M., Van Der Smagt, P., Leib, M.: Layerwise learning for quantum neural networks. Quant. Mach. Intell. **3**(1), 5 (2021)
65. Stein, J., Christ, I., Kraus, N., Mansky, M.B., Müller, R., Linnhoff-Popien, C.: Applying QNLP to sentiment analysis in finance. In: 2023 IEEE International Conference on Quantum Computing and Engineering (QCE), vol. 2, pp. 20–25. IEEE (2023)
66. Sun, Y., Ma, Y., Tresp, V.: Differentiable quantum architecture search for quantum reinforcement learning. In: 2023 IEEE International Conference on Quantum Computing and Engineering (QCE), vol. 2, pp. 15–19. IEEE (2023)
67. Sutton, R.S., Barto, A.G., et al.: Reinforcement Learning: An Introduction, vol. 1. MIT press, Cambridge (1998)
68. Vaswani, A., et al.: Attention is all you need. Adv., Neural Inf. Process. Syst. **30** (2017)
69. Watkins, W., et al.: Quantum privacy aggregation of teacher ensembles (qpate) for privacy preserving quantum machine learning. In: ICASSP 2024-2024 IEEE International Conference on Acoustics, Speech and Signal Processing (ICASSP), pp. 6875–6879. IEEE (2024)
70. Ye, E., Chen, S.Y.C.: Quantum architecture search via continual reinforcement learning. arXiv preprint arXiv:2112.05779 (2021)
71. Zhang, S.X., Hsieh, C.Y., Zhang, S., Yao, H.: Differentiable quantum architecture search. Quant. Sci. Technol. **7**(4), 045023 (2022)

Compact Quantum Modules for Scalable Fine-Tuning in Large Language Models

Emily Jimin Roh and Joongheon Kim(✉)

Korea University, Seoul 02841, Korea
{emilyjroh,joongheon}@korea.ac.kr

Abstract. The rapid growth of large language models (LLMs) has intensified the demand for efficient fine-tuning techniques under constrained computational and memory budgets. Parameter-efficient fine-tuning (PEFT) addresses this challenge, yet existing methods often sacrifice expressive capacity for reduced complexity. This paper introduces Quantum-amplitude embedded (QAE) adaptation, a novel PEFT framework that leverages quantum-amplitude embedding to achieve logarithmic compression of activation vectors while preserving task-relevant information. By integrating parameterized quantum circuits (PQCs) as nonlinear transformation layers, QAE adaptation replaces conventional linear adapters in attention modules with compact quantum-inspired operators. This design provides high expressivity with significantly fewer trainable parameters. Experimental evaluations across benchmark tasks demonstrate that QAE achieves competitive or superior performance compared with state-of-the-art PEFT methods, which highlights its effectiveness for resource-constrained LLM adaptation.

Keywords: Quantum Computing · Parameter-Efficient Fine-Tuning · Large Language Model

1 Introduction

Background and Motivation. Large language models (LLMs) have achieved strong results in tasks such as summarization and question answering, yet their adaptation to new domains still demands additional fine-tuning. Updating billions of parameters for every task is infeasible in practice, which has motivated parameter-efficient fine-tuning (PEFT) [6]. Methods such as LoRA [4] and prefix tuning [5] significantly reduce cost by introducing lightweight modules, but their reliance on linear projections or prompt vectors often restricts representational power. This limitation is particularly evident when operating under strict resource budgets, where efficiency and expressivity directly conflict.

Proposed Approach. To overcome these constraints, we propose quantum-amplitude embedded (QAE) adaptation, which substitutes linear adapters with quantum modules that combine logarithmic amplitude embedding [3] and

V. Aggarwal et al. (Eds.): PQAI 2025, CCIS 2724, pp. 22–27, 2026.
https://doi.org/10.1007/978-981-95-7829-0_2

parameterized quantum circuits (PQCs) [7]. This design compresses activation vectors in log scale while enabling expressive non-linear transformations over entangled states, thus capturing complex dependencies that classical adapters cannot.

Contributions. The contributions are threefold: (i) QAE introduces a quantum-inspired PEFT framework with logarithmic compression and PQC-driven adaptation, (ii) It integrates into transformers without altering pretrained weights, allowing modular deployment, (iii) Empirical evaluation confirms that QAE delivers competitive accuracy with a fraction of parameters compared to classical baselines.

2 LLM Fine-Tuning Framework

Fine-tuning LLMs is costly due to their massive parameter counts, often exceeding billions [2]. Updating all weights for each task is infeasible in multi-task or low-resource scenarios, as memory and training overheads scale dramatically.

Let $\mathcal{Z} = \{(x_i, y_i)\}_{i=1}^{N}$ denote the task dataset, with inputs x_i and outputs y_i. For a pre-trained model $P_\Phi(y \mid x)$ with full parameters Φ, standard fine-tuning maximizes sequence likelihood,

$$\max_{\Phi} \sum_{(x,y)\in\mathcal{Z}} \sum_{t=1}^{|y|} \log P_\Phi(y_t \mid x, y_{<t}). \tag{1}$$

To reduce cost, PEFT updates only a small set of variables while freezing most of Φ [10]. Methods such as LoRA [8] introduce lightweight modules, learning a compact update function $\Delta h(\theta)$ with $|\theta| \ll |\Phi|$, yielding new parameters $\Phi + \Delta h(\theta)$. The objective becomes,

$$\max_{\theta} \sum_{(x,y)\in\mathcal{Z}} \sum_{t=1}^{|y|} \log P_{\Phi+\Delta h(\theta)}(y_t \mid x, y_{<t}). \tag{2}$$

In our framework, $\Delta\Phi(\theta)$ is realized through a quantum-amplitude adapter, which logarithmically compresses activations via amplitude embedding and applies trainable quantum circuits for expressive yet efficient adaptation.

3 Algorithm Design

3.1 QAE Adaptation

We propose the QAE, a PEFT mechanism that substitutes classical adapters with quantum modules. As illustrated in Fig. 1, QAE projects hidden activations into a quantum feature space, which enables non-linear transformations through PQCs while requiring only logarithmic trainable parameters with respect to hidden size d.

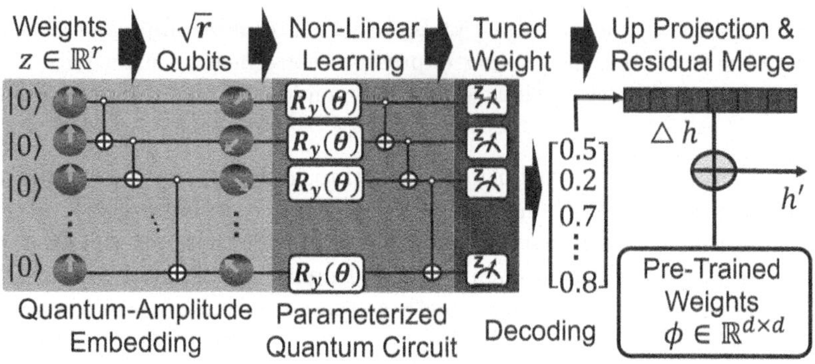

Fig. 1. QAE-based LLM parameter-efficient fine-tuning framework.

Quantum-Amplitude Embedding. Given an activation $x_i \in \mathbb{R}^d$, amplitude embedding maps it into a normalized quantum state over $n = \lceil \log_2 d \rceil$ qubits. After zero-padding or truncation, the vector is normalized as $\tilde{x}_i = x'_i / \|x'_i\|_2$, that produces a valid quantum state $|x_i\rangle = \sum_j \tilde{x}_{i,j} |j\rangle$. This compact encoding enables global, parallel processing of features and reduces resource requirements from d to $\log_2 d$ qubits.

PQC-Based Non-linear Learning. The embedded state $|x_i\rangle$ is processed by a PQC consisting of parameterized single-qubit rotations and CNOT-based entanglement. Formally, the circuit applies a unitary $U_T(\theta)$ over n qubits, where θ are trainable parameters. Rotations introduce local non-linearity, while entanglement captures dependencies across qubits, yielding expressive transformations in Hilbert space.

Decoding and Up Projection. After PQC processing, the output state $|\psi_i\rangle$ is measured via Pauli-Z observables, which generates a compact feature vector $z_i^{(q)} \in \mathbb{R}^n$. A linear up-projection $W_u \in \mathbb{R}^{d \times n}$ maps $z_i^{(q)}$ back to dimension d, producing $\hat{x}_i = W_u z_i^{(q)}$. The final adapted activation is obtained through a residual merge $x_i^{(\text{aug})} = x_i + \hat{x}_i$. This procedure defines $\Delta\Phi(\theta)$, enabling efficient and expressive adaptation with minimal parameters.

4 Performance Evaluation

To examine the effectiveness of the proposed QAE, we conduct experiments on three representative LLM backbones—GPT-NEO [1], TinyLLaMA [12], and Qwen2.5 [11]—using the Alpaca dataset [9]. The evaluation benchmarks QAE against full fine-tuning, LoRA with rank $r = 8$, and prefix tuning. Metrics include BLEU and BERTScore F1, alongside the absolute and relative number of trainable parameters.

Table 1. Comparison of the standard NLG evaluation metrics using different PEFT methods.

Model & Method	#Model Parameters	#Trainable Parameters	Trainable Ratio	Standard BLEU	NLG Metrics BERTScoreF1
GPT-NEO (Full)	125.2M	125.2M	100%	10.62	83.06
GPT-NEO (LoRA)	125.2M	294.9K	0.24%	0.734	77.76
GPT-NEO (Prefix)	125.2M	368.6K	0.29%	0.827	76.95
GPT-NEO (QAE)	125.2M	**184.5K**	**0.14%**	**33.37**	**89.11**
TinyLLaMA (Full)	1.100B	1.100B	100%	24.25	85.44
TinyLLaMA (LoRA)	1.100B	1.126M	0.11%	12.24	85.13
TinyLLaMA (Prefix)	1.100B	**0.225M**	**0.02%**	21.07	87.17
TinyLLaMA (QAE)	1.100B	0.540M	0.05%	**20.66**	**88.17**
Qwen2.5 (Full)	494.0M	494.0M	100%	13.92	86.14
Qwen2.5 (LoRA)	494.0M	0.540M	0.10%	6.549	83.78
Qwen2.5 (Prefix)	494.0M	**0.122M**	**0.02%**	1.715	82.56
Qwen2.5 (QAE)	494.0M	0.215M	0.04%	**12.95**	**87.76**

Parameter Efficiency. Table 1 highlights the drastic reduction of trainable parameters achieved by QAE compared with other PEFT techniques. For GPT-NEO, QAE updates only 184.5K weights (0.14% of the total), whereas LoRA and prefix tuning require 0.24% and 0.29%, respectively. On TinyLLaMA, QAE adapts 0.05% of parameters, which is smaller than LoRA's 0.11% while remaining higher than prefix tuning's 0.02%. For Qwen2.5, QAE modifies 0.04% of parameters, again below LoRA's 0.10%. These observations confirm that QAE achieves adaptation with one of the smallest parameter footprints across all model families.

Generation Quality. Table 1 further shows that QAE not only minimizes parameter usage but also delivers superior text generation quality. On GPT-NEO, QAE reports a BLEU of 33.37 and BERTScore of 89.11, outperforming LoRA (BLEU: 0.734, BERTScore: 77.76) and prefix tuning (BLEU: 0.827, BERTScore: 76.95), while also surpassing full fine-tuning. For TinyLLaMA, QAE reaches a BERTScore of 88.17 with just 0.05% parameters, which exceed LoRA and prefix tuning, and even improving over the full model baseline. On Qwen2.5, QAE attains BLEU 12.95 and BERTScore 87.76, outperforming LoRA (6.549, 83.78) and prefix tuning (1.715, 82.56), while closely approaching the full fine-tuned performance. These outcomes demonstrate that QAE effectively balances compactness and semantic fidelity, consistently outperforming existing PEFT baselines.

5 Conclusion

This work introduces QAE, a quantum-inspired adapter for efficient fine-tuning of LLMs. By combining amplitude embedding with PQC-based transformations,

QAE enables expressive non-linear adaptation while substantially lowering the number of trainable parameters. Unlike conventional PEFT approaches that rely primarily on linear projections or prompt engineering, QAE leverages quantum state representations to achieve logarithmic compression of activation vectors, thereby offering both efficiency and representational richness. Comprehensive experiments on GPT-NEO, TinyLLaMA, and Qwen2.5 demonstrate that QAE consistently achieves generation quality comparable to, and in many cases exceeding, full fine-tuning while requiring only a fraction of the parameters. The empirical results further highlight QAE's robustness across model scales, confirming its suitability for resource-constrained environments where memory and computation budgets are limited. These findings establish QAE as a practical alternative to classical PEFT baselines such as LoRA and prefix tuning, achieving an improved balance between compactness and expressivity.

Looking forward, QAE opens new research directions at the intersection of quantum machine learning and large-scale language modeling. Extensions may include hardware-aware optimizations for near-term quantum processors, integration with multi-modal architectures, and theoretical analysis of quantum expressivity under PEFT constraints. Overall, this study demonstrates that quantum-inspired fine-tuning provides a viable path toward scalable, adaptable, and energy-efficient LLM deployment in real-world applications.

Acknowledgments. The full version of this work was accepted to ACM Conference on Information and Knowledge Management (CIKM), Seoul, South Korea, November 2026. This is the extended abstract for the full version under the permission and guidance of Purdue Quantum AI (PQAI) Symposium. This work was supported by the National Research Foundation of Korea(NRF) grant funded by the Korea government(MSIT)(RS-2025-00561377).

References

1. Black, S., Gao, L., Wang, P., Leahy, C., Biderman, S.: GPT-Neo: large scale autoregressive language modeling with mesh-tensorflow (2021). https://doi.org/10.5281/zenodo.5297715
2. Ding, N., et al.: Parameter-efficient fine-tuning of large-scale pre-trained language models. Nat. Mach. Intell. **5**(3), 220–235 (2023)
3. Gonzalez-Conde, J., Watts, T.W., Rodriguez-Grasa, P., Sanz, M.: Efficient quantum amplitude encoding of polynomial functions. Quantum **8**, 1297 (2024)
4. Hu, E.J., et al.: LoRA: low-rank adaptation of large language models. In: Proceedings of the International Conference on Learning Representations (ICLR). Virtual (2022)
5. Huang, H., Liu, X., Shi, G., Liu, Q.: Event extraction with dynamic prefix tuning and relevance retrieval. IEEE Trans. Knowl. Data Eng. **35**(10), 9946–9958 (2023)
6. Ma, X., Guo, J., Zhang, R., Fan, Y., Cheng, X.: Scattered or connected? An optimized parameter-efficient tuning approach for information retrieval. In: Proceedings of the ACM International Conference on Information and Knowledge Management (CIKM), pp. 1471–1480. Atlanta, Georgia, USA (2022)

7. Mahmud, D., et al.: Integrating LLMs with ITS: recent advances, potentials, challenges, and future directions. IEEE Trans. Intell. Transp. Syst. **26**(5), 5674–5709 (2025)
8. Mao, Y., et al.: A survey on lora of large language models. Front. Comp. Sci. **19**(7), 197605 (2025)
9. Taori, R., et al.: Stanford alpaca: an instruction-following llama model (2023). https://github.com/tatsu-lab/stanford_alpaca . https://github.com/tatsu-lab/stanford_alpaca
10. Usman, M.T., Khan, H., Singh, S.K., Lee, M.Y., Koo, J.: Efficient deepfake detection via layer-frozen assisted dual attention network for consumer imaging devices. IEEE Trans. Consum. Electron. 1–11 ((Early Access) 2024)
11. Xiang, M., Fernando, R., Wang, B.: On-device qwen2. 5: efficient LLM inference with model compression and hardware acceleration. arXiv preprint arXiv:2504.17376 (2025)
12. Zhang, P., Zeng, G., Wang, T., Lu, W.: Tinyllama: an open-source small language model. arXiv preprint arXiv:2401.02385 (2024)

Robustness of Classical and Quantum Inspired Architectures Against Structured Corruptions in Vision Tasks

Sarika Gauba(✉) and Aarav Gauba

Maryland Heights, USA
sarikagauba29@gmail.com

Abstract. Robust performance under distributional shifts and noisy inputs is critical for real world deployment of machine learning models. While Convolutional Neural Networks (CNNs) remain the foundation of vision models, they are notoriously sensitive to corruptions in the input data, such as sensor noise, partial occlusions, or bit level errors. Motivated by the growing interest in quantum machine learning, we investigate whether quantum inspired architectural inductive biases can confer greater resilience to such perturbations.

We conduct a systematic evaluation of three neural architectures Classical CNN, Quantum Convolutional Neural Network (QCNN), and Quantum Multi Head Attention (QMHA) on the CIFAR 10 dataset under a diverse set of corruption regimes. Specifically, we test inference time robustness against Gaussian noise, salt and pepper corruption, Fourier masking, stripe noise, block masking, and bit flip noise, without any retraining or augmentation.

Our results demonstrate that while CNNs degrade significantly under most corruptions, quantum inspired architectures, particularly QMHA, exhibit improved robustness in multiple scenarios. These findings highlight the potential of quantum informed designs in developing resilient vision models and suggest promising directions for future hybrid quantum classical architectures in real world deployment settings.

1 Introduction

Deep learning has achieved state of the art performance in a wide array of computer vision tasks, ranging from image classification to object detection and segmentation. However, a critical limitation of modern neural networks is their fragility to corrupted, noisy, or adversarially perturbed inputs. Even minor distortions such as Gaussian blur, pixel masking, or bit flips can cause significant performance degradation. This vulnerability poses major challenges in deploying these models in safety critical or real world applications, such as autonomous driving, medical imaging, or robotics.

Recent work has explored data augmentation, adversarial training, and input preprocessing as means to improve robustness. However, these methods often

V. Aggarwal et al. (Eds.): PQAI 2025, CCIS 2724, pp. 28–35, 2026.
https://doi.org/10.1007/978-981-95-7829-0_3

increase computational cost and may generalize poorly to unseen corruption types. An emerging alternative is to redesign the model architecture itself to inherently withstand noise without relying on retraining or post hoc defenses.

In this paper, we explore the robustness properties of *quantum inspired neural architectures*, which integrate principles from quantum computing into classical models to induce novel inductive biases. These principles include entanglement like feature mixing, probabilistic measurement via expectation values, and circuit based computation. Such inductive biases may enhance the model's ability to encode global structure, suppress irrelevant noise, and generalize under perturbations.

We evaluate three architectures:

- **CNN:** A standard convolutional neural network with ReLU and pooling layers.
- **QCNN:** A variant of CNN with quantum inspired latent projections and dropout like mechanisms.
- **QMHA:** A novel architecture combining CNN encoders with multi head quantum circuit simulations, where each head computes expectation values from parameterized quantum transformations.

Our contribution is a unified empirical study that tests these models across six distinct corruption types, each representing a different class of structured or unstructured noise. Crucially, all models are trained solely on clean data to isolate architectural robustness from training time exposure. We show that quantum inspired models particularly QMHA exhibit notable resilience across multiple corruption scenarios, supporting their relevance for robust machine learning in noisy environments.

2 Related Work

2.1 Robustness Benchmarks

Robustness in computer vision has become a growing concern, with datasets such as CIFAR 10 C [1] and ImageNet C providing standardized corruption benchmarks. These benchmarks evaluate model performance under corruptions such as noise, blur, compression, and weather artifacts. While many defenses rely on data augmentation, adversarial training, or test time adaptation, recent studies emphasize the importance of architectural robustness—that is, models that are resilient by design, even without retraining.

2.2 Quantum Inspired Neural Networks

Quantum Machine Learning (QML) explores using quantum computing for learning tasks, while quantum inspired neural networks aim to simulate aspects of quantum computation using classical hardware. Models such as the Quantum Convolutional Neural Network (QCNN) [2] mimic quantum feature maps and hierarchical entanglement, while attention based variants such as Quantum Multi Head Attention (QMHA) simulate parallel quantum circuits to compute attention via expectation values. Prior work has focused on performance and expressivity; here, we study their robustness Fig. 1.

2.3 Model Architectures

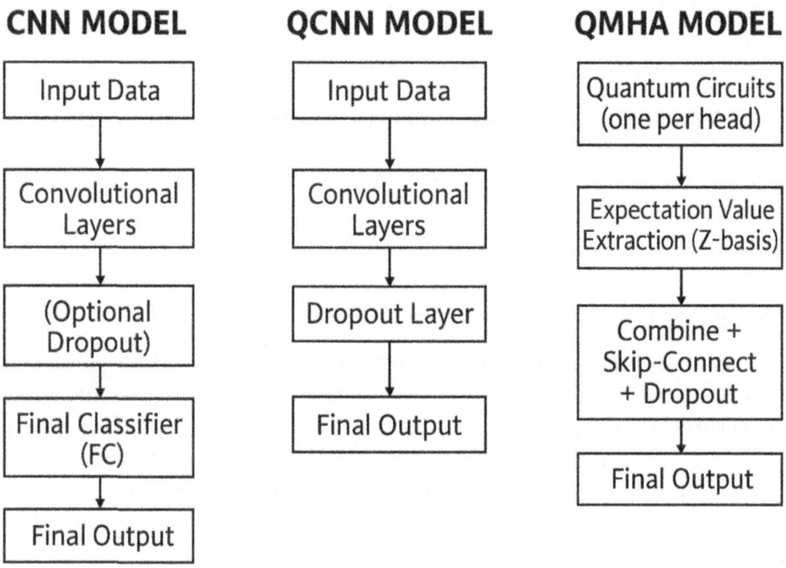

Fig. 1. Comparison of the three neural architectures evaluated in this study. The CNN model uses standard convolutional layers followed by an optional dropout and a final fully connected classifier. The QCNN model extends this by integrating a quantum inspired dropout mechanism and latent projection layers to simulate quantum feature encoding. The QMHA model replaces traditional convolutional processing with parallel quantum inspired attention heads, each simulating a parameterized quantum circuit and extracting Z basis expectation values. Outputs from each head are aggregated via skip connections and regularized using dropout before classification.

To isolate architectural robustness, all three models are trained on clean CIFAR 10 images using identical optimization settings. The primary difference lies in their architectural choices, especially the integration of quantum inspired inductive biases.

CNN: The classical baseline is a lightweight convolutional neural network comprising two convolutional layers with ReLU activations and max pooling, optionally followed by a dropout layer. The resulting features are flattened and passed through a linear classifier. This architecture represents a standard approach in low capacity vision models and serves as a reference for robustness degradation.

QCNN: The Quantum Convolutional Neural Network (QCNN) extends the classical CNN by incorporating two key enhancements inspired by quantum computing principles. First, a latent projection module is introduced after the convolutional backbone, simulating a quantum feature map by applying a high dimensional nonlinear embedding. Second, a dropout mechanism is employed that probabilistically suppresses feature channels in a pattern inspired by quantum decoherence. These modifications aim to introduce redundancy and disentanglement in the latent space, potentially improving robustness to structured noise.

QMHA: The Quantum Multi Head Attention (QMHA) architecture replaces the convolutional feature extractor with a quantum inspired attention mechanism. The encoder consists of a shallow CNN that projects inputs to a latent representation, which is then processed by H parallel attention heads. Each head simulates a parameterized quantum circuit $U_i(x)$, and the output is computed as an expectation value in the Z basis:

$$z_i = \langle 0|U_i(x)^\dagger Z U_i(x)|0\rangle$$

The expectation values from all heads are concatenated and passed through a fusion layer with residual connections and dropout before being classified. This architecture introduces an implicit ensemble of disentangled feature paths, inspired by quantum entanglement, and is hypothesized to confer robustness by enabling the model to extract invariant representations under noise.

These models were designed to reflect a progression from purely classical architectures to increasingly quantum inspired designs, allowing us to directly assess the contribution of quantum inductive biases to robustness under various corruptions.

2.4 Noise Types

To evaluate the intrinsic robustness of each architecture, we apply a diverse set of structured and unstructured noise corruptions to the CIFAR 10 test dataset. These noise types were chosen to reflect common challenges in real world vision systems, including sensor degradation, transmission errors, and environmental distortions. Noise is applied post training, during inference only, without any retraining or augmentation, ensuring that robustness is purely architectural.

- **Gaussian noise:** Additive zero mean Gaussian noise with varying standard deviation $\sigma \in [0.0, 1.0]$, simulating camera sensor noise or low light environments.
- **Salt and pepper noise:** A portion of image pixels are randomly set to either 0 (black) or 1 (white), modeling binary sensor corruption and bit level faults.
- **Fourier masking:** Central frequency components in the Fourier domain are suppressed, reducing structural and low frequency information. This simulates transmission artifacts and lossy compression effects.

- **Stripe noise:** Random vertical columns are zeroed out, producing periodic occlusion patterns similar to column sensor failure or interlaced interference.
- **Block noise:** Square regions of the image are occluded with black patches, mimicking real world occlusions such as raindrops, dust, or partial visibility.
- **Bit flip noise:** Random pixel values are inverted, simulating low level hardware faults and memory corruption at the bit level, particularly relevant in embedded and edge computing scenarios.

Each corruption type is applied at multiple severity levels, and models are evaluated based on classification accuracy across these perturbations.

2.5 Training and Evaluation

All models are trained on clean CIFAR 10 images for 50 epochs using the Adam optimizer, with a learning rate of 1×10^{-3} and a batch size of 32. To ensure fair comparison, all training hyperparameters are kept constant across models.

Importantly, no noise augmentation is used during training—noise is applied exclusively at test time. This ensures that the observed performance degradation stems from architectural robustness rather than exposure to corrupted samples during optimization.

Model evaluation is performed using top-1 classification accuracy on the corrupted test set. For each noise type, we report accuracy at two severity levels (e.g., $\sigma = 0.3$ and $\sigma = 0.5$ for Gaussian noise), along with accuracy under clean (unperturbed) conditions. No finetuning or postprocessing is performed Table 1.

3 Results

Table 1. Accuracy (%) on clean and corrupted CIFAR-10 test data. Abbreviations: g = Gaussian, s = Salt and Pepper.

Model	Clean	g0.3	g0.5	g0.7	s0.3	s0.5	Fourier	Bitflip
CNN	57.46	24.60	15.88	13.09	19.24	14.06	22.39	54.85
QMHA	55.26	31.37	18.07	14.24	23.81	14.68	18.40	54.88
QCNN	57.94	24.97	15.09	12.58	19.17	12.75	19.27	56.52

Observations:

- **QMHA** shows superior robustness under Gaussian and salt and pepper noise, outperforming both CNN and QCNN by a notable margin at medium noise levels ($\sigma = 0.5$).
- **QCNN** slightly edges out the others under bit-flip noise, suggesting that its latent projection layer introduces redundancy beneficial for binary level perturbations.

- **CNN** suffers significant accuracy drops across nearly all corruption types, especially in Fourier masked and block like scenarios where its local feature detectors become unreliable Table 2.

Table 2. Accuracy (%) on clean and corrupted CIFAR-10 test data with dropout.

Model	Clean	g0.5	g0.7	g1.0	s0.5	s0.7	Fourier	Bitflip
CNN	54.51	15.07	13.10	10.65	12.36	10.62	16.08	53.88
QMHA+DO	54.90	17.26	16.01	12.24	14.70	11.82	23.36	54.06
QCNN+DO	56.30	16.51	14.34	11.50	13.31	11.41	20.01	55.07

Additional Observations (with Dropout):

- **QMHA+DO** achieves the best accuracy under Fourier corruption (23.36%), outperforming all other configurations, including the vanilla QMHA and CNN.
- Dropout improves robustness for both QMHA and QCNN models under Gaussian noise ($\sigma = 0.5$ and 0.7), though the gain diminishes at extreme levels ($\sigma = 1.0$).
- **QCNN+DO** continues to perform strongly under bit-flip corruption, maintaining high resilience even after dropout is introduced.
- While dropout does not dramatically improve salt and pepper robustness beyond QMHA's baseline, it does offer small consistent improvements and regularizes performance across moderate noise regimes.

4 Discussion

The results indicate that quantum inspired inductive biases can offer tangible robustness benefits in vision tasks without requiring noise specific training. The QMHA architecture, in particular, maintains stronger performance under unstructured and semi structured corruption types such as Gaussian and salt and pepper noise. This may stem from its use of disentangled attention heads and expectation value based feature extraction, which introduce implicit regularization and redundancy.

Interestingly, the QCNN's resilience under bit flip corruption implies that its latent projection encodes information in a way that tolerates binary level noise, possibly due to redundancy in its intermediate representations. This suggests that even relatively minor architectural changes inspired by quantum processing can improve robustness in specific scenarios.

However, no model demonstrated universal robustness across all corruption types. In fact, QMHA underperforms CNN under Fourier noise, indicating that high frequency or global distortions still challenge quantum inspired structures. This emphasizes the importance of evaluating multiple corruption types and severity levels when assessing robustness.

Limitations.

While QMHA exhibits promising robustness, it is not universally superior. Under high-frequency corruptions such as Fourier masking, it performs worse than the CNN baseline. Additionally, quantum-inspired models incur additional computational complexity, especially with increasing attention heads or deeper circuit analogs. Future work must also validate robustness on larger, real-world datasets and resource-constrained deployments.

5 Conclusion

In this study, we compared the robustness of classical and quantum inspired architectures CNN, QCNN, and QMHA, on the CIFAR 10 dataset subjected to a diverse set of noise corruptions. All models were trained exclusively on clean data, isolating architectural robustness from data driven adaptation.

Our findings demonstrate that quantum inspired architectures, particularly QMHA, offer promising robustness advantages under common corruption types, including Gaussian and salt and pepper noise. These results suggest that incorporating quantum design principles such as circuit inspired attention, expectation value extraction, and entangled feature paths can enhance the resilience of deep learning models in adverse environments.

Future work may extend this analysis to:

- Larger and higher resolution datasets such as CIFAR 100 and ImageNet.
- Real world corruptions (e.g., weather, motion blur, lens distortion).
- Hybrid training regimes that combine robustness with noise aware fine tuning.

Ultimately, this study supports the growing interest in quantum inspired neural architectures not just for performance, but also for their potential to improve reliability and generalization in challenging settings.

References

1. Hendrycks, D., Dietterich, T.: Benchmarking neural network robustness to common corruptions and perturbations. In: International Conference on Learning Representations (ICLR) (2019)
2. Cong, I., Choi, S., Lukin, M.D.: Quantum convolutional neural networks. Nat. Phys. **15**(12), 1273–1278 (2019)
3. Zhang, S., Zhao, W., Yu, Z.: Quantum attention mechanism: a novel architecture for natural language understanding. arXiv preprint arXiv:2012.03807 (2020)
4. Hendrycks, D., et al.: AugMix: a simple data processing method to improve robustness and uncertainty. In: International Conference on Learning Representations (ICLR) (2020)

5. Bhojanapalli, S., Chakrabarti, A., Menon, A., Neyshabur, B.: Understanding robustness of vision transformers. In: International Conference on Machine Learning (ICML), pp. 10117–10125 (2021)
6. Gauba, A.: Quantum-Enhanced multi-head attention for hybrid classical-quantum neural networks. Preprint available at https://doi.org/10.13140/RG.2.2.13590.66880 (2025)

Geometric Origin of Fourier Coefficients in Data-Reuploading Single-Qubit Circuits

Seungcheol Oh[1], Chaemoon Im[1], Soohyun Park[2(✉)], and Joongheon Kim[1(✉)]

[1] Korea University, Seoul 02841, Korea
{seungoh,anscodla0314,joongheon}@korea.ac.kr
[2] Sookmyung Women's University, Seoul 04310, Korea
soohyun.park@sookmyung.ac.kr

Abstract. In this work, we characterize the Fourier structure of single-qubit, data-reuploading parameterized quantum circuits (PQCs). Each data-encoding rotation constrains the Bloch vector to a circle whose radius is fixed by the trainable blocks. We prove the *Bloch Circle Projection Theorem*: for a single layer, projecting this trajectory onto the measurement axis directly yields the circuit's Fourier coefficients. This gives a concrete design rule: choose rotation axes non-collinear with the encoding axis to avoid degenerate projections. The same geometric argument extends to L-layer circuits; permitting layer-wise flexible rotation axes preserves full expressivity, even for targets with rich high-frequency content.

Keywords: Quantum Neural Networks · Parameterized Quantum Circuits

1 Introduction

Background and Motivation. Quantum computing lies at the confluence of physics and computation. By exploiting quantum-mechanical principles such as superposition and entanglement, quantum devices can tackle certain highly complex problems that are otherwise intractable on classical computers, achieving what is termed *quantum supremacy* [1,7,9,18]. However, the label *quantum supremacy* should not be read as an automatic guarantee of superiority; genuine advantages emerge only when algorithms are carefully designed to harness problem-relevant quantum effects. Consequently, realizing quantum advantage in any scientific domain requires a deep understanding of how quantum systems behave under task-specific conditions so that these behaviors can be meaningfully translated into algorithmic design [12]. Quantum methods are now being explored across diverse scientific areas, prominently via quantum neural networks (QNNs) [4,16]. Within this paradigm, parameterized quantum circuits (PQCs) have become a central building block [3], trained in a manner analogous to classical deep neural networks (DNNs) to learn inputâĂŞoutput mappings. Recent

V. Aggarwal et al. (Eds.): PQAI 2025, CCIS 2724, pp. 36–41, 2026.
https://doi.org/10.1007/978-981-95-7829-0_4

studies have applied PQCs to tasks traditionally handled by DNNs [2,5,10]. While these applications are valuable, it is equally important to elucidate the *intrinsic characteristics* of PQCs. As argued in [15], discovering uniquely quantum capabilities can guide us toward problems naturally aligned with those capabilities, thereby pursuing *genuine quantum supremacy* for QNNs.

Related Work. PQCs are typically composed of sequences of unitary gates acting on quantum states. Despite progress in QNNs, PQC design remains more art than theory, with gate sequences often chosen ad hoc. A growing body of work seeks principled foundations for design [13,19]. For instance, [14] introduced the data-reuploading paradigm and established a rigorous analogy between classical DNNs and PQCs. Building on this, [17] demonstrated that data-reuploading PQCs produce outputs that can be expressed as partial Fourier series. More recently, [20] proved that single-qubit QNNs are universal approximators for bounded univariate functions. Nevertheless, a fundamental gap persists in understanding the variational power of PQCs when the target function is a true black box. In particular, [20] constructs circuits under the assumption that the target's Fourier coefficients are known in advance, which limits applicability in realistic learning scenarios. This motivates the central question of how PQCs learn directly through training—without any prior knowledge of the target's Fourier coefficients.

Overview. This work provides both theoretical and empirical analyses explaining how PQC architecture influences the Fourier coefficients of a model's output. By interpreting the qubit's evolution under a PQC as a trajectory on the Bloch sphere, we establish a formal connection between this geometry and the Fourier coefficients of the target function. We show that distinct PQC configurations induce distinct trajectories, which in turn determine the learned coefficients. Leveraging this viewpoint, we derive insights for systematically designing PQC architectures to enhance learning capability. To make the perspective actionable, we further introduce a principled method for constructing data-encoding Hamiltonians with tunable scalers, enabling spectrum selection aligned to desired frequencies and more deliberate control over the output's frequency content.

Contributions. The primary contributions are two-fold. First, we present empirical and mathematical evidence that a single-layer PQC learns a target function by modulating the radius of the circular trajectory traced by the qubit on the Bloch sphere. Moreover, using a Lie-algebraic framework [11], we establish a theorem that explicitly links the trajectory radius, the associated Fourier coefficient, and the measurement axis.

2 Quantum Model and Preliminaries

Qubit. A quantum bit (qubit) is the elementary unit of quantum information. Formally, a qubit is a unit-norm vector in a two-dimensional Hilbert space, $|\psi\rangle = \alpha\,|0\rangle + \beta\,|1\rangle$, where $|0\rangle = \begin{bmatrix}1\\0\end{bmatrix}$ and $|1\rangle = \begin{bmatrix}0\\1\end{bmatrix}$ are the computational basis states, and the complex amplitudes $\alpha, \beta \in \mathbb{C}$ satisfy $|\alpha|^2 + |\beta|^2 = 1$. The *Bloch sphere* provides a geometric representation in which any single-qubit pure state corresponds to a point on the unit sphere in $\mathbb{R}^3$. Equivalently, every state admits the parameterization $|\psi\rangle = \cos(\frac{\theta}{2})|0\rangle + e^{i\phi}\sin(\frac{\theta}{2})|1\rangle$, with polar and azimuthal angles $\theta \in [0, \pi]$ and $\phi \in [0, 2\pi)$, respectively.

Parameterized Quantum Circuit. QNNs are quantum counterparts of classical neural networks that leverage superposition (and related effects) to learn from data [8]. A widely used construction is the PQC. A PQC comprises a sequence of quantum gates whose rotation angles depend on classical inputs and/or tunable parameters optimized during training. Mathematically, a PQC implements a unitary $U(x, \boldsymbol{\theta})$ acting on an initial state $|0\rangle^{\otimes n}$, where $x \in \mathbb{R}^d$ is the input, $\boldsymbol{\theta} = (\theta_1, \ldots, \theta_p) \in \mathbb{R}^p$ the trainable parameters, and n the number of qubits. We adopt an architecture of alternating data-encoding gates $S(x)$ and trainable blocks $W(\boldsymbol{\theta})$ [14], yielding $U(x, \boldsymbol{\theta}) = W^{L+1}(\boldsymbol{\theta})\, S(x)\, W^L(\boldsymbol{\theta}) \cdots S(x)\, W^1(\boldsymbol{\theta})$, where L is the number of reuploading steps. Each W^ℓ typically consists of cascaded parameterized Pauli rotations, $R_X(\theta) = e^{-i\frac{\theta}{2}\sigma_X}$, $R_Y(\theta) = e^{-i\frac{\theta}{2}\sigma_Y}$, $R_Z(\theta) = e^{-i\frac{\theta}{2}\sigma_Z}$, with the Pauli matrices $\sigma_X = \begin{bmatrix} 0 & 1 \\ 1 & 0 \end{bmatrix}$, $\sigma_Y = \begin{bmatrix} 0 & -i \\ i & 0 \end{bmatrix}$, $\sigma_Z = \begin{bmatrix} 1 & 0 \\ 0 & -1 \end{bmatrix}$.

Measurement and Projection. Measurement in the computational basis yields $|0\rangle$ or $|1\rangle$ with probabilities $|\alpha|^2$ and $|\beta|^2$, respectively, and irreversibly collapses the superposition. We typically measure an observable M, most often $M = \sigma_Z$, whose eigenvectors are $|0\rangle$ and $|1\rangle$ with eigenvalues ± 1. The model output is the expectation value of M after the PQC, $f_{\boldsymbol{\theta}}(x) = \langle 0| U^\dagger(x, \boldsymbol{\theta})\, M\, U(x, \boldsymbol{\theta})\, |0\rangle$. For $M = \sigma_Z$, this equals $f_{\boldsymbol{\theta}}(x) = |\alpha|^2 - |\beta|^2$. Geometrically, this is the projection of the Bloch vector onto the measurement axis $\hat{z}$, so the output is bounded in $[-1, 1]$, defining a real-valued map $f_{\boldsymbol{\theta}} : \mathbb{R}^d \to [-1, 1]$.

Partial Fourier Series Analysis. Consider a data-reuploading unitary [14] with encoder $S(x) = e^{-ix\boldsymbol{H}}$, where the Hamiltonian $\boldsymbol{H}$ determines how the input is embedded. The spectrum of $\boldsymbol{H}$ fixes the set of accessible frequencies in the model output. As shown in [17], the resulting function naturally admits a partial Fourier expansion $f(x) = \sum_{\omega \in \Omega} c_\omega\, e^{i\omega x}$, where the accessible set Ω consists of differences of eigenvalues of $\boldsymbol{H}$, and the coefficients c_ω depend on circuit parameters and the measurement choice. Moreover, [20] proved that a single-qubit circuit can universally approximate bounded univariate functions. That work, however, does not provide an intuitive or structural account of how specific PQC configurations shape the resulting Fourier coefficients.

3 Bloch Circle Projection Theorem

Single-qubit gates act on $\mathbb{C}^2$ and form the special unitary group $SU(2)$. Via the standard adjoint map, each $U \in SU(2)$ induces a unique rotation of the Bloch sphere in $\mathbb{R}^3$, i.e., an element of $SO(3)$. We analyze single-qubit learning behavior entirely in this geometric picture.

Consider the single-layer data-reuploading ansatz

$$U(x, \boldsymbol{\theta}, \boldsymbol{\phi}) = R_L(\boldsymbol{\theta})\, R_X(x)\, R_R(\boldsymbol{\phi}), \tag{1}$$

where $R_X(x) = \exp(-ix\sigma_X/2)$ encodes the input and $R_Q(\eta) = \exp(-i\eta Q/2)$ for $Q \in \{\sigma_X, \sigma_Y, \sigma_Z\}$ appear inside the trainable blocks R_L, R_R. Let $\mathcal{R}_L, \mathcal{R}_R \in SO(3)$ be the corresponding Bloch-sphere rotations, and write $\hat{\mathbf{z}} = (0,0,1)^\top$ for the measurement axis. Define the post-encoding axis $\mathbf{u} := \mathcal{R}_R\, \hat{\mathbf{z}}, \mathbf{n} := \hat{\mathbf{x}}$.

Theorem 1 (Bloch Circle Projection). *Let $\mathcal{R}_{\mathbf{n}}(x) \in SO(3)$ denote rotation by angle x about the unit vector $\mathbf{n}$. The model output (the Z-expectation) is*

$$f(x) \;=\; \hat{\mathbf{z}} \cdot \big(\mathcal{R}_L\, \mathcal{R}_{\mathbf{n}}(x)\, \mathbf{u}\big) \;=\; a_0 + a_c \cos x + a_s \sin x, \tag{2}$$

with coefficients

$$\begin{aligned} a_0 &= (\hat{\mathbf{z}} \cdot \mathcal{R}_L \mathbf{n})\,(\mathbf{n} \cdot \mathbf{u}), && (3)\\ a_c &= \hat{\mathbf{z}} \cdot \mathcal{R}_L \mathbf{u} \;-\; (\hat{\mathbf{z}} \cdot \mathcal{R}_L \mathbf{n})\,(\mathbf{n} \cdot \mathbf{u}), && (4)\\ a_s &= \hat{\mathbf{z}} \cdot \mathcal{R}_L (\mathbf{n} \times \mathbf{u}). && (5) \end{aligned}$$

Hence the output is a first-order trigonometric polynomial; it is constant iff $a_c = a_s = 0$.

Proof. By Rodrigues' formula [6],

$$\mathcal{R}_{\mathbf{n}}(x)\mathbf{u} = \mathbf{u} \cos x + (\mathbf{n} \times \mathbf{u}) \sin x + (\mathbf{n} \cdot \mathbf{u})\, \mathbf{n}\, (1 - \cos x).$$

Applying $\mathcal{R}_L$ and projecting onto $\hat{\mathbf{z}}$ gives

$$f(x) = \underbrace{\hat{\mathbf{z}} \cdot \mathcal{R}_L \mathbf{n}}_{d_z}(\mathbf{n} \cdot \mathbf{u}) + \Big(\hat{\mathbf{z}} \cdot \mathcal{R}_L \mathbf{u} - d_z (\mathbf{n} \cdot \mathbf{u})\Big) \cos x + \big(\hat{\mathbf{z}} \cdot \mathcal{R}_L (\mathbf{n} \times \mathbf{u})\big) \sin x, \tag{6}$$

which matches the stated form with the displayed coefficients.

Geometric Interpretation and Design Rule. The state traces a circle around $\mathbf{n}$ with radius $r = \|\mathbf{u} \times \mathbf{n}\| = \sqrt{1 - (\mathbf{u} \cdot \mathbf{n})^2}$, set entirely by the trainable choice of $\mathbf{u} = \mathcal{R}_R \hat{\mathbf{z}}$. The Fourier amplitudes (a_c, a_s) are the $\hat{\mathbf{z}}$-projections of the rotated circle's in-plane directions $\mathcal{R}_L \mathbf{u}$ and $\mathcal{R}_L(\mathbf{n} \times \mathbf{u})$, while a_0 is the projection of the circle's axis. Degeneracy occurs if $r = 0$ (i.e., $\mathbf{u} \parallel \mathbf{n}$) or if both in-plane directions are orthogonal to $\hat{\mathbf{z}}$. Thus, to avoid degenerate projections and preserve expressivity, choose trainable axes so that $\mathbf{u} \nparallel \mathbf{n}$ and $\hat{\mathbf{z}}$ is not orthogonal to the plane spanned by $\mathcal{R}_L \mathbf{u}$ and $\mathcal{R}_L(\mathbf{n} \times \mathbf{u})$.

For

$$U(x, \boldsymbol{\theta}) \;=\; W^{L+1}(\boldsymbol{\theta})\, S(x)\, W^{L}(\boldsymbol{\theta}) \cdots S(x)\, W^{1}(\boldsymbol{\theta}), \tag{7}$$

let $\mathcal{A}_\ell = \mathrm{Ad}_{W^\ell} \in SO(3)$ and write the induced Bloch evolution as

$$f_L(x) \;=\; \hat{\mathbf{z}} \cdot \left(\mathcal{A}_{L+1} \prod_{\ell=1}^{L} \mathcal{R}_{\mathbf{n}_\ell}(x)\, \mathcal{A}_\ell \right) \hat{\mathbf{z}}, \qquad \|\mathbf{n}_\ell\| = 1. \tag{8}$$

If $\boldsymbol{r}(x) \in \text{span}\{1, \sin kx, \cos kx : k \leq m\}$, then $\mathcal{R}_{\mathbf{n}}(x)\boldsymbol{r}(x)$ lies in $\text{span}\{1, \sin kx, \cos kx : k \leq m+1\}$ by product-to-sum identities applied to Rodrigues' form. Inducting over layers yields

$$f_L(x) \in \text{span}\{1, \sin x, \cos x, \ldots, \sin(Lx), \cos(Lx)\}.$$

Allowing layer-wise flexible axes $\{\mathbf{n}_\ell\}$ (set by trainable blocks) therefore sustains full expressivity up to harmonic order L, enabling accurate approximation of targets with rich high-frequency content.

Key Design Insights. We highlight key design insights. **i)** *Choose a non-collinear* $\mathbf{u}$ to $\mathbf{n}$: To avoid degeneracy, $\mathbf{u}$ must be designed such that it *must not be* collinear with the data-encoding axis $\mathbf{n}$. **ii)** *Avoid Degenerate Projections:* If the final rotation leaves the Bloch vector primarily parallel to the xz-plane, then projecting onto the yz-plane degenerates the circle to a line, which loses the trajectory's expressivity. These insights emphasize *flexible* design of rotation axes, which allows robust tilting of the rotating vector the circuit's expressiveness is well preserved.

4 Conclusion

In this paper, we analyze PQCs through the Bloch-sphere trajectories of the output qubit. Our theoretical and empirical results show that the trainable parameters govern the radius and orientation of this trajectory. We establish a rigorous link between the trajectory geometry, the measurement axis, and the resulting Fourier coefficients. Together, these findings connect circuit structure with geometric insight and yield practical guidance for realizing the full expressive potential of PQCs.

Acknowledgments. The complete version of this paper is currently under review at the Physical Review A. This is the extended abstract for the full version under the permission and guidance of the Purdue Quantum AI (PQAI) Symposium. This work was supported by the National Research Foundation of Korea (NRF) grant funded by the Korea government [MSIT (Ministry of Science and ICT (Information and Communications Technology))] (RS-2025-00561377).

References

1. Arute, F., et al.: Quantum supremacy using a programmable superconducting processor. Nature **574**(7779), 505–510 (2019)
2. Bausch, J.: Recurrent quantum neural networks. Adv. Neural. Inf. Process. Syst. **33**, 1368–1379 (2020)
3. Benedetti, M., Lloyd, E., Sack, S., Fiorentini, M.: Parameterized quantum circuits as machine learning models. Quant. Sci. Technol. **4**(4), 043001 (2019)
4. Biamonte, J., et al.: Quantum machine learning. Nature **549**(7671), 195–202 (2017)

5. Chen, S.Y.C., et al.: Variational quantum circuits for deep reinforcement learning. IEEE Access **8**, 141007–141024 (2020)
6. Dai, J.S.: Euler-Rodrigues formula variations, quaternion conjugation and intrinsic connections. Mech. Mach. Theor. **92**, 144–152 (2015)
7. Deutsch, D., Jozsa, R.: Rapid solution of problems by quantum computation. Proc. Roy. Soc. Lond. Ser. A Math. Phys. Sci. **439**(1907), 553–558 (1992)
8. Gawron, P., Lewiński, S.: Multi-spectral image classification with quantum neural network. In: Proceedings of the IEEE International Geoscience and Remote Sensing Symposium, pp. 3513–3516. Waikoloa, HI, USA (2020)
9. Grover, L.K.: A fast quantum mechanical algorithm for database search. In: Proceedings ACM Symposium on Theory of Computing (STOC), pp. 212–219. Philadelphia, PA, USA (1996)
10. Hur, T., Kim, L., Park, D.K.: Quantum convolutional neural network for classical data classification. Quantum Mach. Intell. **4**(1), 3 (2022)
11. Kharchenko, V.: Quantum Lie theory. Lecture Notes in Mathematics (2015)
12. Li, G., Ye, R., Zhao, X., Wang, X.: Concentration of data encoding in parameterized quantum circuits. Adv. Neural. Inf. Process. Syst. **35**, 19456–19469 (2022)
13. Nguyen, Q.T., et al.: Theory for equivariant quantum neural networks. PRX Quantum **5**(2), 020328 (2024)
14. Pérez-Salinas, A., Cervera-Lierta, A., Gil-Fuster, E., Latorre, J.I.: Data re-uploading for a universal quantum classifier. Quantum **4**, 226 (2020)
15. Schuld, M., Killoran, N.: Is quantum advantage the right goal for quantum machine learning? PRX Quantum **3**, 030101 (2022)
16. Schuld, M., Sinayskiy, I., Petruccione, F.: The quest for a quantum neural network. Quantum Inf. Process. **13**, 2567–2586 (2014)
17. Schuld, M., Sweke, R., Meyer, J.J.: Effect of data encoding on the expressive power of variational quantum-machine-learning models. Phys. Rev. A **103**, 032430 (2021)
18. Shor, P.W.: Polynomial-time algorithms for prime factorization and discrete logarithms on a quantum computer. SIAM Rev. **41**(2), 303–332 (1999)
19. Yu, Z., et al.: Non-asymptotic approximation error bounds of parameterized quantum circuits. Proc. Adv. Neural Inf. Process. Syst. 10–15 (2024)
20. Yu, Z., Yao, H., Li, M., Wang, X.: Power and limitations of single-qubit native quantum neural networks. Adv. Neural. Inf. Process. Syst. **35**, 27810–27823 (2022)

Accelerating Tensor Ring VQE with Batched Contraction

Keun Jun Park(✉), Dheeraj Peddireddy, and Vaneet Aggarwal

Purdue University, West Lafayette, IN 47907, USA
{park0925,dpeddire,vaneet}@purdue.edu
https://www.purdue.edu/

Abstract. The Variational Quantum Eigensolver (VQE) is a central method in quantum machine learning, with applications across disciplines such as finance, biostatistics, and related fields. Despite its impact, VQE suffers from limited scalability due to the exponential space complexity $O(2^N)$ required to store quantum states and compute expectation values, where N is the number of qubits. To address this, the Matrix Product State (MPS) VQE was developed, reducing storage complexity to $O(n\chi^2)$, where χ is the bond dimension. However, the bond dimension typically grows with entangling operations such as CNOT or SWAP, leading to large memory demands in highly entangled circuits. The Tensor Ring (TR) VQE improves upon this by fixing the bond dimension and connecting the first and last qubits, thereby avoiding unbounded memory growth and enabling more complex circuits. Nevertheless, TR VQE still requires exponential space complexity $O(2^N)$ to evaluate expectation values, which restricts its scalability. In this work, we propose a space-efficient TR VQE method for combinatorial optimization. Our key contribution is a batched expectation value calculation that leverages MPS-style contraction to evaluate multiple Pauli terms simultaneously. This batching improves efficiency while reducing expectation evaluation space complexity to $O(n\chi^4)$. Moreover, it increases only linear space complexity with respect to the number of Pauli terms, providing a scalable path toward practical quantum optimization.

Keywords: Quantum Machine Learning · Tensor Ring · Combinatorial Optimization · Variational Quantum Eigensolver

1 Introduction

The Variational Quantum Eigensolver (VQE) has emerged as one of the most promising hybrid quantum–classical algorithms for near-term quantum devices. By iteratively preparing parameterized quantum states and evaluating expectation values of a target Hamiltonian [1], VQE has demonstrated applications across diverse disciplines, including finance [2], biostatistics [5]. Its adaptability and relatively shallow circuit requirements have made it a cornerstone of

V. Aggarwal et al. (Eds.): PQAI 2025, CCIS 2724, pp. 42–50, 2026.
https://doi.org/10.1007/978-981-95-7829-0_5

quantum machine learning research. Yet, despite this broad impact, scalability remains a key limitation: classical simulation of VQE requires representing full quantum states of size 2^N for N qubits, making storage and expectation evaluation prohibitively expensive beyond intermediate system sizes. [11]

To address this challenge, tensor network methods provide compressed representations of quantum states. The Matrix Product State (MPS) reduces storage to $O(n\chi^2)$ by exploiting local entanglement, but the bond dimension χ can grow quickly under entangling gates, leading to substantial memory overhead [6]. The Tensor Ring (TR) representation mitigates this issue by fixing χ and connecting the first and last sites to form a closed loop, thereby bounding memory use and increasing expressivity [10]. However, TR VQE introduces a new difficulty: expectation value evaluation, in its naive form, still requires exponential resources $O(2^N)$ [7], limiting its scalability.

In this work, we propose a space-efficient Tensor Ring VQE (TR VQE) with a batched expectation value evaluation scheme for combinatorial optimzation. Our method adapts MPS-style contraction techniques to the TR setting, reducing the cost of expectation evaluation from exponential to polynomial $O(M\chi^4)$, where M is the number of Pauli terms. While this per-term cost is higher than the $O(\chi^3)$ scaling of MPS [6], the TR framework avoids uncontrolled bond growth, ensuring controllable memory requirements [10]. Moreover, batching enables simultaneous evaluation of many Pauli terms, a strategy that maps naturally to GPUs and other tensor-accelerated hardware. This balance of representation efficiency, bounded resources, and parallelizable evaluation makes TR VQE a practical framework for large-scale combinatorial optimization.

The remainder of this paper is organized as follows. Section 2 reviews related work on tensor-network-based VQE approaches. Section 3 details our batched contraction method and analyzes its complexity. Section 4 reports experimental results on the Max-Cut problem, and Sect. 5 concludes with future directions.

2 Related Work

Several extensions of the Variational Quantum Eigensolver (VQE) have been proposed to address its scalability limitations. MPS-based VQE (MPS-VQE) represents quantum states as Matrix Product States, reducing storage to $O(n\chi^2)$ and enabling efficient expectation value calculation with cost $O(n\chi^3)$ [6]. This extends feasible simulations but suffers from rapid bond-dimension growth under entangling operations, leading to large memory costs.

Tensor Ring (TR) representations were first applied in simulating variational quantum classifiers. By fixing the bond dimension and introducing a ring topology, TR avoids uncontrolled bond growth and improves expressivity. When applied to VQE, this results in TR-VQE, which maintains bounded memory but requires exponential $O(2^n)$ cost for expectation value calculation. [7]

The Noisy Tensor Ring approximation [8] extended TR-VQE to combinatorial optimization. It showed that reducing the bond dimension lowers space complexity while inducing approximation error that models NISQ-type noise.

Importantly, Peddireddy et al. also noted that expectation value calculation in TR-VQE could be reduced by adapting the MPS evaluation method established in tensor network literature [6].

In this work, we build directly on that idea by formalizing it into an explicit algorithm for expectation value calculation in TR-VQE. We further introduce batching, which reduces memory usage and enables efficient GPU parallelization. This makes TR-VQE more practical and scalable for combinatorial optimization.

3 Proposed Method

We introduce a space-efficient Tensor Ring Variational Quantum Eigensolver (TR VQE) with batched expectation value calculation. The approach extends the Matrix Product State (MPS) expectation evaluation scheme to the tensor ring representation, with modifications that enable efficient contraction and support batching across multiple Pauli terms. This combination reduces memory usage while maintaining accuracy, making TR VQE more scalable for combinatorial optimization.

Consider a parameterized quantum state $|\psi(\theta)\rangle$ represented as a tensor ring with bond dimension χ. The Hamiltonian is decomposed into Pauli terms [4],

$$H = \sum_{i=1}^{M} c_i P_i,$$

where each P_i is a tensor product of Pauli operators. The goal is to compute the expectation value [9]

$$\langle H \rangle = \langle \psi(\theta)|H|\psi(\theta)\rangle = \sum_{i=1}^{M} c_i \langle \psi(\theta)|P_i|\psi(\theta)\rangle.$$

Direct evaluation of this sum requires exponential cost $O(2^n)$ in the number of qubits [7]. To mitigate this, we employ a contraction procedure inspired by MPS methods, sequentially contracting transfer tensors around the ring. This avoids exponential summation and achieves polynomial scaling in χ. Algorithm 1 summarizes this contraction procedure.

Algorithm 1. Expectation via MPS-style Contraction on a Tensor Ring

Require: TR tensors $A_1, \ldots, A_N$ with $A_i \in \mathbf{C}^{\ell_i \times r_i \times d}$ and d=2; Pauli mask $P \in \{0,1\}^{T \times N}$ (1 = apply Z, 0 = apply I); coefficients $c \in \mathbf{C}^T$

Ensure: $E = \sum_{t=1}^{T} c_t \langle \psi \mid P_t \mid \psi \rangle$

$E_{\text{tot}} \leftarrow 0$
for $t = 1$ to T **do**
  $W \leftarrow$ UNSET
  for $i = 1$ to N **do**
    $O \leftarrow \begin{cases} Z, & \text{if } P_{t,i} = 1 \\ I, & \text{otherwise} \end{cases}$
    $AO \leftarrow \text{einsum}_{lrd,\ dk \to lrk}(A_i, O)$
    $E_i \leftarrow \text{einsum}_{l'r'k,\ lrk \to ll'rr'}(\text{conj}(A_i), AO)$
    if W is UNSET **then**
      $W \leftarrow E_i$
    else
      $W \leftarrow \text{einsum}_{ijab,\ abkl \to ijkl}(W, E_i)$
    end if
  end for
  $v \leftarrow \text{einsum}_{ikik \to}(W)$
  $E_{\text{tot}} \leftarrow E_{\text{tot}} + c_t \cdot v$
end for
return E_{tot}

This construction leads to higher space complexity than MPS. In MPS, the boundary bonds collapse directly to scalars, while in the tensor ring a diagonal summation is required to close the loop. As a result, the accumulated transfer tensor W has dimension χ^4, compared to the $O(\chi^3)$ scaling of MPS [6].

$$v = \sum_{i=1}^{\chi} \sum_{k=1}^{\chi} W_{i\,i\,k\,k} = \text{einsum}_{ikik \to}(\mathrm{W})$$

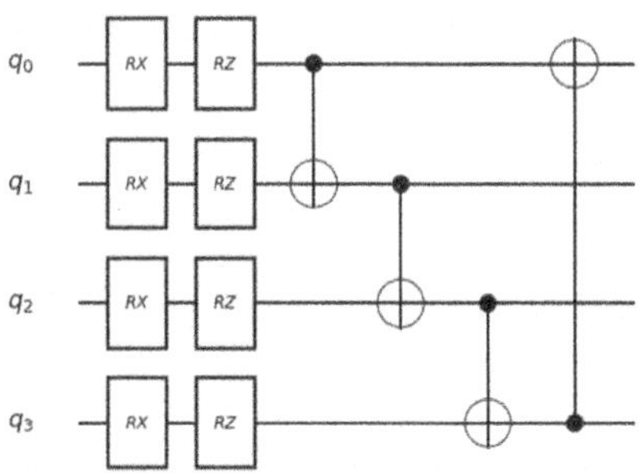

Fig. 1. One-depth instance of the parameterized ansatz used in benchmarks. Each qubit has R_x and R_z rotations, followed by a ring of CNOT gates closed with CX($N-1, 0$).

We next benchmark the naive TR contraction method on states prepared by a parameterized ansatz of depth $D = 5$ acting on N qubits. Each layer consists of alternating R_x and R_z rotations on all qubits followed by a ring of CNOTs closed with $\mathrm{CX}(N-1, 0)$. A one-depth instance of the circuit is shown in Fig. 1, while all experiments use depth $D = 5$. The reference state $|\psi(\theta)\rangle$ is obtained from Qiskit's `Statevector` simulator. To isolate contraction performance, we evaluate expectation values of a single Pauli term.

Figure 2 summarizes runtime and accuracy across four plots. Figure 2(a) plots the average time per expectation evaluation as the number of qubits n increases with bond dimension fixed at $\chi = 10$. Runtime grows approximately linearly with n, and notably, the TR method remains runnable up to $n = 1000$ qubits with an average evaluation time of about 5 seconds per expectation. Figure 2(b) plots runtime versus bond dimension χ at fixed $n = 10$, showing the expected rapid growth with χ due to cubic contraction scaling.

Accuracy is quantified as the absolute relative error of the expectation value with respect to the statevector ground truth,

$$\frac{|\langle H\rangle_{\mathrm{TR}} - \langle H\rangle_{\mathrm{SV}}|}{|\langle H\rangle_{\mathrm{SV}}|}.$$

Figure 2(c) shows this error versus n at $\chi = 10$: the error grows with system size since the fixed bond dimension cannot fully capture increasing entanglement. Figure 2(d) reports error versus χ at fixed $n = 10$: increasing χ systematically reduces the error, and for sufficiently large χ the TR result nearly coincides with the statevector ground truth. This trend is expected because the depth-5 circuit contains about 50 CNOT gates, generating strong entanglement. As a result, limited bond dimension truncates significant information, but convergence improves rapidly as χ grows.

While Algorithm 1 evaluates each Pauli term independently, Algorithm 2 introduces batching to reuse precomputed per-site transfers. This eliminates redundancy and exploits GPU parallelism through batched matrix multiplications, significantly reducing execution time. The performance of this batched scheme will be examined in the following section, where we compare naive TR contraction and batched TR contraction in terms of both runtime and approximation accuracy.

4 Experiment

We evaluate the proposed batched Tensor Ring VQE method on benchmark combinatorial optimization tasks, focusing on the MaxCut problem. All experiments are conducted on an NVIDIA GPU using PyTorch with circuit depth fixed to 1 and bond dimension set to $\chi = 10$.

MaxCut is a canonical NP-hard optimization task and a standard benchmark for variational quantum algorithms [3]. Given a graph $G = (V, E)$, the objective is to partition V into two subsets so as to maximize the number of crossing edges. The corresponding cost Hamiltonian is

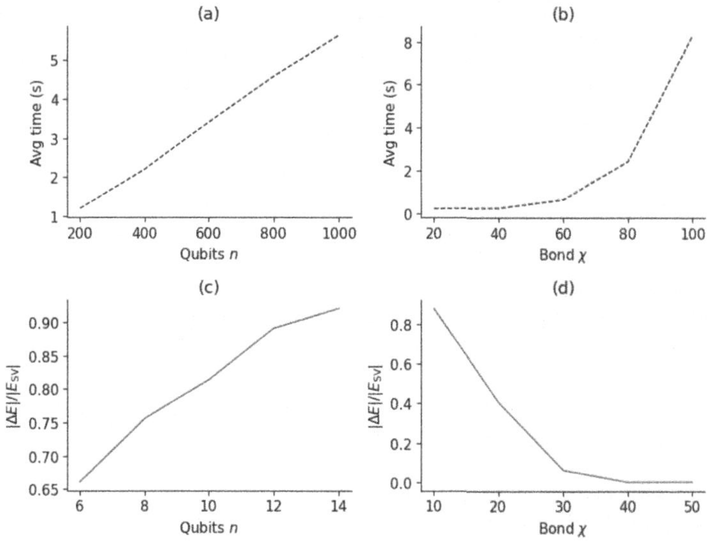

Fig. 2. Benchmark of the naive TR contraction method. (a) Average time per expectation evaluation vs. number of qubits n at fixed χ. (b) Average time per expectation evaluation vs. bond dimension χ at fixed n. (c) Absolute relative expectation error vs. number of qubits n, using the statevector simulation as ground truth. (d) Absolute relative expectation error vs. bond dimension χ, relative to the statevector ground truth.

$$H_{\text{MaxCut}} = \tfrac{1}{2} \sum_{(i,j)\in E} \big(I - Z_i Z_j\big),$$

where each edge (i, j) contributes a two-qubit Pauli term $Z_i Z_j$ [4]. We generate random instances with varying numbers of vertices n and edge densities. For runtime evaluation, we compare the standard TR VQE with naive contraction against the proposed batched TR VQE, both executed on GPU to isolate the batching effect. For accuracy evaluation, both TR variants are benchmarked against a StateVector VQE baseline on small-n instances. Following prior work [8], performance at iteration t is quantified by the approximation ratio

$$\alpha = \frac{M - \langle H(\theta_t) \rangle}{M - m},$$

where M and m denote the maximum and minimum eigenvalues of H. This metric normalizes results between $\alpha = 0$ (worst case) and $\alpha = 1$ (optimal).

Figure 3(a) shows that batching yields consistent speedups, exceeding 4× at $n = 30$ qubits. Figure 3(b) demonstrates that all methods achieve very similar approximation ratios, with Batched TR VQE matching Naive TR VQE accuracy while offering substantially reduced runtime. These results confirm that batching provides significant efficiency gains without loss of accuracy, making TR VQE more scalable to larger problem sizes.

Algorithm 2. Batched TR Expectation via MPS-style Contraction

Require: TR tensors $A_1, \dots, A_N$ with $A_i \in \mathbf{C}^{\ell_i \times r_i \times d}$ and d=2; Pauli mask $P \in \{0,1\}^{T \times N}$ (1 = apply Z, 0 = apply I); coefficients $c \in \mathbf{C}^T$
Ensure: $E = \sum_{t=1}^{T} c_t \langle \psi \mid P_t \mid \psi \rangle$
$E_{\text{tot}} \leftarrow 0$
$A \leftarrow A_1$
$AO_I \leftarrow \text{einsum}_{lrd,\, dk \to lrk}(A, O_I)$; $AO_Z \leftarrow \text{einsum}_{lrd,\, dk \to lrk}(A, O_Z)$
$E_I \leftarrow \text{einsum}_{l'r'k,\, lrk \to l'lr'r}(\text{conj}(A), AO_I)$
$E_Z \leftarrow \text{einsum}_{l'r'k,\, lrk \to l'lr'r}(\text{conj}(A), AO_Z)$
$p \leftarrow \text{reshape}(P[:, 1], T, 1, 1, 1, 1)$
$\mathsf{W} \leftarrow \text{where}(p, E_Z, E_I)$
for $i \leftarrow 2$ **to** N **do**
    $A \leftarrow A_i$
    $AO_I \leftarrow \text{einsum}_{lrd,\, dk \to lrk}(A, O_I)$; $AO_Z \leftarrow \text{einsum}_{lrd,\, dk \to lrk}(A, O_Z)$
    $E_I \leftarrow \text{einsum}_{l'r'k,\, lrk \to l'lr'r}(\text{conj}(A), AO_I)$
    $E_Z \leftarrow \text{einsum}_{l'r'k,\, lrk \to l'lr'r}(\text{conj}(A), AO_Z)$
    $p \leftarrow \text{reshape}(P[:, i], T, 1, 1, 1, 1)$
    $E \leftarrow \text{where}(p, E_Z, E_I)$
    $\mathsf{W} \leftarrow \text{einsum}_{bijpq,\, bpqrs \to bijrs}(\mathsf{W}, E)$
end for
$v \leftarrow \text{einsum}_{bijij \to b}(\mathsf{W})$
$E_{\text{tot}} \leftarrow \sum_{b=1}^{B} c_b\, v_b$
return E_{tot}

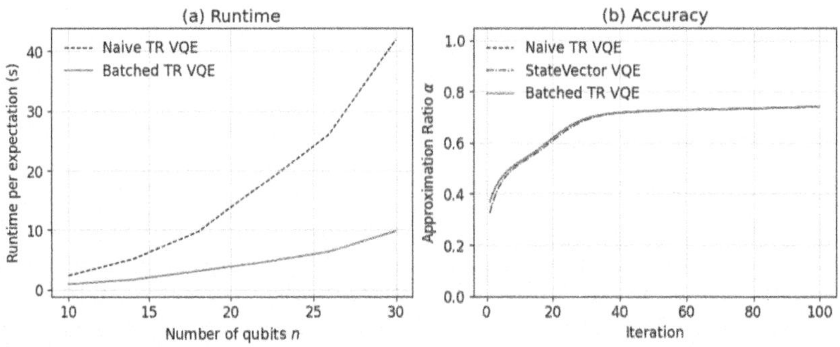

Fig. 3. Results with circuit depth fixed to 1 and bond dimension $\chi = 5$. (a) Runtime per expectation evaluation as a function of qubit number n, comparing Naive TR VQE and Batched TR VQE on GPU. (b) Approximation ratio α versus optimizer iterations, comparing Naive TR VQE, Batched TR VQE, and the StateVector baseline. Batched TR VQE achieves accuracy comparable to StateVector VQE while significantly reducing runtime.

5 Conclusion and Future Work

In this work, we presented a space-efficient Tensor Ring VQE (TR VQE) method with batched expectation value evaluation. By adapting MPS-style contraction techniques to the tensor ring structure, our approach enables multiple Pauli terms to be evaluated simultaneously. This batching strategy reduces redundant computations and effectively leverages GPU acceleration for large-scale tensor operations.

Through experiments on MaxCut instances, we demonstrated that the proposed method delivers significant runtime improvements while preserving accuracy comparable to both the naive TR contraction and the Statevector VQE baseline. These results highlight batching as a practical strategy for scaling hybrid quantum–classical algorithms to larger qubit systems and denser Hamiltonians.

Several promising directions remain for future research. First, extending the batching strategy to cover gradient computations would further accelerate VQE training and optimization. Second, exploring additional Pauli operators beyond $\{I, Z\}$ in the precomputation scheme could broaden applicability to more general Hamiltonians. Finally, applying the approach to other combinatorial optimization problems beyond MaxCut could validate its generality and effectiveness.

Overall, this work contributes a step toward making tensor-network-based VQE methods more scalable and practical, and we hope it will motivate further research at the intersection of quantum optimization, tensor networks, and high-performance computing.

References

1. Cerezo, M., Sharma, K., Arrasmith, A., Coles, P.J.: Variational quantum state eigensolver. npj Quant. Inf. **8**(1), 1–11 (2022). https://doi.org/10.1038/s41534-022-00611-6. https://www.nature.com/articles/s41534-022-00611-6
2. Egger, D.J., et al.: Quantum computing for finance: state-of-the-art and future prospects. IEEE Trans. Quant. Eng. **1**, 1–24 (2020)
3. Glover, F., Kochenberger, G., Hennig, R., Du, Y.: Quantum bridge analytics I: a tutorial on formulating and using QUBO models. Ann. Oper. Res. **314**(1), 141–183 (2022)
4. Lucas, A.: Ising formulations of many NP problems. Front. Phys. **2** (2014). https://doi.org/10.3389/fphy.2014.00005. http://journal.frontiersin.org/article/10.3389/fphy.2014.00005/abstract
5. McWeeney, S., Perciano, T., Susut, C., Chatterjee, L., Fornari, M., Biven, L., Siwy, C.: Quantum Computing for Biomedical Computational and Data Sciences: A Joint DOE-NIH Roundtable. Technical Report. None, 2228574 (2023). https://doi.org/10.2172/2228574. https://www.osti.gov/servlets/purl/2228574/
6. Orús, R.: A practical introduction to tensor networks: matrix product states and projected entangled pair states. Ann. Phys. **349**, 117–158 (2014)
7. Peddireddy, D., Bansal, V., Aggarwal, V.: Classical simulation of variational quantum classifiers using tensor rings. Appl. Soft Comput. **141**, 110308 (2023)
8. Peddireddy, D., Priyam, U., Aggarwal, V.: Noisy tensor-ring approximation for computing gradients of a variational quantum eigensolver for combinatorial optimization. Phys. Rev. A **108**(4), 042429 (2023)
9. Tilly, J., et al.: The variational quantum eigensolver: a review of methods and best practices. Phys. Rep. **986**, 1–128 (2022)

10. Wang, W., Aggarwal, V., Aeron, S.: Efficient low rank tensor ring completion. In: 2017 IEEE International Conference on Computer Vision (ICCV), pp. 5698–5706. IEEE, Venice (Oct 2017). https://doi.org/10.1109/ICCV.2017.607. http://ieeexplore.ieee.org/document/8237869/
11. Zhou, Y., Stoudenmire, E.M., Waintal, X.: What limits the simulation of quantum computers? Phys. Rev. X **10**(4), 041038 (2020)

Satellite Constellation Scheduling with Quantum Reinforcement Learning

Gyu Seon Kim[1], Samuel Yen-Chi Chen[2], Soohyun Park[3], and Joongheon Kim[1(✉)]

[1] Korea University, Seoul 02841, Korea
{kingdom0545,joongheon}@korea.ac.kr
[2] Wells Fargo, New York, NY 10017, USA
yen-chi.chen@wellsfargo.com
[3] Sookmyung Women's University, Seoul 04310, Korea
soohyun.park@sookmyung.ac.kr

Abstract. Reinforcement learning (RL) utilizing classical neural networks (NNs) has achieved substantial progress across numerous application domains. Nevertheless, classical RL faces significant challenges in training when applied to systems with high-dimensional action spaces, such as coordinated satellite networks. In these scenarios, the exponential increase in model parameters imposes heavy computational requirements, thereby constraining scalability and slowing convergence. Quantum reinforcement learning (QRL), which incorporates quantum neural networks (QNNs), provides a promising alternative by exploiting quantum mechanical properties, including superposition and entanglement. Through QNNs, multiple states can be compactly represented using a limited number of quantum bits (qubits), effectively reducing computational overhead. Due to these characteristics—namely, rapid convergence and enhanced scalability—QRL constitutes a suitable substitute for classical RL in coordinated satellite applications. Moreover, the proposed QRL framework mitigates the curse of dimensionality by efficiently leveraging qubit-based representations. In experimental environments characterized by high-dimensional action spaces, the proposed algorithm demonstrates superior training performance compared to conventional RL approaches.

Keywords: Quantum Reinforcement Learning (QRL) · Quantum Neural Network (QNN) · Satellite Communication

1 Introduction

Reinforcement learning (RL) with classical neural networks (NNs) has demonstrated substantial progress in various domains [2,6,11,12,17,24]. However, it encounters significant limitations when applied to high-dimensional environments such as coordinated satellite networks [8,9]. As the dimensionality of state

V. Aggarwal et al. (Eds.): PQAI 2025, CCIS 2724, pp. 51–56, 2026.
https://doi.org/10.1007/978-981-95-7829-0_6

and action spaces increases, the number of trainable parameters expands exponentially, leading to elevated computational demands, data sparsity, and slow convergence—an issue widely recognized as the *curse of dimensionality* [5,23]. Quantum reinforcement learning (QRL) and quantum multi-agent reinforcement learning (QMARL) present a compelling alternative [10,15,22]. By utilizing quantum neural networks (QNNs), quantum AI leverages fundamental principles such as *superposition* and *entanglement*, which enable compact representations of high-dimensional data and efficient model training [16,20]. With relatively few qubits, QNNs can simultaneously encode multiple states, thereby alleviating scalability constraints and expediting convergence [14,21]. These characteristics render QRL and QMARL particularly advantageous for coordinated satellite applications, where an increasing number of agents contribute to the expansion of action dimensions. Through basis measurement techniques, QNNs extend output dimensionality with a minimal number of qubits, offering both *accelerated convergence* and *enhanced scalability*. This paper introduces the structural design of QNNs and their application within QRL and QMARL for satellite scheduling, emphasizing the role of basis measurement in addressing dimensionality challenges and exploring the broader potential of these methods.

2 Quantum Neural Network Architecture

The architecture of the QNN is organized into three principal phases: *i) state encoding*, *ii) parameterized quantum circuit (PQC)*, and *iii) measurement* [3].

State Encoding. In QRL, the phase referred to as *state encoding* entails transforming environmental states—traditionally represented as vectors in classical reinforcement learning—into quantum states suitable for computation on quantum devices. In essence, this process encodes classical state information into the quantum domain. This step is pivotal to realizing the potential computational advantage of quantum systems, as it governs the ability to represent and manipulate complex environments. Effective encoding enables more efficient and precise processing of environmental states, exploiting quantum phenomena such as *entanglement* and *superposition*, thereby facilitating the resolution of problems that exceed the capabilities of classical reinforcement learning algorithms. Among available approaches, angle encoding introduces classical data into qubits through rotational parameters of quantum gates, such as R_X, R_Y, and R_Z [18].

Parameterized Quantum Circuits. The parameterized quantum circuit (PQC) constitutes the central computational component of a QNN in QRL, serving an analogous role to neurons and synaptic connections in classical neural networks [7,16]. The PQC is composed of quantum gates parameterized by adjustable variables, which are optimized during the training process. In QRL, these circuits transform the encoded quantum state into a new quantum representation that approximates policy or value functions relevant to reinforcement learning tasks [13,19]. The learnable parameters within the PQC are equivalent to weights in classical neural networks and are optimized through interaction with the environment to enhance policy performance. PQC design typically integrates both *rotational gates*, such as R_X, R_Y, and R_Z, and *entangling gates*, such as the controlled-NOT (CNOT) operation, which collectively manipulate the quantum state. The effectiveness of QRL training is therefore strongly influenced by the configuration and choice of gates within the PQC [1].

Measurement. The measurement phase in QRL represents the process of translating quantum states, which have been evolved within the PQC, back into classical information. In practice, this corresponds to decoding a quantum state into a classical action distribution. The extracted information is subsequently utilized to determine the *actions* executed within the environment. Measurement is fundamental for converting the results of quantum computation into actionable data for decision-making. Upon measurement, the quantum state collapses into one of its basis states, with the outcome governed by the probability distribution generated by prior quantum operations. The measurement outcome is interpreted as the action or set of actions to be applied, thereby completing the decision-making cycle within the reinforcement learning framework.

3 Design of the Quantum Reinforcement Learning Framework for Large-Scale Satellite Scheduling

The Pauli-Z measurement evaluates *individual* qubits in quantum states through the Pauli-Z operator, defined as, $\begin{bmatrix} 1 & 0 \\ 0 & -1 \end{bmatrix}$, where the columns correspond to the computational basis states $|0\rangle$ and $|1\rangle$. Nevertheless, in an environment involving $\mathcal{W}$ coordinated satellites, a total of $2^{\mathcal{W}}$ qubits would still be required to represent the $2^{\mathcal{W}}$ action dimensions necessary for combinatorial scheduling decisions. As a result, the classical challenge referred to as the "curse of dimensionality" persists, since this form of measurement does not alleviate the exponential complexity that arises with the increasing number of coordinated satellites [4].

By contrast, basis measurement enables computation of the probabilities across all $2^{\mathcal{W}}$ combinations using only $\mathcal{W}$ qubits. This is achieved by performing measurements across the full set of $2^{\mathcal{W}}$ basis states, expressed as

$$\{|\mathbb{P}(\mathcal{M}_k)\rangle\}_{k=1}^{2^{\mathcal{W}}} \triangleq \left\{ \bigotimes_{k=1}^{\mathcal{W}} |\Omega_j^i\rangle \right\}, \tag{1}$$

where the Ω_j^i denotes the selection vector of the i-th control tower for the j-th satellite, with $\forall \Omega_j^i \in \{0,1\}$ and $\forall j \in [1, \mathcal{W}]$. In (1), the $\mathcal{M}_k$ represents the action of the agents. Accordingly, the probability that the i-th control tower selects the k-th action among $2^{\mathcal{W}}$ possible combinations at time t, according to its policy, is given by

$$\pi(\mathcal{M}_k(t)|\varpi_i(t); \boldsymbol{\rho}_i) = \langle\psi|\nu_k\rangle\langle\nu_k|\psi\rangle = |\langle\psi|\nu_k\rangle|^2 = |\alpha_k|^2, \tag{2}$$

where $|\nu_k\rangle\langle\nu_k|$ is the projector associated with the k-th basis, and the complete set of projectors is given by $|\nu_k\rangle\langle\nu_k|_{k=1}^{2^{\mathcal{W}}}$. In (2), the π, $\varpi_i(t)$, $\boldsymbol{\rho}_i$, and α_k denote the policy, state space of i-th agent, training parameter of i-th agent, and probability amplitude, respectively. Because the probability values for each action correspond to distinct outputs and collectively form a normalized distribution, the summation of probabilities over all actions satisfies

$$\sum_{k=1}^{2^{\mathcal{W}}} \pi(\mathcal{M}_k(t)|\varpi_i(t); \boldsymbol{\rho}_i) = 1. \tag{3}$$

4 Conclusion

This study demonstrates that QRL effectively overcomes the limitations of conventional RL when applied to environments characterized by extensive action spaces, such as coordinated satellite networks. The distinctive strengths of QRL, particularly its rapid convergence and high scalability, underscore its suitability for deployment in large-scale and complex operational scenarios. Looking ahead, QMARL presents substantial potential for broader application across diverse systems.

Acknowledgments. The full version of this work was presented at the IEEE International Conference on Acoustics, Speech, and Signal Processing (ICASSP), held in Hyderabad, India, in April 2025. This is the extended abstract for the full version under the permission and guidance of the Purdue Quantum AI (PQAI) Symposium. This research was supported by the MSIT (Ministry of Science and ICT), Korea, under the ITRC (Information Technology Research Center) support program (IITP-2024-RS-2024-00436887) supervised by the IITP (Institute for Information & Communications Technology Planning & Evaluation); and also by IITP grant funded by MSIT (RS-2024-00439803, SW Star Lab). The corresponding authors of this paper are Prof. Soohyun Park and Prof. Joongheon Kim (e-mail: soohyun.park@sookmyung.ac.kr, joongheon@korea.ac.kr).

References

1. Abbas, A., Sutter, D., Zoufal, C., Lucchi, A., Figalli, A., Woerner, S.: The power of quantum neural networks. Nat. Comput. Sci. **1**(6), 403–409 (2021)
2. Cao, Z., Wong, K., Lin, C.T.: Weak human preference supervision for deep reinforcement learning. IEEE Trans. Neural Netw. Learn. Syst. **32**(12), 5369–5378 (2021)
3. Chu, C., Jiang, L., Swany, M., Chen, F.: Qtrojan: a circuit backdoor against quantum neural networks. In: Proc. IEEE International Conference on Acoustics, Speech and Signal Processing (ICASSP), pp. 1–5. Rhodes Island, Greece (2023)

4. Cicek, C.T.: A reinforcement learning algorithm for data collection in UAV-aided IoT networks with uncertain time windows. In: Proc. IEEE International Conference on Communications Workshops (ICC Workshops), pp. 1–6. Montreal (2021)
5. Du, W., Ding, S.: A survey on multi-agent deep reinforcement learning: from the perspective of challenges and applications. Artif. Intell. Rev. **54**(5), 3215–3238 (2020)
6. He, Z., Li, J., Wu, F., Shi, H., Hwang, K.S.: DeRL: coupling decomposition in action space for reinforcement learning task. IEEE Trans. Emerging Top. Comput. Intell. **8**(1), 1030–1043 (2024)
7. Killoran, N., Bromley, T.R., Arrazola, J.M., Schuld, M., Quesada, N., Lloyd, S.: Continuous-variable quantum neural networks. Phys. Rev. Res. **1**(3), 33063–33084 (2019)
8. Kim, G.S., et al.: Quantum multi-agent reinforcement learning for cooperative mobile access in space-air-ground integrated networks. IEEE Trans. Mobile Comput., 1–18 (2025)
9. Kim, G.S., Cho, Y., Park, S., Jung, S., Kim, J.: Quantum multi-agent reinforcement learning for joint cube-satellites and high-altitude long-endurance aerial vehicles in SAGIN. IEEE Trans. Aerosp. Electron. Syst. **61**(4), 9490–9510 (2025)
10. Kim, G.S., Chung, J., Park, S.: Realizing stabilized landing for computation-limited reusable rockets: a quantum reinforcement learning approach. IEEE Trans. Veh. Technol. **73**(8), 12252–12257 (2024)
11. Kiran, B.R., et al.: Deep reinforcement learning for autonomous driving: a survey. IEEE Trans. Intell. Transp. Syst. **23**(6), 4909–4926 (2022)
12. Lee, D., Kwon, M.: Episodic future thinking with offline reinforcement learning for autonomous driving. IEEE Internet Things J. **12**(11), 17012–17023 (2025)
13. Lockwood, O., Si, M.: Reinforcement learning with quantum variational circuit. In: Proc. AAAI Conference on Artificial Intelligence and interactive digital entertainment, pp. 245–251. Virtual (2020)
14. Ma, H., Dong, D., Ding, S.X., Chen, C.: Curriculum-based deep reinforcement learning for quantum control. IEEE Trans. Neural Netw. Learn. Syst. **34**(11), 8852–8865 (2023)
15. Meyer, N., Ufrecht, C., Periyasamy, M., Scherer, D.D., Plinge, A., Mutschler, C.: A survey on quantum reinforcement learning. arXiv preprint arXiv:2211.03464 (2022)
16. Mitarai, K., Negoro, M., Kitagawa, M., Fujii, K.: Quantum circuit learning. Phys. Rev. A **98**(3), 32309–32314 (2018)
17. Park, S., Kim, G.S., Jung, S., Kim, J.: Markov decision policies for distributed angular routing in LEO mobile satellite constellation networks. IEEE Internet Things J. **11**(23), 38744–38754 (2024)
18. Shindi, O., Yu, Q., Girdhar, P., Dong, D.: Model-free quantum gate design and calibration using deep reinforcement learning. IEEE Trans. Artifi. Intell. **5**(1), 346–357 (2024)
19. Skolik, A., Jerbi, S., Dunjko, V.: Quantum agents in the gym: a variational quantum algorithm for deep Q-learning. Quantum **6**, 720–745 (2022)
20. Wang, T., Tseng, H., Yoo, S.: Quantum federated learning with quantum networks. In: Proc. IEEE International Conference on Acoustics, Speech and Signal Processing (ICASSP), pp. 13401–13405. Seoul, Republic of Korea (2024)
21. Yu, H., Zhao, X.: Deep reinforcement learning with reward design for quantum control. IEEE Trans. Artif. Intell. **5**(3), 1087–1101 (2024)
22. Yun, W.J., et al.: Quantum multi-agent reinforcement learning via variational quantum circuit design. In: Proc. IEEE International Conference on Distributed Computing Systems (ICDCS), pp. 1332–1335. Bologna, Italy (2022)

23. Zhu, L., Cui, Y., Takami, G., Kanokogi, H., Matsubara, T.: Scalable reinforcement learning for plant-wide control of vinyl acetate monomer process. Control. Eng. Pract. **97**, 104331–104340 (2020)
24. Zhu, Z., Lin, K., Jain, A.K., Zhou, J.: Transfer learning in deep reinforcement learning: a survey. IEEE Trans. Pattern Anal. Mach. Intell. **45**(11), 13344–13362 (2023)

Accelerating Quantum Reinforcement Learning with a Quantum Natural Policy Gradient Based Approach

Yang Xu(✉) and Vaneet Aggarwal

Purdue University,West Lafayette, IN, USA
{xu1720,vaneet}@purdue.edu

Abstract. This extended abstract summarizes [10], and cites it as the primary source. We present a quantum-compatible reformulation of Natural Policy Gradient (NPG) using deterministic, truncation-based estimators for both the policy gradient and Fisher information, enabling coherent evaluation with standard quantum environment and policy oracles. Leveraging quantum mean estimation and variance reduction within a classical–quantum double-loop scheme, the resulting Quantum NPG (QNPG) achieves a sample complexity of $\tilde{\mathcal{O}}(\epsilon^{-1.5})$ to reach an ϵ-optimal policy, improving upon the classical $\Omega(\epsilon^{-2})$ lower bound for policy-gradient methods. We outline the required assumptions (smooth score function, Fisher non-degeneracy, bounded compatible approximation error), highlight the exponentially decaying truncation bias, and sketch the convergence analysis that yields the overall complexity. This manuscript is intended as a concise, approximately six-page overview. We refer the full details, proofs, and extended discussion to [10].

Keywords: Quantum reinforcement learning · Natural policy gradient · Quantum mean estimation · Sample complexity

1 Introduction

Reinforcement learning (RL) is a powerful framework for sequential decision-making with applications ranging from robotics to healthcare. Among policy optimization methods, policy gradient (PG) algorithms, and in particular the Natural Policy Gradient (NPG), are well known for their stability and sample efficiency. However, the sample complexity of NPG is lower bounded by $\Omega(\epsilon^{-2})$ [2] for achieving an ϵ-optimal policy in the classical setting of infinite-horizon discounted setting. As RL continues to scale to large, complex environments, the possibility of breaking the theoretical classical lower bound and reducing this complexity have become a challenge.

Quantum computing offers a promising route for accelerating RL. Quantum mean estimation, amplitude amplification, and related primitives have yielded speedups in multi-armed bandits [9], stochastic optimization [8], and tabular

V. Aggarwal et al. (Eds.): PQAI 2025, CCIS 2724, pp. 57–63, 2026.
https://doi.org/10.1007/978-981-95-7829-0_7

RL [4,6]. Yet, extending these benefits to RL with parameterized policies has remained elusive, leaving open the question of whether quantum advantages can be realized for model-free algorithms in general parameterized MDPs.

This work develops the first quantum-compatible version of NPG, called *Quantum Natural Policy Gradient (QNPG)*. Classical NPG estimators depend on random-length trajectories drawn from a geometric distribution, which are infeasible to encode coherently in quantum systems. To address this, we introduce a deterministic truncation-based procedure that yields quantum-compatible estimators with exponentially decaying bias. Combined with quantum variance reduction via mean estimation, this leads to a provable improvement in query complexity from $\tilde{\mathcal{O}}(\epsilon^{-2})$ to $\tilde{\mathcal{O}}(\epsilon^{-1.5})$.

1.1 Contributions

The novelties of this work, distinguishing it from both classical RL and prior quantum RL results, are:

- **Quantum-compatible NPG:** First construction that embeds NPG updates into quantum states using only standard environment and policy oracles, enabling coherent quantum subroutines.
- **Deterministic truncation estimators:** Replacement of random-length rollouts by fixed-length truncations, ensuring implementability on quantum devices while guaranteeing exponentially decaying bias.
- **Quantum variance reduction:** Incorporation of variance-reduced estimators via quantum mean estimation, giving a quadratic improvement over classical averaging.
- **Provable quantum speedup:** QNPG achieves $\tilde{\mathcal{O}}(\epsilon^{-1.5})$ query complexity, surpassing the $\tilde{\mathcal{O}}(\epsilon^{-2})$ classical lower bound for policy gradient methods.

2 Background

We consolidate the main preliminaries on quantum oracles, policy gradients, and the challenges of quantum RL aiming to provide foundation to follow the method and analysis without reproducing full technical details.

2.1 Policy Gradients and Natural Gradient

An infinite-horizon discounted MDP $M = (\mathcal{S}, \mathcal{A}, r, P, \gamma, \rho)$ is defined by state space $\mathcal{S}$, action space $\mathcal{A}$, deterministic reward function r, transition kernel P, discount factor γ, and initial state distribution ρ. For a parameterized policy π_θ with policy parameter $\theta \in \Theta$, objective the objective is defined as

$$J_\rho(\theta) = \mathbb{E}_{s\sim\rho}[V^{\pi_\theta}(s)] \quad \text{where} \quad V^{\pi_\theta}(s) = \mathbb{E}\left[\sum_{t=0}^{\infty} \gamma^t r(s_t, a_t) \middle| s_0 = s\right]$$

Define the Q function and the advantage function as follows:

$$Q^{\pi_\theta}(s,a) = \mathbb{E}\left[\sum_{t=0}^{\infty} \gamma^t r(s_t,a_t)\middle| s_0 = s, a_0 = a\right], \; A^{\pi_\theta}(s,a) - Q^{\pi_\theta}(s,a) = V^{\pi_\theta}(s)$$

Then the policy gradient is expressed as

$$\nabla_\theta J_\rho(\theta) = \frac{1}{1-\gamma} \mathbb{E}_{(s,a)\sim\nu_{\pi_\theta,\rho}} \left[A^{\pi_\theta}(s,a)\nabla_\theta \log \pi_\theta(a|s)\right],$$

where $\nu_{\pi_\theta,\rho}$ is the discounted stateâĂŞaction occupancy and A^{π_θ} is the advantage function. NPG preconditions this gradient with the Fisher information matrix

$$F_\rho(\theta) = \mathbb{E}_{(s,a)\sim\nu_{\pi_\theta,\rho}}\big[\nabla_\theta \log \pi_\theta(a|s) \otimes \nabla_\theta \log \pi_\theta(a|s)\big],$$

yielding the update $\theta_{k+1} = \theta_k + \eta F_\rho(\theta_k)^\dagger \nabla_\theta J_\rho(\theta_k)$.

2.2 Quantum Oracles and Quantum Mean Estimation

We assume access to two standard quantum environment oracles:

- The *quantum transition oracle* U_P, which given state s and action a returns a coherent superposition over next states with amplitudes $\sqrt{P(s'|s,a)}$.
- The *quantum initial state oracle* U_ρ, which prepares a superposition over initial states sampled from ρ.

A parameterized policy π_θ can likewise be encoded in a unitary Π, producing a superposition of actions $\sqrt{\pi_\theta(a|s)}$ for a given state s. These oracles allow us to generate quantum superpositions of fixed-length trajectories, which are essential for our algorithm.

We then utilize quantum mean estimation [3,8] to perform variance reduction, which improves the convergence rate of sample averages from the classical $\mathcal{O}(1/\sqrt{n})$ to $\mathcal{O}(1/n)$ via amplitude amplification. This quadratic gain is crucial for variance reduction in QNPG.

2.3 Challenges for Quantum RL

To summarize, classical NPG algorithms rely on unbiased estimators constructed from random-length trajectories, but this randomness cannot be directly encoded into quantum systems. Three obstacles arise:

1. **Trajectory encoding:** Random-length rollouts are infeasible to represent coherently.
2. **Bias control:** Deterministic truncation introduces bias that must be shown to decay exponentially with horizon length.
3. **Variance reduction:** To exploit quantum advantages, one must reduce estimator variance without relying on classical averaging.

Our QNPG algorithm is explicitly designed to address the above challenges.

3 Quantum Natural Policy Gradient Method

We now present our Quantum Natural Policy Gradient (QNPG) algorithm. The key idea is to replace the classical use of stochastic trajectory lengths with deterministic truncation, making the estimators quantum-compatible, and then apply quantum mean estimation for variance reduction.

3.1 Deterministic Truncation Estimators

Classical NPG algorithms [1,5,7] estimate the policy gradient and Fisher information matrix by sampling trajectories whose lengths follow a geometric distribution with mean $(1-\gamma)^{-1}$. While this ensures unbiasedness, it is not feasible in quantum settings since random trajectory lengths cannot be coherently encoded.

To address this, we fix a truncation horizon N and define the estimators:

$$\hat{g}_\rho(\tau_N|\theta) = \sum_{n=0}^{N-1} \Big(\sum_{t=0}^{n} \nabla_\theta \log \pi_\theta(a_t|s_t) \Big) \big(\gamma^n r(s_n, a_n)\big), \tag{1}$$

$$\hat{F}_\rho(\tau_N|\theta) = (1-\gamma) \sum_{n=0}^{N-1} \gamma^n \nabla_\theta \log \pi_\theta(a_n|s_n) \otimes \nabla_\theta \log \pi_\theta(a_n|s_n). \tag{2}$$

Here $\tau_N = (s_0, a_0, \ldots, s_{N-1}, a_{N-1})$ is a trajectory of fixed length N sampled using the quantum oracles U_P and U_ρ together with the policy unitary Π.

These truncated estimators are biased relative to their infinite-horizon counterparts, but the bias decays exponentially as $\mathcal{O}(\gamma^N)$. Choosing $N = \mathcal{O}(\log(1/\epsilon))$ ensures the bias remains below the target accuracy.

3.2 Quantum Variance Reduction

Even with truncation, estimators may suffer from high variance. We leverage *quantum mean estimation* to reduce variance quadratically compared to classical averaging. In particular, given access to the quantum states encoding $\hat{g}_\rho(\tau_N|\theta)$ and $\hat{F}_\rho(\tau_N|\theta)$ across superpositions of trajectories, quantum mean estimation yields variance-reduced versions $\tilde{g}$ and $\tilde{F}$ with only $\tilde{\mathcal{O}}(1/\epsilon)$ queries, as opposed to the classical $\Omega(\epsilon^{-2})$. This is essential for achieving the overall sample complexity that breaks the classical barrier of $\Omega(\epsilon^{-2})$.

3.3 Algorithm Structure

QNPG combines the above ingredients into a double-loop scheme: a classical outer loop over policy updates and a quantum inner loop implementing variance-reduced stochastic gradient descent. The outer loop ensures convergence at the same rate as classical NPG, while the inner loop provides the quantum speedup.

Algorithm 1. Quantum Natural Policy Gradient (QNPG)

Input: Initial parameter θ_0, horizon N, steps K, H, learning rates η, α
for $k = 0$ to $K - 1$ **do**
  Initialize $\omega_0 \leftarrow 0$
  for $h = 0$ to $H - 1$ **do**
    Sample quantum superpositions of τ_N via U_P, U_ρ, Π
    Compute truncated estimators $\hat{g}, \hat{F}$ as in (1)–(2)
    Apply quantum variance reduction to obtain $\tilde{g}, \tilde{F}$
    Update inner variable: $\omega_{h+1} \leftarrow \omega_h - \alpha(\tilde{F}\omega_h - \tilde{g})$
  end for
  Outer update: $\theta_{k+1} \leftarrow \theta_k + \eta\,\omega_H$
end for

3.4 Discussion

The algorithm can be interpreted as a quantum-accelerated analogue of classical NPG:

- The outer loop mirrors classical policy updates and requires no quantum speedup.
- The mini-batch in the inner loop is where quantum mean estimation yields a quadratic improvement in variance reduction, cutting the overall query complexity.
- Truncation bias decays exponentially in N, and setting $N = \mathcal{O}(\log(1/\epsilon))$ balances bias and variance.

Together, these elements yield the first quantum-compatible NPG algorithm with provable complexity guarantees.

4 Convergence and Complexity Analysis

We now summarize the theoretical guarantees of QNPG. Full proofs are given in [10]; here we focus on the key assumptions, the main result, and its intuition.

4.1 Assumptions

We adopt standard conditions used in policy gradient analysis:

1. **Smooth score function:** $\nabla_\theta \log \pi_\theta(a|s)$ is G-Lipschitz and B-smooth in θ.
2. **Fisher non-degeneracy:** The Fisher matrix $F_\rho(\theta)$ has minimum eigenvalue at least $\mu > 0$ on the relevant subspace.
3. **Bounded approximation error:** The compatible function approximation error is uniformly bounded by δ.

These ensure that the policy gradient is well defined, updates are stable, and NPG iterations converge to a stationary point.

4.2 Main Theorem

Theorem 1 (Informal). *Under the above assumptions, with truncation horizon $N = \mathcal{O}(\log(1/\epsilon))$, QNPG achieves an ϵ-optimal policy with high probability using*

$$\tilde{\mathcal{O}}(\epsilon^{-1.5})$$

oracle queries to the environment and policy, improving over the classical $\tilde{\mathcal{O}}(\epsilon^{-2})$ lower bound for policy gradient methods.

Proof Sketch. The argument combines three ingredients:

- **Bias control:** Truncation introduces bias of order $\mathcal{O}(\gamma^N)$, which is negligible for $N = \mathcal{O}(\log(1/\epsilon))$.
- **Variance reduction:** Quantum mean estimation reduces the variance of gradient and Fisher estimators quadratically compared to classical sampling.
- **Convergence rate:** With variance reduced estimators, stochastic natural gradient descent converges in $\tilde{\mathcal{O}}(\epsilon^{-1.5})$ oracle queries.

Intuition. Classical NPG requires $\tilde{\mathcal{O}}(1/\epsilon^2)$ samples to ensure both bias and variance are sufficiently small. By introducing deterministic truncation, we trade a negligible $\log(1/\epsilon)$ factor for the ability to embed trajectories into quantum states. Once this is achieved, quantum mean estimation reduces variance so effectively that the sample complexity improves to $\tilde{\mathcal{O}}(1/\epsilon^{1.5})$.

4.3 Comparison

This result provides, to our knowledge, the first quantum speedup for model-free RL with parameterized policies. Prior work on quantum RL has focused on tabular models or bandits, where complexity improvements arise from simpler problem structure. In contrast, QNPG handles general parameterizations and infinite-horizon MDPs, filling an important gap.

5 Conclusion and Outlook

We presented the Quantum Natural Policy Gradient (QNPG) algorithm, the first quantum-compatible reformulation of natural policy gradients with provable guarantees. By combining deterministic truncation with quantum variance reduction, QNPG achieves an $\tilde{\mathcal{O}}(\epsilon^{-1.5})$ query complexity, improving over the classical $\tilde{\mathcal{O}}(\epsilon^{-2})$ bound.

This framework opens several directions for future work: extending to quantum actor–critic algorithms, incorporating constraints, and exploring connections with quantum linear algebra methods for policy evaluation. More broadly, our results suggest that carefully designed deterministic estimators, together with quantum primitives, can bring meaningful speedups to reinforcement learning beyond tabular settings.

References

1. Agarwal, A., Kakade, S.M., Lee, J.D., Mahajan, G.: On the theory of policy gradient methods: optimality, approximation, and distribution shift. J. Mach. Learn. Res. **22**(98), 1–76 (2021)
2. Auer, P., Jaksch, T., Ortner, R.: Near-optimal regret bounds for reinforcement learning. Adv. Neural Inf. Process. Syst. **21** (2008)
3. Cornelissen, A., Hamoudi, Y., Jerbi, S.: Near-optimal quantum algorithms for multivariate mean estimation. In: Proceedings of the 54th Annual ACM SIGACT Symposium on Theory of Computing, pp. 33–43 (2022)
4. Ganguly, B., Xu, Y., Aggarwal, V.: Quantum speedups in regret analysis of infinite horizon average-reward Markov decision processes. arXiv preprint arXiv:2310.11684 (2023)
5. Liu, Y., Zhang, K., Basar, T., Yin, W.: An improved analysis of (variance-reduced) policy gradient and natural policy gradient methods. Adv. Neural. Inf. Process. Syst. **33**, 7624–7636 (2020)
6. Luo, B., Huang, Y., Allcock, J., Lin, X., Zhang, S., Lui, J.: Quantum algorithms for finite-horizon Markov decision processes. In: Forty-second International Conference on Machine Learning (2025)
7. Mondal, W.U., Aggarwal, V.: Improved sample complexity analysis of natural policy gradient algorithm with general parameterization for infinite horizon discounted reward Markov decision processes. In: International Conference on Artificial Intelligence and Statistics, pp. 3097–3105. PMLR (2024)
8. Sidford, A., Zhang, C.: Quantum speedups for stochastic optimization. Adv. Neural Inf. Process. Syst. **36** (2024)
9. Wu, Y., Guan, C., Aggarwal, V., Wang, D.: Quantum heavy-tailed bandits. arXiv preprint arXiv:2301.09680 (2023)
10. Xu, Y., Aggarwal, V.: Accelerating quantum reinforcement learning with a quantum natural policy gradient based approach. In: Forty-second International Conference on Machine Learning (2025)

Quantum Kernel Methods for Industrial Anomaly Detection

Takao Tomono(✉)

Keio University, Yokohama, Kanagawa 212-0032, Japan
takao.tomono@ieee.org

Abstract. Anomaly detection plays a vital role in industrial quality control and manufacturing processes. Traditional machine learning methods face significant challenges when training data is limited, particularly in specialized manufacturing contexts. Quantum machine learning (QML) offers a promising approach to enhance anomaly detection capabilities under these constraints. This paper provides a comprehensive review of QML applications in industrial anomaly detection, with particular focus on image-based inspection systems, presenting our novel contributions. This paper classifies various types of anomalies encountered in industrial environments and provides a detailed review of classical and quantum anomaly detection approaches. In addition, we present the latest advances in quantum kernel methods in image-based anomaly detection. The analysis includes experimental results showing that quantum kernels outperform classical methods in certain industrial applications. In shipment inspection using an imbalanced dataset of 400 samples (300 normal, 100 anomalous), QSVM achieved an F1 score of 0.990 compared to 0.964 for classical SVM and 0.958 for ResNet, representing a 2.7% improvement in detection performance. We also discuss the implementation of quantum support vector machines (QSVM) with quantum kernels and their performance on quantum simulators and actual quantum hardware. Hardware validation reveals that quantum circuits with depths ≤ 32 maintain consistent performance between simulators and actual quantum devices, while circuits with depths > 273 suffer significant degradation (AUC: $0.89 \rightarrow 0.59$) due to noise accumulation. These findings establish practical guidelines for deploying quantum machine learning in industrial settings and provide a roadmap for future quantum-enhanced manufacturing systems.

Keywords: Quantum machine learning · Anomaly detection · Quantum kernels · Support vector machines · Industrial inspection · Image processing

1 Introduction

The manufacturing industry faces increasing demands for efficient and accurate quality control systems. This challenge is particularly acute in modern production scenarios characterized by small-batch, high-mix manufacturing. Traditional machine learning approaches for anomaly detection typically require large training datasets, which can be difficult to obtain in specialized manufacturing contexts where defective samples are rare

V. Aggarwal et al. (Eds.): PQAI 2025, CCIS 2724, pp. 64–85, 2026.
https://doi.org/10.1007/978-981-95-7829-0_8

and costly to generate. Quantum machine learning (QML) has emerged as a promising paradigm that leverages quantum mechanical principles to potentially overcome these limitations [1–3].

Quantum computing leverages phenomena such as superposition and entanglement to process information fundamentally differently than classical computers. While quantum speedup is often emphasized, we adopt a different perspective on the advantages of quantum computing. In artificial intelligence and machine learning, quantum algorithms can explore exponentially large feature spaces more efficiently than their classical counterparts, potentially leading to improved classification performance with smaller training datasets. This capability is particularly relevant for industrial anomaly detection, where obtaining large numbers of defective samples for training can be impractical or costly.

The integration of quantum machine learning in industrial settings represents a significant step toward what has been termed the "sustainable smart factory with quantum technology". This concept encompasses not only improved efficiency in manufacturing processes but also reduced computational energy consumption through reversible quantum computation. As we progress through the NISQ era, practical applications of quantum machine learning are becoming increasingly feasible, with several successful demonstrations in industrial trial [4–6].

This paper provides a comprehensive review of quantum machine learning applications for industrial anomaly detection [7–9]. We begin by examining the various types of anomalies encountered in industrial settings (Sect. 2), followed by a detailed review of both classical and quantum approaches to anomaly detection (Sect. 3). Section 4 presents our recent advances in image-based anomaly detection using quantum kernels, including experimental results from industrial applications. Finally, Sect. 5 discusses future perspectives and challenges in implementing quantum machine learning for industrial quality control. Within this review, we present the results of our work in Sects. 3.3 and 4.

2 Types of Anomaly Detection

2.1 Classification of Anomalies in Manufacturing

Anomaly detection in industrial contexts encompasses a wide range of defect types and detection scenarios. Understanding these different categories is essential for selecting appropriate detection methods and evaluating the potential benefits of quantum approaches. We examine point anomalies [10–12], contextual anomalies [13–18], and collective anomalies [19–22].

Point Anomalies:
These represent individual data instances that deviate significantly from normal patterns. Point anomalies typically appear as statistical outliers that show significant deviation from expected data trends. For instance, if a credit card holder suddenly makes an unusually large purchase, this triggers investigation for potential fraud.

In the manufacturing, point anomalies are often the most straightforward to detect but can have significant implications:

Manufacturing Examples: A single defective product with a crack in an otherwise perfect batch, sudden spike in temperature readings during a stable process, individual measurement exceeding tolerance limits.

Detection Characteristics: Point anomalies typically manifest as outliers in statistical distributions. They can be detected using threshold-based methods, statistical process control charts, or distance-based approaches.

Challenges: Distinguishing between true anomalies and measurement noise, setting appropriate thresholds that balance sensitivity and specificity.

Contextual Anomalies: Also known as conditional anomalies, these occur when data instances are anomalous in specific contexts but normal in others. This type occurs when data point behavior that is normal in one context becomes anomalous in another. A common example is when a retail site experiences a large increase in traffic and sales on Black Friday, the busiest shopping day of the year. These spikes would be abnormal at other times of the year, so we set special parameters for this day.

Manufacturing Examples: High power consumption during machine startup (normal) vs. during idle state (anomalous), dimensional variations acceptable for prototype products but not for mass production, seasonal variations in environmental conditions affecting product quality.

Detection Requirements: Contextual anomaly detection requires understanding of operational states, temporal patterns, and environmental conditions. Methods must incorporate contextual features such as time of day, production phase, or product type.

Technical Approaches: Conditional probability models, time-series segmentation, multi-modal analysis combining sensor data with operational logs.

Collective Anomalies: These involve collections of related data instances that are anomalous when considered together. This is when a group of data points together exhibit abnormal behavior, even though each individual data point may appear normal. This anomaly can be identified by observing associations or patterns between multiple data points. A DDoS attack is an example of a collective anomaly, as it generates traffic from multiple sources that differ from the normal traffic pattern.

Manufacturing Examples: Gradual tool wear causing progressive quality degradation, coordinated drift in multiple process parameters indicating systemic issues, batch effects where entire production runs show subtle deviations.

Detection Complexity: Individual instances may appear normal, requiring analysis of patterns, trends, and correlations across multiple data points.

Detection Methods: Sequential pattern mining, change point detection, multivariate statistical process control.

2.2 Domain-Specific Anomaly Types

Industrial anomalies are classified into product anomalies (surface defects, dimensional variations, structural issues) and process anomalies (equipment degradation, parameter drift, environmental factors) [23–27]. This work addresses both standardized industrial products [28–30] and non-standardized agricultural products [26, 31, 32], as demonstrated in our shipping inspection and apple quality control applications.

These are detailed as follows.
Industrial Product:

- Surface defects: scratches, dents, discoloration
- Dimensional anomalies: size variations, misalignment
- Structural defects: cracks, voids, inclusions
- Assembly errors: missing components, incorrect assembly

Agricultural Product:

- Internal defects: rot, hollow areas, internal browning
- External blemishes: bruises, insect damage, disease spots
- Shape irregularities: deformation, size variations
- Ripeness variations: over-ripe or under-ripe products

Manufacturing (Growth) Process:

- Equipment degradation: bearing wear, alignment issues
- Process parameter drift: temperature, pressure, speed variations
- Environmental anomalies: contamination, humidity effects

This detailed classification of content helps clarify the specific anomaly types and their detection requirements in each industry sector and allows the selection of appropriate detection methods.

2.3 Challenges in Industrial Anomaly Detection

Industrial anomaly detection presents several unique challenges that make quantum approaches particularly attractive as well as classical approaches. We describe the following challenges: Data Scarcity and Imbalance [33–36], High Dimensionality and Multimodality [37–40], Real-time Requirements [41–44], Dynamic and Evolving Patterns [45–49], and Cost Sensitivity [50, 51].

Data Scarcity and Imbalance:
High-quality production processes naturally result in low defect rates, making anomalous samples rare and challenging to collect for model training.

Statistical Challenge: In the high-quality manufacturing, defect rates often fall below 0.1%, creating severe class imbalance.

Economic Impact: Cost of producing defective samples for training purposes.

Solution Approaches: Data augmentation, synthetic defect generation, transfer learning, few-shot learning techniques.

High Dimensionality and Multi-modality:
Modern inspection systems generate high-resolution images and multi-sensor data. As a results, they generate multiple data types.

Visual: RGB images, hyperspectral data, thermal imaging
Physical: Force, torque, vibration measurements
Chemical: Spectroscopy, gas chromatography

Acoustic: Ultrasonic signals, acoustic emissions

Real-Time Requirements:
Production line speeds demand rapid detection and classification.

Production Speeds: Modern production lines operate at speeds requiring sub-second decision making.

Latency Constraints: Detection, classification, and action must occur within tight time windows.

Edge Computing: Need for onsite processing capabilities rather than cloud based solutions.

Dynamic and Evolving Patterns:
Natural variations in materials and processes require robust detection methods.

Concept Drift: Normal patterns change over time due to tool wear, seasonal variations, or process improvements.

New Defect Types: Previously unseen anomaly patterns emerge with process or material changes.

Adaptive Requirements: Detection systems must continuously learn and update.

Cost Sensitivity:
False positives and false negatives both carry significant economic impacts.

2.4 Anomaly Detection Metrics and Evaluation

When performing anomaly detection, it is sometimes difficult to know what evaluation metrics to use. Here, we will explain two important metrics.

Performance metrics that can be detected include true positive rate (sensitivity), false positive rate (1-specificity), precision, and F1 score. Classification Performance Metrics are formally defined as:

- True Positive Rate (TPR/Sensitivity/Recall): TPR = TP/(TP + FN)
- False Positive Rate (FPR): FPR = FP/(FP + TN)
- Precision: Precision = TP/(TP + FP)
- F1-score: F1 = 2 × (Precision × Recall)/(Precision + Recall)

where TP, TN, FP, and FN represent true positive, true negative, false positive, and false negative classifications, respectively.

In addition, economic metrics include the cost of false positives (unnecessary exclusions) and the cost of false negatives (shipment of defective products). Meanwhile, operational metrics include the impact on throughput, inspection time, and computational resources.

As an evaluation challenge, there is the establishment of ground truthing techniques. This is due to the difficulty of obtaining accurate labels for all anomaly types. In addition, as a time aspect, there is a trade-off between early detection and accurate classification. Finally, as a multi-class problem, different anomaly types require different actions.

2.5 Related Work

Classical Anomaly Detection in Industrial Setting:
Traditional industrial anomaly detection has evolved from statistical control charts [52] to machine learning approaches, with SVMs becoming popular for small datasets. Recent surveys highlight the dominance of deep learning, particularly autoencoder-based methods for unsupervised anomaly detection[4, 5]. However, these methods typically require substantial training data, limiting applicability in specialized manufacturing contexts with limited defective samples.

Quantum Machine Learning for Pattern Recognition:
Quantum machine learning addresses classical limitations through theoretical speedups. Rebentrost et al. [53] demonstrated quantum advantages for SVMs, while Havlíček et al. [2] provided experimental validation using quantum feature maps. Quantum kernel methods show particular promise for NISQ devices [54], with Huang et al.[55] proving that quantum kernels can achieve lower generalization error than optimal classical methods. However, most results assume ideal conditions, raising questions about practical applicability under noise.

Quantum Anomaly Detection:
Direct quantum applications to anomaly detection remain limited. Liu and Rebentrost [56] introduced quantum algorithms for anomaly detection with logarithmic scaling advantages. Corli et al. [1] surveyed quantum anomaly detection across supervised, unsupervised, and reinforcement learning paradigms. Industrial applications are largely unexplored, with Bhowmik and Thapliyal [57] discussing consumer electronics potential.

NISQ-Era Limitation and Error Mitigation:
NISQ devices face significant noise challenges [58]. Wang et al. [54] demonstrated that quantum advantages can vanish under realistic noise, with circuit depth as a critical factor. Some commercial solutions, including those involving error suppression, offer practical approaches to noise mitigation in near-term quantum devices.

Positioning of This Work:
This work addresses key gaps by tackling real-world industrial applications rather than synthetic datasets. Our systematic investigation of quantum kernel architectures (QK0–QK10) and comparative analysis of simulator versus hardware performance provides practical insights for NISQ-era deployment. The demonstrated quantum advantage in small-data regimes and applications across industrial and agricultural domains establish empirical evidence for quantum benefits in the resource-constrained manufacturing environments. This work makes several distinctive contributions: (1) systematic investigation of quantum kernel architectures (QK0-QK10) with comparative performance analysis, (2) empirical validation across both standardized industrial products and non-standardized agricultural products, and (3) comprehensive hardware validation demonstrating the practical constraints of NISQ-era quantum devices for industrial deployment.

3 Methods for Anomaly Detection

3.1 Classical Machine Learning Approach

Traditional machine learning methods have been extensively applied to anomaly detection in industrial settings. We provide a comprehensive analysis of each approach with mathematical foundations and practical considerations. We describe statistical method [8, 52] and machine learning Algorithm [7, 59, 60] as following item.

Statistical Method:

- Gaussian distribution modeling assumes normal data follows a Gaussian distribution, flagging instances beyond certain standard deviations as anomalies
- Regression analysis identifies anomalies as instances with high prediction errors
- Time series analysis methods like ARIMA detect anomalies in temporal data patterns

Machine Learning Algorithms:
Support Vector Machines (SVM):

SVMs construct optimal hyperplanes to separate normal and anomalous classes. The kernel trick enables non-linear classification by mapping data to higher-dimensional spaces. Key advantages include effectiveness with small datasets and robust performance in high-dimensional spaces.

Neural Networks:

Deep learning approaches, particularly Convolutional Neural Networks (CNNs), have shown remarkable success in image-based anomaly detection. Autoencoders learn compressed representations of normal data, identifying anomalies through high reconstruction errors. However, these methods typically require large training datasets and significant computational resources.

Ensemble Methods:

Random Forests and Gradient Boosting combine multiple weak learners to create robust anomaly detectors. Isolation Forests specifically designed for anomaly detection, isolate anomalies using fewer random partitions than normal instances.

Clustering-Based Methods:

K-means, DBSCAN, and other clustering algorithms identify anomalies as instances that don't belong to any cluster or form small, isolated clusters.

3.2 Quantum Machine Learning Approaches

Quantum machine learning leverages quantum mechanical principles to potentially achieve computational advantages over classical methods. Several quantum approaches have been developed for anomaly detection. Here, we describe an interface between PCA-quantum interface [61, 62], QSVM [53, 63, 64], VQA [65–68], and quantum neural networks [69–71]to perform image quality control.

PCA-Quantum Interface:
The number of principal components directly determines the required number of qubits in quantum circuits. This creates a natural bridge between classical preprocessing and quantum processing, where PCA not only reduces computational complexity but also enables quantum encoding of classical industrial data.

Quantum Support Vector Machines (QSVM):
QSVMs utilize quantum kernels to compute inner products in exponentially large feature spaces. The quantum kernel function is expressed as:

$$\kappa\left(x_i, x_j\right) = \left|\langle\phi(x_i)^\dagger|\phi\left(x_j\right)\rangle\right|^2 = \left|\langle 0|U_k{}^\dagger(x_i)U_k\left(x_j\right)|0\rangle\right|^2$$

where $\phi(x)$ represents the quantum feature map, and $\kappa\left(x_i, x_j\right)$ is the quantum kernel function computing the inner product between quantum states $|\phi(x_i)\rangle$ and $|\phi\left(x_j\right)\rangle$. This approach has shown particular promise for problems with limited training data.

$\varphi(x)$: $\mathbb{R}^n \to \mathscr{H}$ represents the quantum feature map that encodes classical data into quantum Hilbert space $\mathscr{H}$, and the kernel computes the squared overlap between corresponding quantum states.

Variational Quantum Algorithms (VQA):
Variational Quantum Circuits (VQCs) combine quantum and classical processing, using parameterized quantum circuits optimized through classical optimization. The Quantum Approximate Optimization Algorithm (QAOA) can be adapted for anomaly detection tasks.

Quantum Neural Networks:
Quantum versions of neural networks leverage quantum gates as neurons and quantum states for information processing. These include Quantum Convolutional Neural Networks (QCNNs) for image processing and Quantum Autoencoders for unsupervised anomaly detection.

3.3 Comparative Analysis

Classical and quantum machine learning differ in two key aspects. First, quantum approaches can build effective models with smaller datasets [72–74]. Second, the quantum learning process itself differs fundamentally from classical approaches and is particularly effective for anomaly detection [75].

The first point is that being able to build a machine learning model with a small data set is very useful for the manufacturing small quantities of a wide variety of products in the manufacturing sites. Figure 1 illustrate performance of Quantum Kernel-SVM (Q-SVM) compared to classical Kernel-SVM (C-SVM) and Resnet: a kind of Convolutional Neural Network (CNN) [74]. Quantum kernel is QK0 on Fig. 4. The comparative performance of classical and quantum approaches is illustrated on across different metrics. Recent studies have demonstrated that quantum methods can achieve superior performance with smaller training datasets, making them particularly suitable for industrial applications

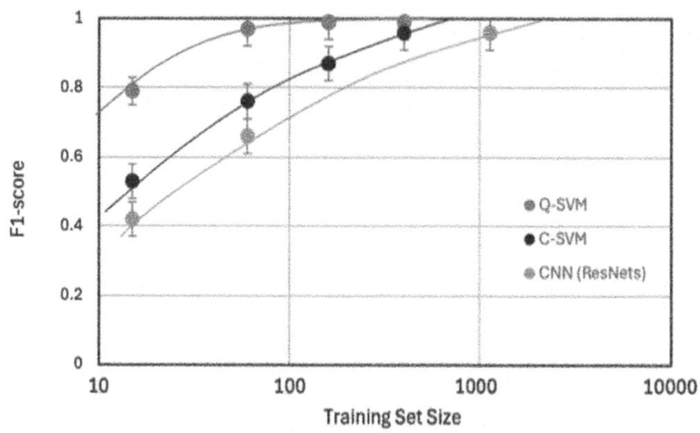

Fig. 1. Performance comparison demonstrating quantum advantage in small-data regimes. F1-score performance across varying training set sizes for industrial product classification. Experimental conditions: Real industrial product images, binary classification (normal vs anomaly), 15 principal components. Methods compared: Quantum SVM (QK0 kernel, red circles), Classical SVM (RBF kernel, black circles), CNN ResNet-18 (green circles). Error bars represent means ± standard deviation across 3 independent runs. Experimental conditions: Real industrial product images, binary classification (normal vs anomaly), 15 principal components (The cumulative contribution rate is 0.82). Methods: Red circles: Quantum SVM (QK0 kernel), Black circles: Classical SVM (RBF kernel, $\gamma = 0.1$), Green circles: CNN (ResNet-18). Key finding: Quantum SVM demonstrates superior performance for training sets < 1000 samples, with F1-score advantage diminishing as dataset size increases. Statistical significance: Error bars represent means ± standard deviation across 3 independent runs. This figure shows the number of training sessions in logarithmic scale by adding some data to the numerical data in ref. [74].

where anomalous samples are rare. Methods for Anomaly Detection Natural variations in materials and processes require robust detection methods.

Additionally, the quantum learning process demonstrates unique characteristics that are particularly beneficial for anomaly detection tasks. Figure 2 shows the FPR and TPR values plotted in the ROC space using the Hearts disease and Iris_2 datasets. Here, we created a two-class dataset Iris_2 (versicolor and Virginia) with attribute 4 from the original Iris dataset. Figures (a) and (b) plot the training set size (ts) from 8 to 200 and 6 to 60 for hearts disease and Iris_2. The red and gray arrows show the hypotheses about the quantum and classical learning processes in the ROC space.

The gray line is how classical machine learning progresses and represents a random machine learning model. It is an ideal learning method that starts from TPR = 0.5, FPR = 0.5 and eventually progresses in the direction of TPR = 1, FPR = 1.

On the other hand, in the case of quantum machine learning, we observed that the learning process starts at a high TPR = 1 and near FPR = 1. If you maintain a high TPR, the FPR will decrease as the learning progresses (orange arrow in the figure). This means that the TPR is always 1, the inspection result is always a good product, and suspicious products are judged as defective. Therefore, no defective products are shipped to the market [75].

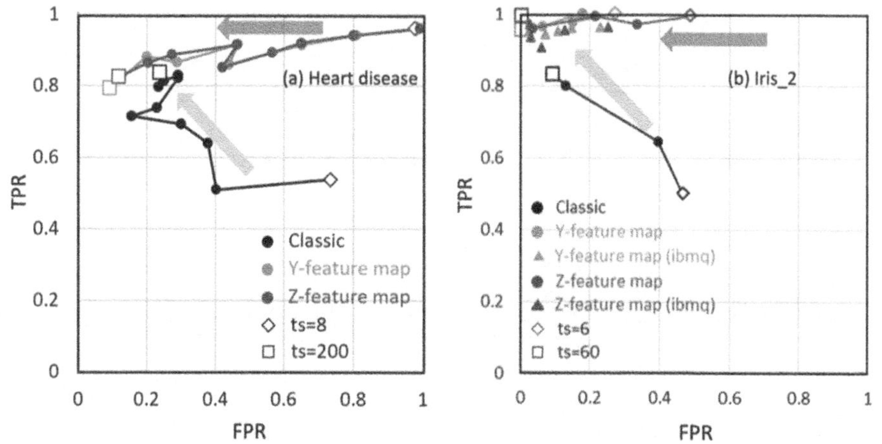

Fig. 2. The plotting onto ROC space. Open diamond and square shape stand for training size of 8 and 200 on heart disease dataset, 6 and 60 on Iris_2 dataset. The red and gray directing arrow indicate hypothesis of learning progress. Ibm_Bogota is actual quantum computer [75]. TPR (True Positive Rate) and FPR (False Positive Rate) trajectories show different learning behaviors between classical and quantum approaches.

4 Image-Based Anomaly Detection Using Quantum Machine Learning

4.1 Quantum Kernel Design for Image Processing

The design of effective quantum kernels is crucial for image-based anomaly detection. Our research has explored various quantum circuit architectures, each offering different advantages.

Figure 3 shows the basic structure of a quantum kernel. It is represented by the inner product of $U_k(x_i)$ and its dagger. This calculated inner product is embedded as the kernel of the SVM. And then a machine learning model is constructed. Figure 4 shows the detailed circuit configuration of $U_k(x_i)$. We designed QK0 to QK10. The design concept is shown below [76].

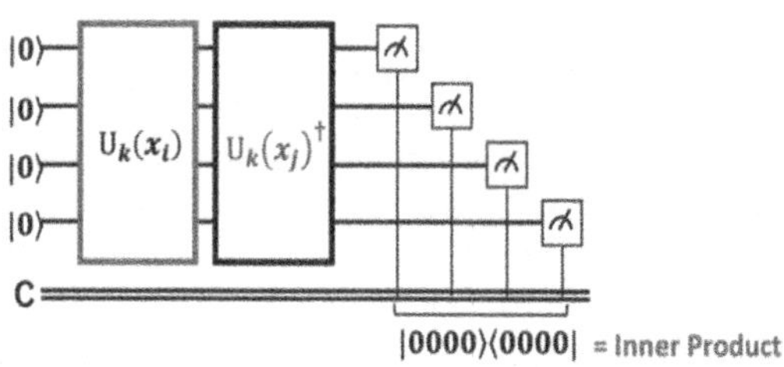

Fig. 3. Quantum kernel circuits diagram in the 4qubits.

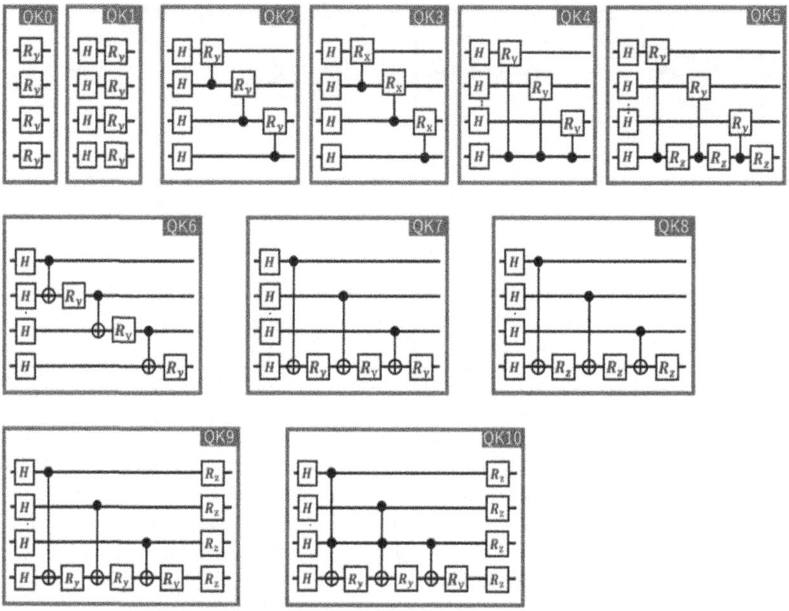

Fig. 4. Quantum circuit architectures for different kernel designs. QK0 and QK1 are superposition only. QK2, QK3, QK4 and QK5 include control rotation gate. QK6, QK7, QK8, QK9, and QK10 include CNOTs and rotation gates.

Basic Rotation Gates (QK0, QK1): Simple circuits using Hadamard gates and single-qubit rotations provide baseline quantum kernels. While these show improvements over classical kernels, their expressivity is limited.

Controlled Rotation Gates (QK2-QK5): Circuits incorporating controlled rotation gates (C-R_y, C-R_x) create entanglement between qubits, enabling more complex feature relationships. These kernels show significant improvements in discriminative power.

CNOT-based Architectures (QK6-QK10): Advanced circuits using *CNOT* gates and combinations with rotation gates achieve the highest performance. The QK9 architecture, featuring *CNOT* gates connecting each qubit to a bottom qubit with Ry rotations between *CNOT*s, has proven particularly effective.

4.2 Industrial Application: Shipping Inspection

Machine learning is now widely used for shipping inspection at production sites. Products undergo quality control using 3σ statistical limits to minimize basic defect occurrence. However, since defective products are extremely rare, machine learning models typically employ unsupervised learning approaches.

Like many manufacturing facilities, TOPPAN's factories employ unsupervised learning approaches. To prevent defective products from reaching customers, the shipping inspection system implements a three-stage model: an image processing model, a discrimination model, and a generative model to prevent the outflow of anomaly products.

This image-based shipping inspection process is shown in Fig. 5. Products are shipped after passing through these three stages of inspection.

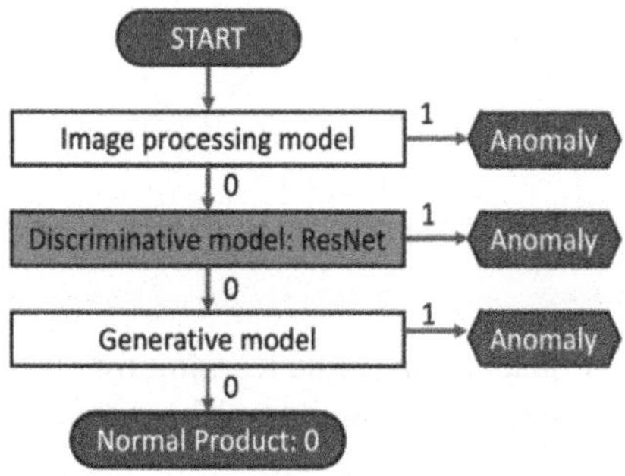

Fig. 5. Overview of shipping inspection using classical machine learning at an actual production site. We use three model described in the figure actually [74]. Here, 0 and 1 stand for normal and anomaly.

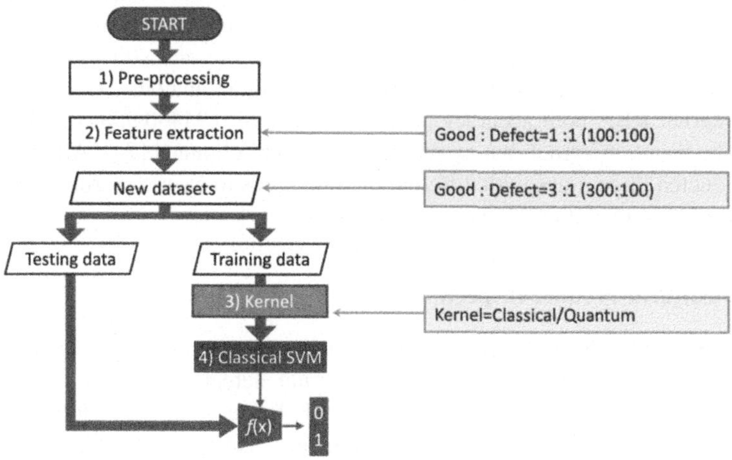

Fig. 6. Application of supervised ML on shipping inspection in actual production sites. ML means machine learning [74].

Table 1. Performance in the shipping inspection.

Model	Note	Datasets size	Accuracy	F1-score
ResNets	Unsupervised learning	1132	0.940 ± 0.047	0.958 ± 0.048

(continued)

Table 1. *(continued)*

Model	Note	Datasets size	Accuracy	F1-score
C-SVM	supervised learning	400	0.944 ± 0.047	0.964 ± 0.048
Q-SVM	supervised learning	400	0.988 ± 0.049	0.990 ± 0.050

Results represent mean ± standard deviation across 3 independent experiments. Original data is based on ref. [74].

Our implementation of quantum machine learning for shipping inspection demonstrates practical advantages in real-world settings. The system processes product images through several stages.

1. Image Preprocessing: Binarization and noise reduction.
2. Feature Extraction: Principal Component Analysis (PCA) to reduce dimensionality.
3. Quantum Kernel Generation: Creating quantum kernels based on selected features.
4. Classification: SVM with embedded quantum kernels.

Results show in the table 1. ResNet using unsupervised learning achieved an F1-score of 0.958 with nearly 1,000 training samples, while our quantum approach achieved superior results with an accuracy of 0.988 and of F1 Score of 0.990 with the quantum kernel using an imbalanced data set (300 normal, 100 anomalous). Even when using classical kernel RBF, both accuracy and F1 score were 0.944–0.958, which is not as good as the quantum kernel, but excellent results were obtained [74]. All experiments were conducted three times with different random seeds to ensure reproducibility, with variations remaining within 5% across all metrics.

4.3 Agricultural Product Inspection

Extending beyond industrial products, we applied quantum anomaly detection to agricultural products, specifically apples with internal defects [76]. While research on quantum technology applications in agriculture is being conducted globally, agricultural applications [72, 77] present unique challenges. This application presents unique challenges:

- Non-standardized product shapes and sizes
- Internal defects invisible from external inspection
- Natural variations requiring robust classification

Figure 7 shows the experimental setup for creating the dataset. Internal defects cannot be detected from external inspection. Therefore, we created a device that irradiates the apple with an LED from the top side and takes a photo from the bottom side. The resolution was 4032 × 3024. The photo taken from the top shifts to orange, but it is converted to a black and white image by binarization. Only the vine is captured, but if there is nectar, a spotted pattern appears. On the other hand, if there is a crack in the vine, it is not visible, but when enlarged, a black spreading area appears around the vine. This enlarged area was 120 x 80. In both cases, by cutting the apple in half and comparing the photo with the binarized image, it is possible to determine whether the apple is filled with nectar or has cracks.

Here, the apple we chose is Fuji. Fuji is the most widely produced apple variety in Japan and is characterized by a good balance of sweetness and sourness, and a good taste with nectar. For this reason, the appearance is also important, and internal defects lead to customer dissatisfaction, so it is best not to ship the product if possible. On the other hand, apples full of honey (which at first glance appear to be browning) are prized as they add to the flavor. However, cracks on the rind result in complaints.

Figure 8 shows the relationship between the F1 scores of the principal components of each quantum kernel used to identify internal vine cracks in apples, compared to classical kernel RBF. Figure 8(a) shows the results of the conventional kernel RBF and QK0-QK5. The horizontal axis is the feature corresponding to the cumulative contribution of the principal components. The vertical axis is the F1 score. When the feature is 3, it means the cumulative contribution of the first to third principal components. When the feature is 7, it means the cumulative contribution of the first to seventh principal components.

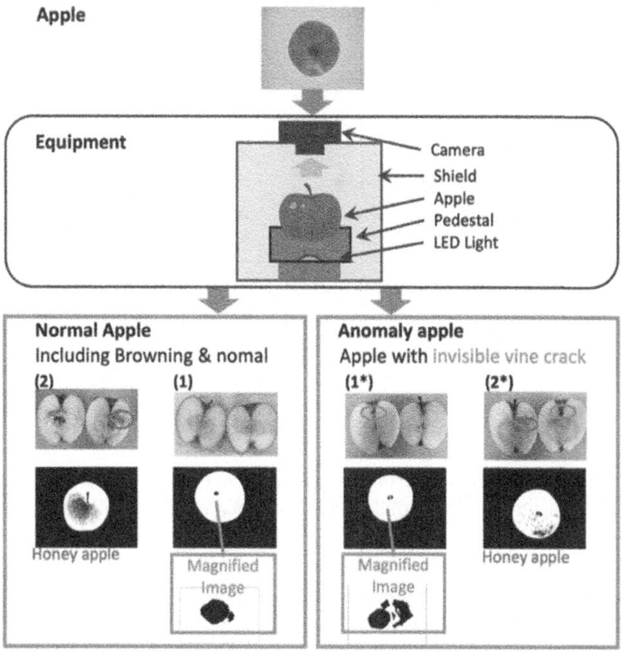

Fig. 7. Experimental setup for discriminating internal vine crack in the apple. Internal defects can be visualized using LEDs. Original image is 4032 × 3024. Magnified image is 120 × 80.

First, we compare the F1 scores of each kernel when the feature is 3 and then compare the trend of the F1 score as the features increases. The classical RBF increased by about 0.2 from the feature of 3 to 6, but there was almost no change when the feature size increased from 6 to 7. QK0 and QK1 are quantum circuits with one H gate, one H gate, and one rotation gate Ry for each qubit. When the features increased from 3 to 7, the F1 score value increased by more than 0.5. The F1 scores of QK2, QK3, QK4, and QK5 increase slightly as the feature value increases from 3 to 7, however, remain almost constant.

Figure 8(b) shows the results of QK6 to QK10 and the classical kernel RBF. When the features is 3, the F1 scores of QK7 and QK8 are more than 0.15 larger than RBF. QK6 shows almost the same trajectory as QK1 and QK2 (1), but QK7 and QK8 increase by 0.1 to 0.15 larger than RBF, even when the feature value increases, and by 0.15 larger. Furthermore, the F1 scores of QK9 and QK10 are more than 0.3 larger than the classical kernel RBF when the feature value is 3, and the F1 scores also increase as the feature value increases.

From the above, it is thought that quantum circuits with entanglement can form more complex separation interfaces than superposition and rotation control gates. In particular, quantum circuits with an Rz rotation gate at the end are thought to be more effective. The F1 scores of QK9 and QK10 are the highest among these quantum kernels, and they are considered promising quantum kernel candidates.

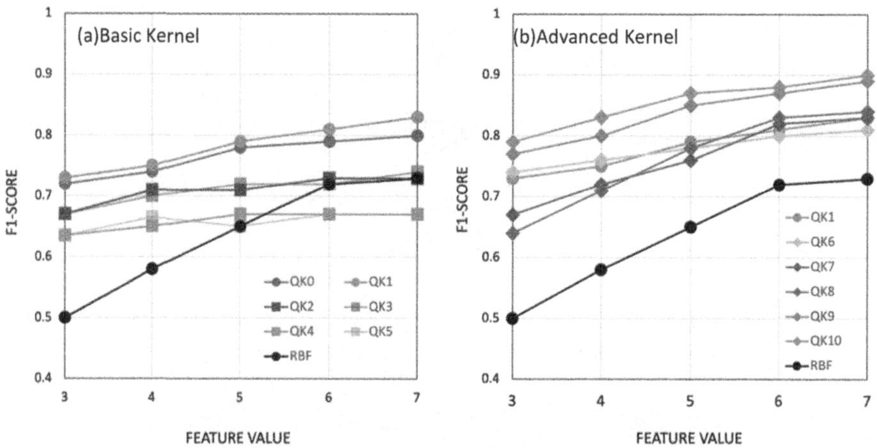

Fig. 8. Quantum kernel performance comparison for apple internal defect detection (vine crack). FEATURE VALUE means numbers of features (number of principal components). (a) Basic quantum kernels: QK0-QK5 vs classical RBF kernel. QK0/QK1 use simple rotation gates, QK2-QK5 incorporate controlled rotations. (b) Advanced quantum kernels: QK6-QK10 with CNOT-based architectures. QK9 and QK10 demonstrate substantial improvement over classical RBF, particularly with 3–7 principal components. Key observation: Quantum kernels with entanglement (QK6-QK10) significantly outperform both classical methods and simple quantum circuits.

4.4 Performance Analysis on Quantum Hardware

In the current analysis, it is better to evaluate the performance with a quantum simulator (ideal, noiseless simulation) and actual quantum hardware (IBM superconducting quantum processor with realistic noise). However, the availability of the quantum simulator depends on the environment, the user, and the conditions, and is subject to various constraints. In addition, the only way to use the quantum computer without error correction is to use it via the cloud. Basically, the data from the factory cannot be taken out of the factory. Therefore, the experiment with the quantum computer was only used for the shipping inspection of agricultural products.

Figure 9 shows the ROC curves comparing the quantum simulator and the quantum computer for the quantum kernels QK9 and QK10, which have 4 qubits and an AUC of 0.9 or more. Here, QC and QS mean the quantum computer and the simulator. The left figure shows the ROC-AUC curves of RBF, QS, and QC for QK9, and the right figure shows the ROC-AUC curves of RBF, QS, and QC for QK10. For reference, we also plot the positions of the random model (black dashed line) and the ideal machine learning model (red dashed line). The dashed line from False Positive Rate (FPR) = 0, True Positive Rate (TPR) = 0 to FPR = 1, TPR = 1 shows the random model. The axis of FPR = 0 and TPR = 1 shows the ideal machine learning model. The AUC of the classical RBF is drawn close to the random model, and the numerical data is 0.62. In QK9, the behavior of the AUC curve on the quantum computer was the same as that of the quantum simulator. The numerical data of the AUC at that time was 0.90 for both, as shown in the figure. The difference in behavior between QK9 and QK10 is thought to be due to the difference in depth. The depths of QK9 are 32, while the depths of QK10 are 273, and it is thought that noise became unacceptable between these two.

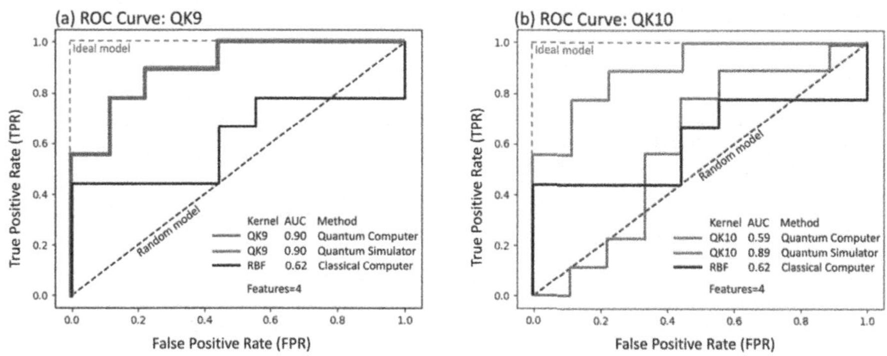

Fig. 9. ROC curve. Hardware validation: Quantum simulator vs actual quantum computer performance. Experimental setup: QK9 (left) and QK10 (right) kernels, 4 qubits, agricultural dataset. Hardware: IBM quantum computer vs ideal simulator. Performance metrics: QK9: AUC = 0.90 (both simulator and hardware) QK10: AUC = 0.89 (simulator) → 0.59 (hardware). Circuit depth impact: QK9 (depths = 32) maintains performance, QK10 (depths = 273) suffers noise degradation. Reference lines: Black dashed = random classifier, Red dashed = ideal classifier. Measurement: 1024 shots, 3 independent runs for hardware validation.

5 Future Perspectives

Figure 10 shows the industrial application roadmap of quantum technology in the manufacturing field. This figure is based on my experience in the quantum computer-related research and development, including references [78–80] and other literature and books.

The quantum technology roadmap for industrial applications spans three phases: 1) Near-term (2025–2030), NISQ-optimized hybrid algorithms and specialized quantum kernels; 2) Medium-term (2030–2040) - on-premises quantum processors and real-time anomaly detection; 3) Long-term (2040+) - fault-tolerant quantum computers enabling fully quantum industrial control systems.

I am thinking of the overall picture of smart manufacturing using quantum technology in the future, as shown in Fig. 11. System and product failure prediction, health management, process optimization, energy management, production line, inspection system, and supply chain are centered on the quantum computers and connected on-promise with classical computers.

I am thinking that integrating quantum processors with classical systems could improve processes by 10% to 20% and potentially make them 30%–50% more energy efficient.

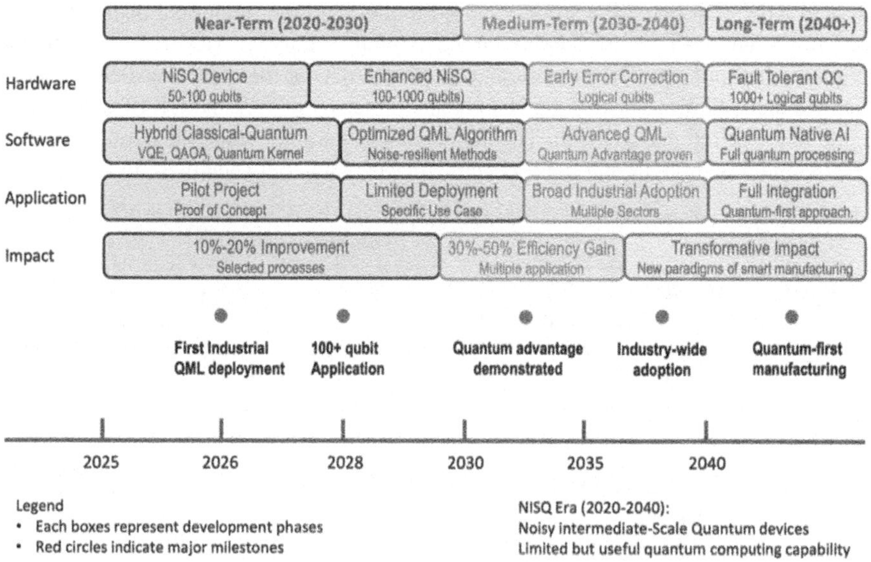

Fig. 10. Roadmap for smart manufacturing using quantum technology.

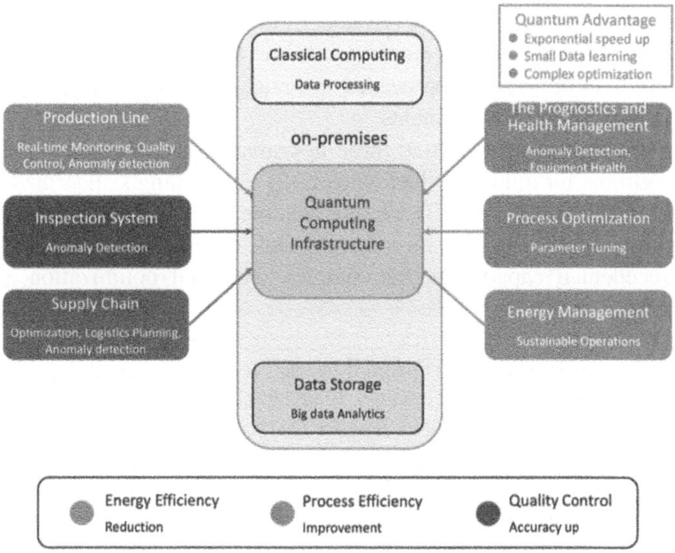

Fig. 11. Smart manufacturing using quantum technology in the future.

6 Conclusion

This study demonstrates the significant potential of quantum machine learning for industrial anomaly detection, establishing both theoretical foundations and practical validation for quantum-enhanced manufacturing quality control systems. Through systematic investigation of quantum kernel methods, we have shown that quantum approaches can achieve superior performance in small-data regimes typical of industrial settings. Our research provides several critical contributions to the field. We have demonstrated quantum advantage in practical industrial scenarios, with quantum SVMs achieving F1-scores of 0.990 compared to 0.964 for classical SVM using 400 training samples. The systematic design and evaluation of quantum kernel architectures (QK0-QK10) revealed that CNOT-based circuits with controlled rotations significantly outperform simple rotation-based designs, providing concrete guidelines for quantum circuit design in industrial applications.

Empirical validation on actual quantum hardware provided crucial insights into NISQ-era limitations. Our analysis revealed that circuit depths ≤ 32 maintain consistent performance between simulators and hardware, while depths > 273 suffer significant degradation due to noise accumulation. This finding establishes practical constraints for current quantum implementations and highlights the importance of error mitigation strategies. Deployment in real-world applications has achieved an F1 score of 0.990 (quantum simulator) for industrial product outgoing inspection and 0.90 (AUC: the quantum computer) for agricultural product quality control. As shown above, we have demonstrated robust performance across a range of defect types, demonstrating the practical feasibility of quantum anomaly detection in manufacturing environments. It is expected that this research will establish quantum machine learning as a transformative technology for industrial quality control.

Current NISQ devices show clear advantages in small-scale data processing but are limited in scalability and circuit complexity. In the future, on-premises infrastructure will be required, so efforts will be needed to break down the barriers to adoption in traditional manufacturing environments.

Future research should prioritize three key areas: (1) developing noise-resilient quantum algorithms for industrial deployment, (2) establishing standardized quantum-classical hybrid frameworks, and (3) creating scalable quantum infrastructure for manufacturing environments. The transition toward fault-tolerant quantum computing will unlock unprecedented capabilities for complex industrial optimization and quality control systems.

Acknowledgments. This work was supported by the New Energy and Industrial Technology Development Organization (NEDO) [Grant No. JPNP23003] and by the Center of Innovations for Sustainable Quantum AI (JST) [Grant No. JPMJPF2221].

Disclosure of Interests. The authors have no competing interests.

References

1. Corli, S., Moro, L., Dragoni, D., Dispenza, M., Prati, E.: Quantum machine learning algorithms for anomaly detection: a review. Futur. Gener. Comput. Syst. **166**, 107632 (2025)
2. Havlíček, V., et al.: Supervised learning with quantum-enhanced feature spaces. Nature **567**(7747) (2019)
3. Tscharke, K., Issel, S., Debus, P.: Semisupervised anomaly detection using support vector regression with quantum kernel. In: Proceedings - 2023 IEEE International Conference on Quantum Computing and Engineering, QCE 2023 (2023)
4. Diers, J., Pigorsch, C.: A survey of methods for automated quality control based on images. Int. J. Comput. Vis. **131**(10) (2023)
5. Zipfel, J., Verworner, F., Fischer, M., Wieland, U., Kraus, M., Zschech, P.: Anomaly detection for industrial quality assurance: a comparative evaluation of unsupervised deep learning models. Comput. Ind. Eng. **177** (2023)
6. Maggipinto, M., Beghi, A., Susto, G.A.: A deep convolutional autoencoder-based approach for anomaly detection with industrial, non-images, 2-dimensional data: a semiconductor manufacturing case study. IEEE Trans. Autom. Sci. Eng. **19**(3) (2022)
7. Pimentel, M.A.F., Clifton, D.A., Clifton, L., Tarassenko, L.: A review of novelty detection. Signal Process. **99** (2014)
8. Woodall, W.H., Montgomery, D.C.: Research issues and ideas in statistical process control. J. Qual. Technol. **31**(4) (1999)
9. Jezek, S., Jonak, M., Burget, R., Dvorak, P., Skotak, M.: Deep learning-based defect detection of metal parts: Evaluating current methods in complex conditions. In: International Congress on Ultra Modern Telecommunications and Control Systems and Workshops, vol. **2021-October** (2021)
10. Fisch, A.T.M., Eckley, I.A., Fearnhead, P.: Subset multivariate collective and point anomaly detection. J. Comput. Graph. Stat. **31**(2) (2022)
11. Fisch, A.T.M., Eckley, I.A., Fearnhead, P.: A linear time method for the detection of collective and point anomalies. Stat. Anal. Data Min. **15**(4) (2022)
12. Li, Z., Xiang, Z., Gong, W., Wang, H.: Unified model for collective and point anomaly detection using stacked temporal convolution networks. Appl. Intell. **52**(3) (2022)

13. Leach, M.J.V., Sparks, E.P., Robertson, N.M.: Contextual anomaly detection in crowded surveillance scenes. Pattern Recognit. Lett. **44** (2014)
14. Kim, H., Kim, H.: Contextual anomaly detection for multivariate time series data. Qual. Eng. **35**(4) (2023)
15. Yu, X., Het al.: An adaptive method based on contextual anomaly detection in Internet of Things through wireless sensor networks. Int. J. Distrib. Sens. Netw. **16**(5) (2020)
16. Mahapatra, A., Srivastava, N., Srivastava, J.: Contextual anomaly detection in text data. Algorithms **5**(4) (2012)
17. Hayes, M.A., Capretz, M.A.: Contextual anomaly detection framework for big sensor data. J. Big Data **2**(1) (2015)
18. Pasini, K., Khouadjia, M., Samé, A., Trépanier, M., Oukhellou, L.: Contextual anomaly detection on time series: a case study of metro ridership analysis. Neural Comput. Appl. **34**(2) (2022)
19. Weng, Y., Liu, L.: A collective anomaly detection approach for multidimensional streams in mobile service security. IEEE Access **7** (2019)
20. Wang, C., et al.: Network traffic analysis over clustering-based collective anomaly detection. Comput. Netw. **205** (2022)
21. Ahmed, M., Mahmood, A.N.: Novel approach for network traffic pattern analysis using clustering-based collective anomaly detection. Ann. Data Sci. **2**(1) (2015)
22. Ahmed, M., Pathan, A.S.K.: Deep learning for collective anomaly detection. Int. J. Comput. Sci. Eng. **21**(1) (2020)
23. Xames, M.D., Torsha, F.K., Sarwar, F.: A systematic literature review on recent trends of machine learning applications in additive manufacturing. J. Intell. Manuf. **34**(6) (2023)
24. Rajesh, A.S., Prabhuswamy, M.S., Krishnasamy, S.: Smart manufacturing through machine learning: a review, perspective, and future directions to the machining industry. J. Eng. (U.K.) **2022** (2022)
25. Jung, H., Jeon, J., Choi, D., Park, A.J.Y.: Application of machine learning techniques in injection molding quality prediction: implications on sustainable manufacturing industry. Sustain. (Switz.) **13**(8) (2021)
26. Javaid, M., Haleem, A., Khan, I.H., Suman, R.: Understanding the potential applications of artificial intelligence in agriculture sector. Adv. Agrochem. **2**(1) (2023)
27. Sharma, A., Jain, A., Gupta, P., Chowdary, V.: Machine learning applications for precision agriculture: a comprehensive review. IEEE Access **9** (2021)
28. Catillo, M., Pecchia, A., Villano, U.: AutoLog: anomaly detection by deep autoencoding of system logs. Expert Syst. Appl. **191** (2022)
29. Liu, Y., Hu, C., Zhan, W., Sun, C., Murch, B., Ma, L.: Identifying industrial heat sources using time-series of the VIIRS Nightfire product with an object-oriented approach. Remote Sens. Environ. **204** (2018)
30. Sahar, T., et al.: Anomaly detection in laser powder bed fusion using machine learning: a review. Results Eng. **17** (2023)
31. Tripathi, M.K., Maktedar, D.D.: A role of computer vision in fruits and vegetables among various horticulture products of agriculture fields: a survey. Inf. Process. Agric. **7**(2) (2020)
32. Zhou, Y., Xia, Q., Zhang, Z., Quan, M., Li, H.: Artificial intelligence and machine learning for the green development of agriculture in the emerging manufacturing industry in the IoT platform. Acta Agric. Scand. B Soil Plant Sci. **72**(1) (2022)
33. Zhou, , Li, G., Wang, R., Chen, R., Luo, S.: A novel contrastive self-supervised learning framework for solving data imbalance in solder joint defect detection. Entropy **25**(2) (2023)
34. Wang, Y., Sun, L., Subramani, S.: CAB: classifying arrhythmias based on imbalanced sensor data. KSII Trans. Internet Inf. Syst. **15**(7) (2021)
35. Liu, F., et al.: DAGAD: data augmentation for graph anomaly detection. In: Proceedings - IEEE International Conference on Data Mining, ICDM, vol. 2022-November (2022)

36. Xie, Y., Wang, M., Zhong, Y., Deng, L., Zhang, J.: Label-free anomaly detection using distributed optical fiber acoustic sensing. Sensors **23**(8) (2023)
37. Viswanath, S.E., Tiwari, P., Lee, G., Madabhushi, A.: Dimensionality reduction-based fusion approaches for imaging and non-imaging biomedical data: Concepts, workflow, and use-cases. BMC Med. Imaging **17**(1) (2017)
38. Ghoniem, R.M., Algarni, A.D., Shaalan, K.: Multi-modal emotion aware system based on fusion of speech and brain information. Inf. (Switz.) **10**(7) (2019)
39. Oyelade, O.N., Irunokhai, E.A., Wang, H.: A twin convolutional neural network with hybrid binary optimizer for multimodal breast cancer digital image classification. Sci. Rep. **14**(1) (2024)
40. Kang, L., Jiang, J., Huang, J., Zhang, T.: Identifying early mild cognitive impairment by multi-modality MRI-based deep learning. Front. Aging Neurosci. **12** (2020)
41. Rakic, S., Pavlovic, M., Marjanovic, U.: A precondition of sustainability: Industry 4.0 readiness. Sustain. (Switz.) **13**(12) (2021)
42. Hussain, M.: YOLO-v1 to YOLO-v8, the rise of YOLO and Its complementary nature toward digital manufacturing and industrial defect detection. Machines **11**(7) (2023)
43. Hung, Y.F., Huang, C.C., Yeh, Y.: Real-time capacity requirement planning for make-to-order manufacturing with variable time-window orders. Comput. Ind. Eng. **64**(2) (2013)
44. Stonebraker, M., Çetintemel, U., Zdonik, S.: The 8 requirements of real-time stream processing. SIGMOD Rec. **34**(4) (2005)
45. Giannakis, G.B., Shen, Y., Karanikolas, G.V.: Topology identification and learning over graphs: accounting for nonlinearities and dynamics. Proc. IEEE **106**(5) (2018)
46. Robinson, J.W., Hartemink, A.J.: Learning non-stationary dynamic Bayesian networks. J. Mach. Learn. Res. **11** (2010)
47. Schnaack, O.H., Peliti, L., Nourmohammad, A.: Learning and organization of memory for evolving patterns. Phys. Rev. X **12**(2) (2022)
48. Cui, Y., Kara, S., Chan, K.C.: Manufacturing big data ecosystem: a systematic literature review. Robot. Comput. Integr. Manuf. **62** (2020)
49. Li, X., Wan, J., Dai, H.N., Imran, M., Xia, M., Celesti, A.: A hybrid computing solution and resource scheduling strategy for edge computing in smart manufacturing. IEEE Trans. Industr. Inform. **15**(7) (2019)
50. Cakir, M., Guvenc, M.A., Mistikoglu, S.: The experimental application of popular machine learning algorithms on predictive maintenance and the design of IIoT based condition monitoring system. Comput. Ind. Eng. **151** (2021)
51. Mushlin, A.I., Kouides, R.W., Shapiro, D.E.: Estimating the accuracy of screening mammography: a meta-analysis. Am. J. Prev. Med. **14**(2) (1998)
52. Montgomery, D.C.: Introduction to Statistical Quality Control, 6th edn. (2009)
53. Rebentrost, P., Mohseni, M., Lloyd, S.: Quantum support vector machine for big data classification. Phys Rev Lett **113**(3) (2014)
54. Wang, X., Du, Y., Luo, Y., Tao, D.: Towards understanding the power of quantum kernels in the NISQ era. Quantum **5**, 531 (2021)
55. Huang, H.Y.,et al.: Power of data in quantum machine learning. Nat. Commun. **12**(1) (2021)
56. Liu, N., Rebentrost, P.: Quantum machine learning for quantum anomaly detection. Phys. Rev. A (Coll. Park.) **97**(4), 042315 (2018)
57. Bhowmik, S., Thapliyal, H.: Quantum machine learning for anomaly detection in consumer electronics. In: 2024 IEEE Computer Society Annual Symposium on VLSI (ISVLSI), pp. 544–550. IEEE (2024)
58. Preskill, J.: Quantum computing in the NISQ era and beyond. Quantum **2** (2018)
59. Chandola, V., Banerjee, A., Kumar, V.: Anomaly detection: a survey. ACM Comput. Surv. **41**(3) (2009)

60. Schölkopf, B., Platt, J.C., Shawe-Taylor, J., Smola, A.J., Williamson, R.C.: Estimating the support of a high-dimensional distribution. Neural Comput. **13**(7) (2001)
61. Cui, Z., Geng, X.: Product service system configuration based on a PCA-QPSO-SVM model. Sustain. (Switz.) **13**(16) (2021)
62. Köse, E., Erkan Köse, M., Güneşdoğdu Sağdınç, S.: Principal component analysis of quantum mechanical descriptors data to reveal the pharmacological activities of oxindole derivatives. Results Chem. **5** (2023)
63. Ding, C., Bao, T.Y., Huang, H.L.: Quantum-inspired support vector machine. IEEE Trans. Neural Netw. Learn. Syst. **33**(12) (2022)
64. Li, Z., Liu, X., Xu, N., Du, J.: Experimental realization of a quantum support vector machine. Phys. Rev. Lett. **114**(14) (2015)
65. McClean, J.R., Romero, J., Babbush, R., Aspuru-Guzik, A.: The theory of variational hybrid quantum-classical algorithms. New J. Phys. **18**(2), 23023 (2016)
66. Wang, D., Higgott, O., Brierley, S.: Accelerated variational quantum eigensolver. Phys. Rev. Lett. **122**(14), 140504 (2019)
67. Moseley, B., Osborne, M., Benjamin, S.: Bayesian optimisation for variational quantum eigensolvers (2018)
68. Wecker, D., Hastings, M.B., Troyer, M.: Progress towards practical quantum variational algorithms. Phys. Rev. A **92**(4), 42303 (2015)
69. Henderson, M., Shakya, S., Pradhan, S., Cook, T.: Quanvolutional neural networks: powering image recognition with quantum circuits. Quantum Mach. Intell. **2**(1) (2020)
70. Cong, I., Choi, S., Lukin, M.D.: Quantum convolutional neural networks. Nat. Phys. **15**(12) (2019)
71. Chen, G., Chen, Q., Long, S., Zhu, W., Yuan, Z., Wu, Y.: Quantum convolutional neural network for image classification. Pattern Anal. Appl. **26**(2) (2023)
72. Caro, M.C., et al.: Generalization in quantum machine learning from few training data. Nat. Commun. **13**(1) (2022)
73. Schuld, M., Bocharov, A., Svore, K.M., Wiebe, N.: Circuit-centric quantum classifiers. Phys. Rev. A (Coll. Park) **101**(3) (2020)
74. Tomono, T., Natsubori, S.: Shipping inspection trial of quantum machine learning toward sustainable quantum factory. In: PHM Society Asia-Pacific Conference, vol. 4, no. 1 (2023)
75. Tomono, T., Natsubori, S.: Performance of quantum kernel on initial learning process. EPJ Quantum Technol. **9**(1) (2022)
76. Tomono, T., Tsujimura, K.: The discriminative ability on anomaly detection using quantum kernels for shipping inspection. EPJ. Quantum Technol. **12**(1), 36 (2025)
77. Velesaca, H.O., Suárez, P.L., Mira, R., Sappa, A.D.: Computer vision based food grain classification: a comprehensive survey. Comput. Electron. Agric. **187** (2021)
78. Kusiak, A.: Smart manufacturing. Int. J. Prod. Res. **56**(1–2), 508–517 (2018)
79. Ghobakhloo, M.: Industry 4.0, digitization, and opportunities for sustainability. J. Clean. Prod. **252** (2020)
80. Da Xu, L., Xu, E.L., Li, L.: Industry 4.0: state of the art and future trends. Int. J. Prod. Res. **56**(8) (2018)

Quantum Optimization and Linear Systems

An Exclusive-Sum-of-Products Pipeline for QAOA

Matthew Brunet[1,3], Shilpi Shah[2], Mostafa Atallah[3,4], Anthony Wilkie[3], and Rebekah Herrman[3(✉)]

[1] Winthrop University, Rock Hill, SC 29733, USA
[2] Rutgers University, New Brunswick, NJ 08901, USA
[3] University of Tennessee Knoxville, Knoxville, TN 37996, USA
rherrma2@utk.edu
[4] Cairo University, Giza 12613, Egypt

Abstract. The quantum approximate optimization algorithm is commonly used to solve combinatorial optimization problems. While unconstrained problems map naturally into the algorithm, incorporating constraints typically requires penalizing constraint violations in the objective function. In this work, we propose an alternative approach that encodes constraints as Boolean expressions in exclusive-sum-of-products (ESOP) form before penalization. We test this method on the maximum independent set problem using graphs with 3 to 20 vertices and find that ESOP constraint formulations achieve higher approximation ratios than standard constraint penalization methods, with percent increases of up to 30.3%. Furthermore, ESOP constraint formulations result in higher approximation ratios than standard QAOA penalization approaches after one layer of the algorithm on approximately 64% of the tested graphs.

Keywords: Quantum approximate optimization algorithm · Exclusive-sum-of-product · Maximum independent set

1 Introduction

The quantum approximate optimization algorithm (QAOA) [12] is a variational quantum algorithm that has the potential to provide a speed-up over classical algorithms for solving combinatorial optimization problems [26,39,42]. Recent QAOA research has resulted in numerous variations of the algorithm that seek to either decrease the circuit depth or problem size [32,48], allow for more expressible circuits [17,43,44,46], or reduce the space of feasible solutions [13,41,45].

While unconstrained combinatorial optimization problems such as MaxCut can easily be mapped into QAOA [28], solving constrained combinatorial optimization problems, such as the Maximum Independent Set (MIS) problem, with the QAOA is more challenging. One of the main methods for handling constraints is to penalize them in the objective function, turning the problem into an unconstrained one that can be naturally mapped to QAOA [18,21]. These

V. Aggarwal et al. (Eds.): PQAI 2025, CCIS 2724, pp. 89–104, 2026.
https://doi.org/10.1007/978-981-95-7829-0_9

approaches can require ancillary qubits and can involve squaring sums of constraints, leading to products of several variables. This penalization can lead to complex cost Hamiltonians whose optimization landscape is difficult to navigate due to barren plateaus and local minima [3,19,20]. Another method of handling constraints is to encode the constraints directly into the initial state of the QAOA circuit, reducing the search space to only the feasible states [6,13,14,41,44,45]. The drawback to this method is the difficulty in creating an efficient (in circuit depth or number of two-qubit gates) state preparation circuit to encode the feasible states. For single constraints, such as linear equality constraints [7,44], the state preparation can be efficient, however it is not clear if this is easy for general sets of constraints, specifically constraints that share large amounts of variables. Other approaches for constrained optimization for QAOA try to avoid penalization altogether [4,10].

The foundational work by Hadfield et al. [16] first established systematic methods for encoding Boolean constraints into phase oracles and Hamiltonians for quantum algorithms. Building on this foundation, recent approaches such as Boolean-Hamiltonian-Transformation QAOA (BHT-QAOA) [2] have extended these techniques to generate QAOA cost Hamiltonians by marking bit-strings that do not satisfy Boolean constraint expressions. In contrast to other penalization methods, BHT-QAOA formulates each constraint as an exclusive-sum-of-product (ESOP) Boolean expression. Interestingly, ESOP expressions have been used in quantum circuit compilation algorithms to reduce T gate count [25,38] and in quantum walk assisted QAOA (QW-QAOA) to solve constrained combinatorial optimization problems [23], though the QW-QAOA approach updates the mixing Hamiltonian to search only over feasible solutions.

In this work, we first introduce background on QAOA, MIS, and ESOPs in Sect. 2. Then, we describe how to use ESOP expressions to encode constraints into QAOA, using MIS as a test problem, in Sect. 3. In Sect. 4, we compare the ESOP constraint approach to standard QAOA penalty approaches to solve MIS on over 3000 graphs with between 3 and 20 vertices and find that encoding constraints using the ESOP formulation results in up to a 30.3% percent increase in approximation ratio over the standard QAOA constraint encoding approach. Finally, we conclude with a discussion in Sect. 5.

2 Background

In this section, we provide background information for QAOA, the MIS problem, SOPs, and ESOPs.

2.1 QAOA

QAOA is a variational quantum algorithm in which two unitaries, a cost unitary $U(C,\gamma) = e^{-iC\gamma}$ and a mixing unitary $U(B,\beta) = e^{-iB\beta}$, are applied to an initial state $|s\rangle$. The cost Hamiltonian, C, is a diagonal matrix that encodes the combinatorial optimization problem $f(x)$ being solved. The construction of C for the MIS problem will be described in more detail later. The mixer Hamiltonian B allows for probability amplitude to be transferred between the computational basis states. In order for QAOA to converge to the optimal solution to the combinatorial optimization problem, the initial state $|s\rangle$ should be an eigenstate of B. B is often chosen to be the transverse field mixer, $B = \sum_i \sigma_i^x$, and $|s\rangle = |+\rangle^{\otimes n}$. The final state after p iterations of the algorithm is

$$|\gamma, \beta\rangle = U(B, \beta_p)U(C, \gamma_p) \cdots U(B, \beta_1)U(C, \gamma_1)|s\rangle.$$

The real-valued parameters β and γ, often called angles, are selected to minimize the expected value of the cost Hamiltonian C,

$$\gamma^*, \beta^* = \underset{\gamma,\beta}{\operatorname{argmin}} \langle C\rangle = \underset{\gamma,\beta}{\operatorname{argmin}} \langle \gamma, \beta|C|\gamma, \beta\rangle.$$

Classical optimization routines such as BFGS [34] and COBYLA [33] are commonly used to find the angles.

A common measure of success for QAOA is the approximation ratio (AR), which for constrained optimization, is defined as

$$\frac{\langle C\rangle - C_{\max}}{C_{\min} - C_{\max}}$$

where $C_{\max}$ ($C_{\min}$) is the largest (smallest) eigenvalue of C. The AR is a real number between 0 and 1, with values closer to 1 indicating better solutions to the combinatorial optimization problem.

2.2 MIS

The goal of the MIS problem on a simple graph $G = (V, E)$ is to find the largest subset of vertices $I \subset V$ such that for all $i, j \in I$, $ij \notin E$. For an example, see Fig. 1. We can formulate the MIS problem as an integer program

$$\max_{x \in \{0,1\}^{|V|}} \sum_{i \in V} x_i$$
$$\text{s.t. } x_i + x_j \leq 1, \quad \forall (i, j) \in E.$$

The MIS problem is commonly written as a quadratic unconstrained binary optimization (QUBO) problem of the form [8,11,31,47]

$$\min \quad -\sum_{i \in V} x_i + J \sum_{ij \in E} x_i x_j,$$

which is mapped to H_C using the substitution $x_i \to \frac{I - Z_i}{2}$ [15]. Note that this formulation recasts MIS as a minimization problem by negating the objective function. Throughout this work, we shall refer to this encoding as the "standard QAOA constraint encoding" method.

Fig. 1. Path on four vertices, P_4. This has three unique maximum independent sets: $\{x_1, x_2\}$, $\{x_2, x_3\}$, and $\{x_3, x_4\}$.

2.3 SOPs and ESOPs

Boolean expressions are commonly written in a SOP formulation [1,5,22], in which products of Boolean variables (ANDs) are added together (ORs). Any Boolean expression can be converted to an equivalent ESOP, which is a Boolean expression consisting solely of exclusive sums of products of literals that can be negated. One can convert a Boolean expression to an ESOP expression using the identity

$$a \vee b = a\bar{b} \oplus b = a \oplus b \oplus ab.$$

ESOP formulations of Boolean expressions have applications in machine learning [30] and Toffoli gate synthesis [37]. While converting Boolean expressions to ESOP expressions is hard, there exist parallel algorithms for converting an arbitrary Boolean expression to an ESOP formulation [29]. Furthermore, there exist efficient algorithms for minimizing ESOP expressions [40]. The author of [15] introduced rules for converting Boolean expressions into sums and products of Pauli-Z operators, which are summarized in Table 1.

Table 1. The table from [15] that shows how to write a Boolean expression $f(x)$ as a Hamiltonian H_f.

$f(x)$	H_f	f(x)	H_f
x	$\frac{1}{2}I - \frac{1}{2}Z$	$\overline{x}$	$\frac{1}{2}I + \frac{1}{2}Z$
$x_1 \oplus x_2$	$\frac{1}{2}I - \frac{1}{2}Z_1 Z_2$	$\bigoplus_{j=1}^{k} x_j$	$\frac{1}{2}I - \frac{1}{2}Z_1 Z_2 \cdots Z_k$
$x_1 \wedge x_2$	$\frac{1}{4}I - \frac{1}{4}(Z_1 + Z_2 - Z_1 Z_2)$	$\bigwedge_{j=1}^{k} x_j$	$\frac{1}{2^k}\prod_j (I - Z_j)$
$x_1 \vee x_2$	$\frac{3}{4}I - \frac{1}{4}(Z_1 + Z_2 + Z_1 Z_2)$	$\bigvee_{j=1}^{k} x_j$	$I - \frac{1}{2^k}\prod_j (I + Z_j)$
$\overline{x_1 x_2}$	$\frac{3}{4}I + \frac{1}{4}(Z_1 + Z_2 - Z_1 Z_2)$	$x_1 \implies x_2$	$\frac{3}{4}I + \frac{1}{4}(Z_1 - Z_2 + Z_1 Z_2)$

Another identity within [15] that will be commonly used in this work is that for two Boolean expressions f and g represented by Hamiltonians H_f and H_g respectively, then

$$H_{f \oplus g} = H_f + H_g - 2H_f H_g. \tag{1}$$

Note in particular that for a Boolean expression of the form $(x_i \wedge x_j) \oplus (\overline{x_j} \wedge x_k)$, where $f = (x_i \wedge x_j)$ and $g = (\overline{x_j} \wedge x_k)$, then $H_f = \frac{1}{4}(I - Z_i)(I - Z_j)$ and $H_g = \frac{1}{4}(I + Z_j)(I - Z_k)$. Substituting these into Eq. 1 yields

$$
\begin{aligned}
H_{f\oplus g} &= \frac{1}{4}(I - Z_i)(I - Z_j) + \frac{1}{4}(I + Z_j)(I - Z_k) - 2\frac{1}{4}(I - Z_i)(I - Z_j)\frac{1}{4}(I + Z_j)(I - Z_k) \\
&= \frac{1}{4}(I - Z_i)(I - Z_j) + \frac{1}{4}(I + Z_j)(I - Z_k) - \frac{1}{8}(I - Z_i)(I - Z_j)(I + Z_j)(I - Z_k) \\
&= \frac{1}{4}(I - Z_i)(I - Z_j) + \frac{1}{4}(I + Z_j)(I - Z_k) \\
&= H_f + H_g
\end{aligned}
$$

as $(I - Z_j)(I + Z_j) = 0$. Thus, one can eliminate higher-order terms from the Hamiltonian formulation of $f \oplus g$ if there exists a variable of the form x_j in one expression and $\overline{x_j}$ in the other. In general, suppose we have a set of Boolean expressions $\{f_i\}_i$ where $f_i = \bigwedge_{u\in V_i} \ell_{\sigma(u)}(x_u)^i$, $V_i \subset V$, $\sigma(u) \in \{0, 1\}$, and

$$
\ell^i_{\sigma(u)} = \begin{cases} x_u & \sigma(u) = 0, \\ \overline{x_u} & \sigma(u) = 1 \end{cases}.
$$

If $|\{u \in V : \exists i, j \text{ s.t. } \ell^i_{\sigma(u)}(x_u) \neq \ell^j_{\sigma(u)}(x_u)\}| > 1$, that is, if there is at least one vertex for which x_u is in f_i and $\overline{x_u}$ is in f_j, then

$$
H_{\bigoplus_i f_i} = \sum_i H_{f_i}. \tag{2}
$$

3 ESOP Pipeline

The following sections describe the framework for generating the QAOA cost Hamiltonian, H_C, for the MIS problem. In this framework, we first convert each constraint in the constrained optimization problem to a Boolean expression. Then, we rewrite the Boolean expression into an ESOP formulation. The ESOP formulation is then multiplied by a large penalty parameter and added to the objective function of the optimization problem. We shall refer to this as the "ESOP pipeline". The objective function is then converted to a cost Hamiltonian, H_C, via the rules from [15]. Once H_C has been generated, we then use it in QAOA with the standard Pauli-X mixer. To automate and test the procedure, we developed a Python library (found at [9]) which takes a Networkx graph as input and produces a QAOA circuit for the MIS problem on said graph. These circuits are then simulated using the Qiskit Aer simulator [35] in Qiskit version 1.3.2 [36]. Classical optimization of angles γ and β is employed using COBYLA [33].

Note that the methods in this section apply to any constrained optimization problem, however we use MIS as an illustrative example. The rest of this section will apply the ESOP pipeline to MIS problems.

3.1 MIS Boolean Encoding

The first step of the ESOP pipeline for a given graph G is to generate a Boolean function $v_{\text{IS}}(\overrightarrow{x_G}) : \{0,1\}^n \rightarrow \{0,1\}$ satisfying

$$v_{\text{IS}}(\overrightarrow{x_G}) = \begin{cases} 1 & \text{if } \{k : \overrightarrow{x_G}[k] = 1\} \text{ form an independet set in G} \\ 0 & \text{otherwise.} \end{cases}$$

To do this, we will require a Boolean function that compares incident vertices. The independence constraints for the MIS problem can be written as

$$\bigwedge_{ij \in E}^{|E|} \overline{x_i \wedge x_j},$$

since if both x_i and x_j are 1, the expression is equal to 0, signaling that both i and j cannot be included in $\overrightarrow{x_G}$. Using De'Morgans law, this is equivalent to

$$v_{\text{IS}}(\overrightarrow{x_G}) = \overline{\bigvee_{ij \in E}^{|E|} x_i \wedge x_j}. \tag{3}$$

This formulation of the constraints can be refactored into an ESOP using algebraic techniques and then added to the objective function.

3.2 Converting Boolean Constraints to ESOP

By inverting both sides of Eq. (3), notice that

$$\overline{v_{\text{IS}}(\overrightarrow{x_G})} = \bigvee_{ij \in E} x_i \wedge x_j.$$

For simplicity, we rewrite each $x_i \wedge x_j$ as some e_k for $0 \leq k \leq |E| - 1$. Then for all edges $e_k \in E$,

$$\begin{aligned} \overline{v_{\text{IS}}(\overrightarrow{x_G})} &= \bigvee_{k=0}^{|E|-1} e_k \\ &= e_0 \vee e_2 \vee \cdots \vee e_{|E|-1}. \end{aligned}$$

Applying the identity $a \vee b = (a \wedge \neg b) \oplus b$ three times yields

$$\begin{aligned} \overline{v_{\text{MIS}}(\overrightarrow{x_G})} &= ((e_0 \wedge \overline{e_1}) \oplus e_1) \vee e_2 \vee e_3 \vee \cdots \vee e_{|E|-1} \\ &= (e_0 \wedge \overline{e_1} \wedge \overline{e_2}) \oplus (e_1 \wedge \overline{e_2}) \oplus (e_2) \vee e_3 \\ &\quad \vee \cdots \vee \ e_{|E|-1} \\ &= (e_0 \wedge \overline{e_1} \wedge \overline{e_2} \wedge \overline{e_3}) \oplus (e_1 \wedge \overline{e_2} \wedge \overline{e_3}) \oplus (e_2 \wedge \overline{e_3}) \\ &\quad \vee \cdots \vee \ e_{|E|-1}. \end{aligned}$$

It is clear that continuing this process until there are no more OR operations will yield an equivalent expression of $v_{\text{IS}}(\overrightarrow{x_G})$, written only using the XOR, AND, NOT, operations:

$$\begin{aligned}\overline{v_{\text{IS}}(\overrightarrow{x_G})} = e_0 &\wedge \overline{e_1} \wedge \overline{e_2} \wedge \overline{e_3} \wedge \overline{e_4} \wedge \cdots \wedge \overline{e_{|E|-1}} \\ \oplus\ e_1 &\wedge \overline{e_2} \wedge \overline{e_3} \wedge \overline{e_4} \wedge \cdots \wedge \overline{e_{|E|-1}} \\ \oplus\ e_2 &\wedge \overline{e_3} \wedge \overline{e_4} \wedge \cdots \wedge \overline{e_{|E|-1}} \\ \oplus\ e_3 &\wedge \overline{e_4} \wedge \cdots \wedge \overline{e_{|E|-1}} \\ \oplus \cdots \oplus\ & e_{|E|-2} \wedge \overline{e_{|E|-1}} \oplus e_{|E|-1}\end{aligned}$$

This expression can be written more succinctly as

$$\begin{aligned}\overline{v_{\text{IS}}(\overrightarrow{x_G})} &= \left[\bigoplus_{k=0}^{|E|-2} e_k \bigwedge_{r=k+1}^{|E|-1} \overline{e_r}\right] \oplus e_{|E|-1} \\ &= \left[\bigoplus_{k=0}^{|E|-2} (x_{i_k} \wedge x_{j_k}) \bigwedge_{r=k+1}^{|E|-1} \overline{(x_{i_r} \wedge x_{j_r})}\right] \oplus (x_{i_{|E|-1}} \wedge x_{j_{|E|-1}}) \\ &= \left[\bigoplus_{k=0}^{|E|-2} (x_{i_k} \wedge x_{j_k}) \bigwedge_{r=k+1}^{|E|-1} (\overline{x_{i_r}} \vee \overline{x_{j_r}})\right] \oplus (x_{i_{|E|-1}} \wedge x_{j_{|E|-1}}) \\ &= \left[\bigoplus_{k=0}^{|E|-2} (x_{i_k} \wedge x_{j_k}) \bigwedge_{r=k+1}^{|E|-1} [(\overline{x_{i_r}} \wedge x_{j_r}) \oplus \overline{x_{j_r}}]\right] \oplus (x_{i_{|E|-1}} \wedge x_{j_{|E|-1}}).\end{aligned}$$

The above process transforms the individual products into even longer terms of conjunctions and negated conjunctions that can be simplified with De'Morgan's law. See Sect. 3.4 for an example of the ESOP derived for the path on four vertices.

3.3 Transformation from Boolean Expression to a Hamiltonian

Once the constraints have been converted into the ESOP formulation $\bigoplus_i c_i$, where $c_i = \bigwedge_j x_j$ with some x_j being negated, we will convert each term into a Hamiltonian using Table 1. Specifically, each variable $x_j \to \frac{1}{2}(I - Z_j)$ and each $\overline{x_j} \to \frac{1}{2}(I + Z_j)$. Where we diverge from Table 1 is how we handle $\bigwedge_{j=1}^{k} x_j$. Similar to [2], we will multiply every second variable in the term c_i by -1, resulting in

$$c_i = \bigwedge_{j=1}^{k} x_j \to H_{c_i} = \frac{1}{2^k} \prod_j (-1)^{j+1}(I - Z_j).$$

Since there exist x_{j_r} and $\overline{x_{j_r}}$ terms in the ESOP formulation, we can use Eq. 2 to write the constraint penalty Hamiltonian as

$$H_c := H_{\oplus_i c_i} = \sum_i H_{c_i}.$$

With the MIS objective Hamiltonian $H_{\mathrm{MAX}} = \sum_{j \in V} \frac{1}{2}(I - Z_j)$, the full cost Hamiltonian is given by

$$H_C = -H_{\mathrm{MAX}} - 2\,V|H_c = -H_{\mathrm{MAX}} - 2\,V|\sum_i H_{c_i}.$$

Note that normally when minimizing an objective function, one adds the Hamiltonians representing constraints to the objective. However, since we derived $\overline{v_{\mathrm{IS}}(\overrightarrow{x_G})}$ above, we subtract it from the objective instead of adding it. Also note that we use $2|V|$ as a penalty in our simulations, but any sufficiently large penalty parameter suffices.

3.4 Example: P_4

Consider the path graph on four vertices, denoted P_4, in Fig. 1. We will apply the ESOP pipeline to this graph to find a cost Hamiltonian that encodes MIS on this problem. Then, we simulate QAOA solving this problem using the Qiskit Aer simulator. It is clear that there are three solutions to this problem: $\{x_1, x_2\}$, $\{x_2, x_3\}$, and $\{x_3, x_4\}$. Furthermore, the objective of this problem is $\max \sum_{i=1}^{4} x_i$, or equivalently, $\min\ -\sum_{i=1}^{4} x_i$.

The ESOP that encodes the MIS constraints is

$$\begin{aligned}
\overline{v_{\mathrm{IS}}(\overrightarrow{x_{P_4}})} &= (x_2 \wedge x_4) \vee (x_4 \wedge x_1) \vee (x_1 \wedge x_3) \\
&= e_1 \vee e_2 \vee e_3 \\
&= (e_1 \wedge \overline{e_2} \wedge \overline{e_3}) \oplus (e_2 \wedge \overline{e_3}) \oplus (e_3) \\
&= (x_2 \wedge x_4 \wedge \overline{x_4 \wedge x_1} \wedge \overline{x_1 \wedge x_3}) \oplus (x_4 \wedge x_1 \wedge \overline{x_1 \wedge x_3}) \oplus (x_1 \wedge x_3) \\
&= (\overline{x_1} \wedge x_2 \wedge x_4) \oplus (x_1 \wedge \overline{x_3} \wedge x_4) \oplus (x_1 \wedge x_3).
\end{aligned}$$

We shall refer to $(\overline{x_1} \wedge x_2 \wedge x_4)$ as c_1, $(x_1 \wedge \overline{x_3} \wedge x_4)$ as c_2, and $(x_1 \wedge x_3)$ as c_3. Note that by Eq. 2, $H_{c_1 \oplus c_2 \oplus c_3} = H_{c_1} + H_{c_2} + H_{c_3}$. Thus,

$$H_C = -H_{\mathrm{MAX}} - 8\sum_{i=1}^{3} H_{c_i}$$

where

$$\begin{aligned}
H_{c_1} &= \frac{1}{8}(I + Z_1)(-I + Z_2)(I - Z_4) \\
H_{c_2} &= \frac{1}{8}(I - Z_1)(-I - Z_3)(I - Z_4) \\
H_{c_3} &= -\frac{1}{4}I + \frac{1}{4}(Z_1 + Z_3 - Z_1 Z_3).
\end{aligned}$$

Figure 2 shows the counts when simulating QAOA with this cost Hamiltonian. Note that the optimal solutions of $\{x_1, x_2\}$, $\{x_2, x_3\}$, and $\{x_3, x_4\}$ correspond to bitstrings 1100, 0110, and 0011 respectively.

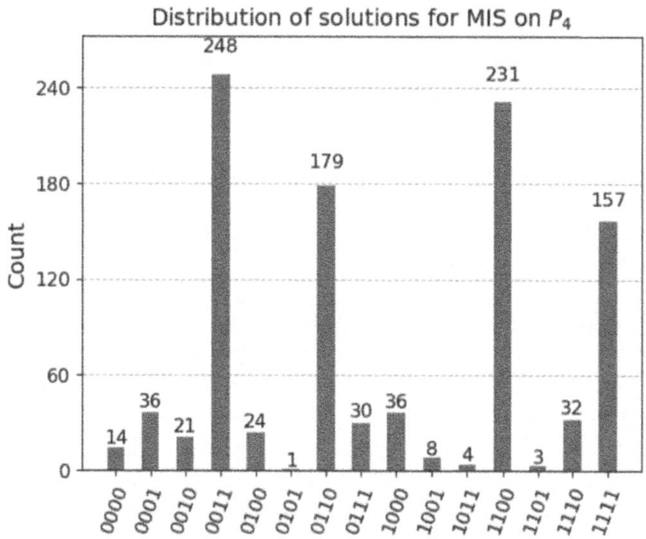

Fig. 2. Measurements for solving MIS on P_4 with only one layer of QAOA and constraints encoded as ESOPs using 1024 shots. The rightmost qubit corresponds to the value of x_1 and the leftmost corresponds to x_4.

4 Results

We implement the ESOP pipeline for penalizing constraints with QAOA on over 3000 graphs: all 3-vertex (2 instances), 4-vertex (6 instances), 5-vertex (21 instances), and 6-vertex graphs graphs (112 instances), 700 randomly sampled 7-vertex graphs, 500 randomly generated graphs for all vertex counts with 8 vertices and between 11 and 14 vertices, and 50 randomly generated graphs for all vertex counts between 15 and 20 vertices. All tested graphs are connected and nonisomorphic. Note that graphs with less than 8 vertices were generated using the g6 files from [24]. We calculate the approximation ratio for each instance and then compare the results to the standard QAOA penalization technique described in Sect. 2.2.

The approximation ratios for tested graphs with between 3 and 14 vertices is found in Fig. 3, while the approximation ratios for graphs with between 15 and 20 vertices is found in Fig. 4. Note that when $p = 1$, the ESOP constraint encoding approach results in higher average approximation ratios across each graph size except $n = 3$, which is a dataset consisting of only two graphs. At $p = 2$ and $p = 3$, the ESOP constraint encoding method results in higher approximation ratios than QAOA for all size graphs except $n = 7$. The average approximation ratios of both approaches are listed in Table 2, as well as the percent change between the standard MIS constraint encoding for QAOA and the ESOP constraint encoding.

When $p = 1$, the ESOP constraint formulations result in higher approximation ratios than standard QAOA penalization approaches on approximately 64% of the tested graphs.

Table 2. Average approximation ratios comparing standard QAOA penalization with ESOP-QAOA across different circuit QAOA layers ($p = 1, 2, 3$) and graph sizes. Percentage changes show ESOP performance relative to standard QAOA, with improvements in blue and negative percent changes in red.

Vertices	QAOA p=1			QAOA p=2			QAOA p=3		
	Standard	ESOP	(% change)	Standard	ESOP	(% change)	Standard	ESOP	(% change)
3	0.660	0.655	(-0.8)	0.749	0.792	(+5.7)	0.835	0.910	(+9.0)
4	0.535	0.560	(+4.7)	0.600	0.782	(+30.3)	0.719	0.787	(+9.5)
5	0.470	0.580	(+23.4)	0.592	0.597	(+0.8)	0.655	0.658	(+0.5)
6	0.411	0.468	(+13.9)	0.563	0.570	(+1.2)	0.621	0.654	(+5.3)
7	0.350	0.384	(+9.7)	0.450	0.441	(-2.0)	0.625	0.586	(-6.2)
8	0.279	0.359	(+28.7)	0.392	0.398	(+1.5)	0.526	0.599	(+13.9)
11	0.194	0.233	(+20.1)	0.231	0.250	(+8.2)	0.292	0.350	(+19.9)
12	0.169	0.215	(+27.2)	0.222	0.230	(+3.6)	0.295	0.303	(+2.7)
13	0.161	0.182	(+13.0)	0.195	0.219	(+12.3)	0.254	0.280	(+10.2)
14	0.156	0.191	(+22.4)	0.194	0.211	(+8.8)	0.245	0.255	(+4.1)
15	0.149	0.172	(+15.4)	0.189	0.216	(+14.3)	0.215	0.252	(+17.2)
16	0.118	0.124	(+5.1)	0.154	0.166	(+7.8)	0.191	0.208	(+8.9)
17	0.098	0.112	(+14.3)	0.130	0.141	(+8.5)	0.160	0.190	(+18.8)
18	0.077	0.089	(+15.6)	0.110	0.123	(+11.8)	0.132	0.159	(+20.5)
19	0.060	0.064	(+6.7)	0.089	0.096	(+7.9)	0.110	0.129	(+17.3)
20	0.036	0.041	(+13.9)	0.077	0.092	(+19.5)	0.095	0.100	(+5.3)

5 Discussion

In this work, we develop an ESOP pipeline for encoding constrained optimization problems into QAOA. In this pipeline, we first write all constraints as Boolean expressions, and then write the Boolean expressions into an ESOP formulation, in contrast to other QAOA constraint encoding methods. Then, we multiply these terms by a large penalty parameter and add it to the objective function. Our experimental results demonstrate that rewriting constraints as ESOP statements and penalizing them in the QAOA cost Hamiltonian results in significantly higher approximation ratios than standard constraint penalization methods. Out of 54 test configurations spanning different graph sizes (3–20 vertices) and QAOA layer depths ($p = 1, 2, 3$), ESOP-QAOA achieves higher average approximation ratios in 51 cases. The method shows particularly large approximation ratio improvements on smaller graphs, with notable increases of 30.3% for 4-vertex

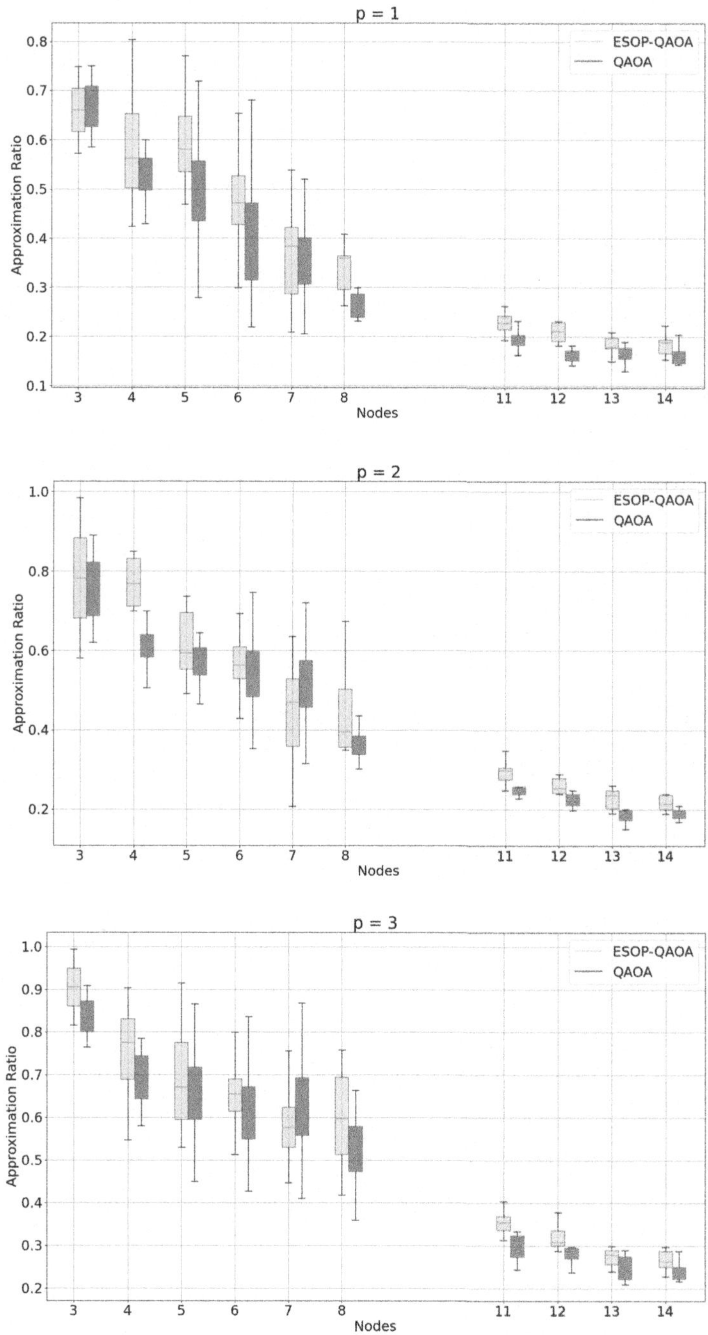

Fig. 3. Comparison of approximation ratios between regular QAOA and our ESOP-QAOA for the MIS problem graphs of size 3 up to 14. Note we did not test our method of graphs of size 9 or 10 vertices.

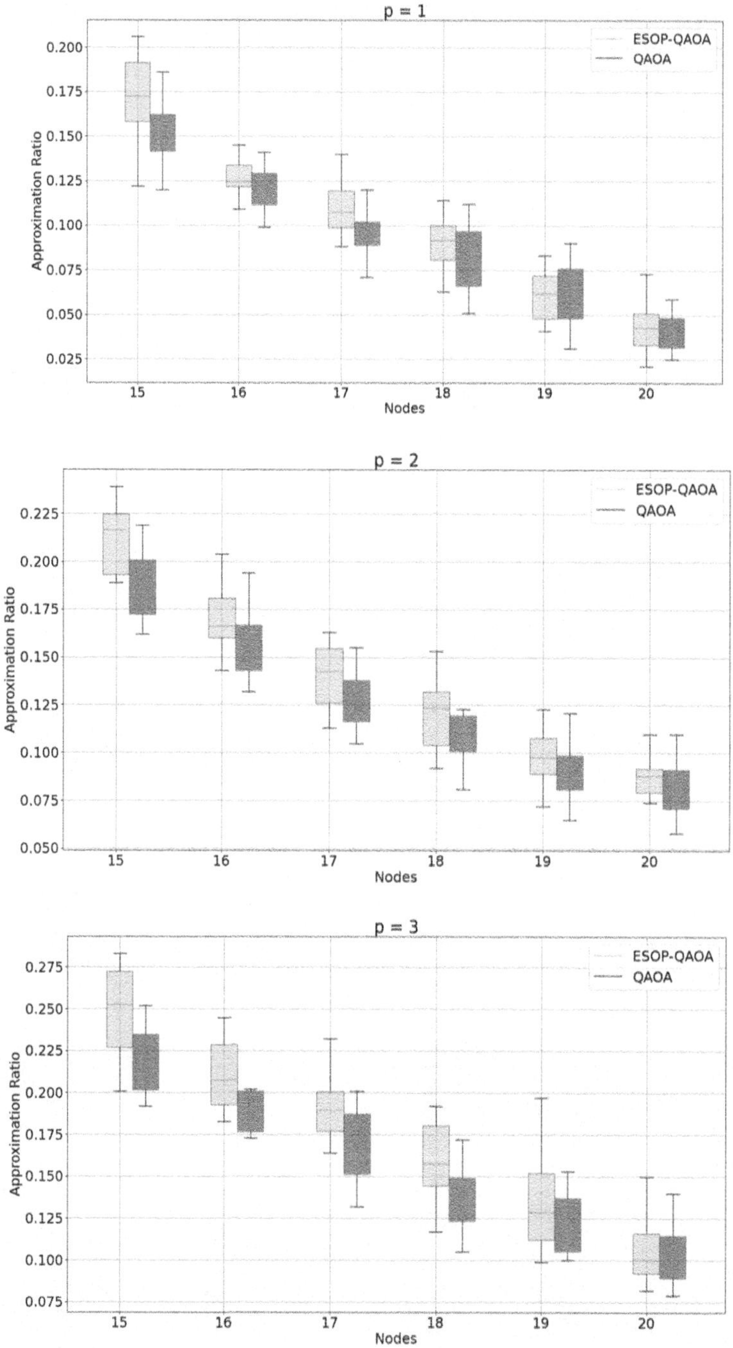

Fig. 4. Comparison of approximation ratios between regular QAOA and our ESOP-QAOA for the MIS problem graphs of size 15 up to 20.

graphs at $p = 2$, 28.7% for 8-vertex graphs at $p = 1$, and 27.2% for 12-vertex graphs at $p = 1$. Even for larger problem instances (15–20 vertices), the ESOP constraint encoding method shows positive increases in approximation ratios, suggesting scalability.

This work highlights the importance of problem encoding in the quantum approximate optimization algorithm, as the ESOP formulation of constraints implements the same logic as the usual Boolean implementation. The ESOP formulation appears to work well for MIS because of the existence of x_i and $\overline{x_i}$ terms in the Boolean expression, which allows for constraint Hamiltonian simplifications. Future interesting research avenues include developing problem-based ESOP simplification methods, similar to the higher order terms canceling in MIS, for other constrained optimization problems. Other future avenues of research include identifying problem classes where ESOP constraints guarantee simpler implementations with shorter circuit depth and fewer ancilla qubits than standard penalty methods, and developing parameter transfer techniques [27] to further enhance optimization efficiency. Finally, the ESOP constraint formulations result in higher approximation ratios than standard constraint encoding methods on approximately 64% of the tested graphs. Characterizing these graphs would allow one to select the best encoding method for a particular MIS instance.

Acknowledgments. M.B., S.S., and R.H. acknowledge NSF CNS-2244512. A.W. and R.H. acknowledge NSF CCF 2210063.

Data Availibility Statement. The code and data for this research can be found at https://github.com/mxttbrunet/Quantum-Walk-Project.

Disclosure of Interests. The authors have no competing interests to declare that are relevant to the content of this article.

References

1. Abhyankar, S.: Minimal "sum of products of sums" expressions of boolean functions. IRE Trans. Electron. Comput. **4**, 268–276 (2009)
2. Al-Bayaty, A., Perkowski, M.: BHT-QAOA: the generalization of quantum approximate optimization algorithm to solve arbitrary boolean problems as hamiltonians. Entropy **26**(10), 843 (2024)
3. Allcock, J., Santha, M., Yuan, P., Zhang, S.: On the dynamical lie algebras of quantum approximate optimization algorithms (2024). https://doi.org/10.48550/ARXIV.2407.12587, https://arxiv.org/abs/2407.12587
4. Angara, P.P., Lykov, D., Stege, U., Alexeev, Y., Müller, H.: The art of avoiding constraints: a penalty-free approach to constrained combinatorial optimization with QAOA. arXiv preprint arXiv:2503.10077 (2025)
5. Arevalo, B.: A method to simplify a Boolean function into a near minimal sum-of-products for programmable logic arrays. IEEE Trans. Comput. **100**(11), 1028–1039 (1978)
6. Bärtschi, A., Eidenbenz, S.: Grover mixers for QAOA: shifting complexity from mixer design to state preparation. In: 2020 IEEE International Conference on Quantum Computing and Engineering (QCE), pp. 72–82. IEEE (2020)

7. Bartschi, A., Eidenbenz, S.: Short-depth circuits for dicke state preparation. In: 2022 IEEE International Conference on Quantum Computing and Engineering (QCE), pp. 87–96. IEEE (Sep 2022). https://doi.org/10.1109/qce53715.2022.00027
8. Brady, L.T., Hadfield, S.: Iterative quantum algorithms for maximum independent set. Phys. Rev. A **110**(5), 052435 (2024)
9. Brunet, M., Shah, S., Atallah, M.: Quantum-walk-project. https://github.com/mxttbrunet/Quantum-Walk-Project
10. Bucher, D., Stein, J., Feld, S., Linnhoff-Popien, C.: If-QAOA: a penalty-free approach to accelerating constrained quantum optimization. arXiv preprint arXiv:2504.08663 (2025)
11. Ebadi, S., et al.: Quantum optimization of maximum independent set using RYDBERG atom arrays. Science **376**(6598), 1209–1215 (2022)
12. Farhi, E., Goldstone, J., Gutmann, S.: A quantum approximate optimization algorithm (2014). https://doi.org/10.48550/ARXIV.1411.4028, https://arxiv.org/abs/1411.4028
13. Golden, J., Bärtschi, A., O'Malley, D., Eidenbenz, S.: Threshold-based quantum optimization. In: 2021 IEEE International Conference on Quantum Computing and Engineering (QCE), pp. 137–147. IEEE (2021)
14. Goldstein-Gelb, B., Lotshaw, P.C.: Convergence guarantee for linearly-constrained combinatorial optimization with a quantum alternating operator ansatz. arXiv preprint arXiv:2409.18829 (2024)
15. Hadfield, S.: On the representation of Boolean and real functions as Hamiltonians for quantum computing. ACM Trans. Quantum Comput. **2**(4), 1–21 (2021)
16. Hadfield, S., Wang, Z., O'gorman, B., Rieffel, E.G., Venturelli, D., Biswas, R.: From the quantum approximate optimization algorithm to a quantum alternating operator ansatz. Algorithms **12**(2), 34 (2019)
17. Herrman, R., Lotshaw, P.C., Ostrowski, J., Humble, T.S., Siopsis, G.: Multi-angle quantum approximate optimization algorithm. Sci. Rep. **12**(1), 1–10 (2022)
18. Herrman, R., Treffert, L., Ostrowski, J., Lotshaw, P.C., Humble, T.S., Siopsis, G.: Globally optimizing qaoa circuit depth for constrained optimization problems. Algorithms **14**(10), 294 (2021)
19. Kazi, S., Larocca, M., Farinati, M., Coles, P.J., Cerezo, M., Zeier, R.: Analyzing the quantum approximate optimization algorithm: ansätze, symmetries, and lie algebras (2024). https://arxiv.org/abs/2410.05187
20. Larocca, M., et al.: Barren plateaus in variational quantum computing. Nat. Rev. Phys. **7**(4), 174–189 (2025). https://doi.org/10.1038/s42254-025-00813-9
21. Lucas, A.: Ising formulations of many np problems. Front. Phys. **2**, 74887 (2014)
22. Luccio, F., Pagli, L.: On a new boolean function with applications. IEEE Trans. Comput. **48**(3), 296–310 (2002)
23. Marsh, S., Wang, J.: A quantum walk-assisted approximate algorithm for bounded np optimisation problems. Quantum Inf. Process. **18**(3), 61 (2019)
24. McKay, B.: Graphs [data set], found at http://users.cecs.anu.edu.au/~bdm/data/graphs.html
25. Meuli, G., Schmitt, B., Ehlers, R., Riener, H., De Micheli, G.: Evaluating ESOP optimization methods in quantum compilation flows. In: International Conference on Reversible Computation, pp. 191–206. Springer (2019)
26. Montanez-Barrera, J., Michielsen, K.: Towards a universal QAOA protocol: evidence of quantum advantage in solving combinatorial optimization problems. arXiv preprint arXiv:2405.09169 (2024)

27. Montanez-Barrera, J., Willsch, D., Michielsen, K.: Transfer learning of optimal QAOA parameters in combinatorial optimization. Quantum Inf. Process. **24**(5), 129 (2025)
28. Ozaeta, A., van Dam, W., McMahon, P.L.: Expectation values from the single-layer quantum approximate optimization algorithm on ising problems. Quantum Sci. Technol. **7**(4), 045036 (2022). https://doi.org/10.1088/2058-9565/ac9013
29. Papakonstantinou, G.: A parallel algorithm for minimizing ESOP expressions. J. Circ. Syst. Comput. **23**(01), 1450015 (2014)
30. Perkowski, M., Ross, T., Gadd, D., Goldman, J.A., Song, N.: Application of ESOP minimization in machine learning and knowledge discovery. In: Proc. Reed Muller. vol. 95 (1995)
31. Pichler, H., Wang, S.T., Zhou, L., Choi, S., Lukin, M.D.: Quantum optimization for maximum independent set using rydberg atom arrays. arXiv preprint arXiv:1808.10816 (2018)
32. Ponce, M., et al.: Graph decomposition techniques for solving combinatorial optimization problems with variational quantum algorithms. Quantum Inf. Process. **24**(2), 60 (2025)
33. Powell, M.J.: A direct search optimization method that models the objective and constraint functions by linear interpolation. In: Advances in Optimization and Numerical Analysis, pp. 51–67. Springer (1994)
34. Press, W.H., Flannery, B.P., Teukolsky, S.A.: Numerical Recipes in Fortran 77: The Art of Scientific Computing. Cambridge University Press, 2nd edn. (1993). https://people.sc.fsu.edu/inavon/5420a/DFP.pdf
35. Qiskit: Qiskit 0.16 release notes. https://quantum.cloud.ibm.com/docs/en/api/qiskit/release-notes/0.16
36. Qiskit: Qiskit sdk 1.3 release notes. https://quantum.cloud.ibm.com/docs/en/api/qiskit/release-notes/1.3
37. Sanaee, Y., Dueck, G.W.: Esop-based toffoli network generation with transformations. In: 2010 40th IEEE International Symposium on Multiple-Valued Logic, pp. 276–281. IEEE (2010)
38. Schmitt, B., Soeken, M., De Micheli, G., Mishchenko, A.: Scaling-up ESOP synthesis for quantum compilation. In: 2019 IEEE 49th International Symposium on Multiple-Valued Logic (ISMVL), pp. 13–18. IEEE (2019)
39. Shaydulin, R., et al.: Evidence of scaling advantage for the quantum approximate optimization algorithm on a classically intractable problem. Sci. Adv. **10**(22) (2024)
40. Stergiou, S., Daskalakis, K., Papakonstantinou, G.: A fast and efficient heuristic ESOP minimization algorithm. In: Proceedings of the 14th ACM Great Lakes Symposium on VLSI, pp. 78–81 (2004)
41. Tate, R., Farhadi, M., Herold, C., Mohler, G., Gupta, S.: Bridging classical and quantum with SDP initialized warm-starts for QAOA. arXiv preprint arXiv:2010.14021 (2020)
42. Tate, R., Moondra, J., Gard, B., Mohler, G., Gupta, S.: Warm-started qaoa with custom mixers provably converges and computationally beats goemans-williamson's max-cut at low circuit depths. Quantum **7**, 1121 (2023)
43. Vijendran, V., Das, A., Koh, D.E., Assad, S.M., Lam, P.K.: An expressive ansatz for low-depth quantum approximate optimisation. Quantum Sci. Technol. **9**(2), 025010 (2024)
44. Wang, Z., Rubin, N.C., Dominy, J.M., Rieffel, E.G.: Xy mixers: analytical and numerical results for the quantum alternating operator ansatz. Phys. Rev. A **101**(1), 012320 (2020)

45. Wilkie, A., DeLise, A., Del Real, A., Herrman, R., Ostrowski, J.: Learning feasible quantum states for quadratic constrained binary optimization problems. arXiv preprint arXiv:2508.02590 (2025)
46. Wilkie, A., Gaidai, I., Ostrowski, J., Herrman, R.: Quantum approximate optimization algorithm with random and subgraph phase operators. Phys. Rev. A **110**(2), 022441 (2024)
47. Xu, H., Liu, X., Pothen, A., Safro, I.: Qaoa parameter transferability for maximum independent set using graph attention networks. arXiv preprint arXiv:2504.21135 (2025)
48. Zhou, Z., Du, Y., Tian, X., Tao, D.: QAOA-in-QAOA: solving large-scale maxcut problems on small quantum machines. Phys. Rev. Appl. **19**(2), 024027 (2023)

A Performance Comparison of Variable Encoding Techniques for QUIO and QUBO Problems

Pedro Maciel Xavier[1], Yirang Park[1], Diego Noble[2], Italo Santana[2], Miguel Paredes[2], and David E. Bernal Neira[1](✉)

[1] Davidson School of Chemical Engineering, Purdue University, 480 Stadium Mall Drive, West Lafayette, IN 47907, USA
dbernaln@purdue.edu

[2] Dell Research, Dell Technologies, Rio de Janeiro, Brazil
{diego.noble,italo.santana,miguel.paredes}@dell.com

Abstract. In this paper, we explore the difficulty of encoding Quadratic Integer Optimization (QIP) problems into Quadratic Unconstrained Integer Optimization (QUIO) formulations with a target integer basis. These formulations are suitable for solving on quantum computers based on qudits, which natively extend the integer representation of their qubit-based counterparts. For qubit-based computers, the standard framework for representing discrete optimization problems is the Quadratic Unconstrained Binary Optimization (QUBO) formulation. Decision variables can, for QUBOs, assume only binary values, while for QUIOs, they can represent integer values from zero up to a machine-dependent maximum. One advantage of working with a larger domain of decision variables is that it enables QIP problems to be cast as QUIO formulations. As our results highlight, these formulations use fewer decision variables than, for example, Quadratic Unconstrained Binary Optimization (QUBO)-based formulations. We selected various candidate problems to empirically verify our reformulations: Quadratic Facility Location Problem, Quadratic Inventory Management Problem, Quadratic Vehicle Routing Problem, and Quadratic Knapsack Problem. Problem instances for these problems are diverse in terms of data distribution and size, and are reformulated into QUBO and QUIO. We compare these formulations, characterizing them using metrics associated with solution performance. Moreover, using qubit- and qudit-based entropy quantum machines, we compared the performance of the resulting formulations for instances amenable to these architectures. Our primary goal is to conduct computational experiments to verify the impacts of each encoding type on these problems, aiming to find insights that could potentially generalize to similar optimization problems. Moreover, we also provide guidelines to accelerate the encoding process, ensuring that the potential quantum advantage offered by qudit quantum computers is not lost during classical pre-processing and problem reformulation. Finally, we provide open-source software to perform the reformulations and communicate them with qudit-based entropy quantum devices, allowing others to map and solve QIP problems using QUIO reformulations.

V. Aggarwal et al. (Eds.): PQAI 2025, CCIS 2724, pp. 105–122, 2026.
https://doi.org/10.1007/978-981-95-7829-0_10

Keywords: QIP · QUIO · QUBO · variable encoding comparison

1 Introduction

Quadratic Integer Problems (QIPs) belong to a class of computational problems in which the objective is a quadratic function and the constraints are linear. Problems of this class can be found in many fields, including engineering, finance, and logistics [12]. Although immensely versatile for modeling, QIP is in the worst-case complexity class of NP [9]. Beyond traditional solvers, such as branch-and-bound algorithms or heuristics executed on classical computers, QIPs can also be addressed by quantum computers via reformulating the problem into a Quadratic Unconstrained Integer Optimization (QUIO) problem or into a Quadratic Unconstrained Binary Optimization (QUBO) problem. Quadratic nonlinearity in the objective aligns well with native operations in many quantum algorithms, making QIPs particularly amenable to quantum formulations.

Mapping constrained optimization problems into formulations amenable to quantum computation is essential for leveraging emerging (quantum) hardware. This mapping is not unique and might reveal strengths associated with a given quantum solver, strengths that often diverge from classical optimization heuristics. Although progress has been made in mapping constrained problems, such as Quadratic Unconstrained Binary Optimization (QUBO) methods [14], traditional quantum architectures and algorithms are not inherently tied to the QUBO formalism.

In general terms, a QUIO problem encodes decision variables using an integer basis, whereas a QUBO problem encodes the same variables using a binary basis, thus making QUIO a more general form of QUBO. For both, the resulting formulation is derived from a QIP problem by introducing slack variables $s \in \mathbb{Z}^m$ for QUIO and $s \in \{0,1\}^m$ for QUBO, and penalizing infeasibility in the objective function instead of enforcing $Ax \leq b$ directly.

Recently, a class of quantum computers capable of encoding information in an integer basis has emerged, with Quantum Computing Inc.'s `Dirac-3` serving as a notable example [8]. These machines are based on qudits, which are a generalization of qubits that allow for a dimensional extension of the computational basis of qubits. In other words, qudits can represent more than two states, contrary to (0 and 1), as natively represented using qubits. From the perspective where a QIP problem has decision variables with a broad domain larger than binary, a reformulation of a QIP into QUIO can be more suitable, as the QUIO can be solved directly on these new devices rather than encoding the integer variables into a binary (e.g., one-hot encoding), as seen in QUBO form. It is essential to highlight that a qudit is not range-free due to physical and measurement limitations. Additionally, problems formulated as QUBOs can be directly solved on qubit-based quantum systems (e.g., quantum annealers), which are more mature technologies than qudit-based devices and typically have a larger number of qubits. This trade-off, therefore, implies that a QIP recast as QUBO may be solved better than a QUIO, even considering the required number of qubits to represent this problem. This, of course, needs to account for all forms of pre- and post-processing tasks (e.g., problem compilation, solution decoding, etc.) when considering either a QUIO or QUBO formulation.

In this study, our objective was to investigate the different encoding types for QUIO and QUBO and their impact on space utilization both before and after solving the optimization. In detail, we highlight the following contributions that are covered in this paper:

1. Define four classes of QIPs and describe the procedure for generating their problem instances.
2. Conduct computational experiments over these problem instances using variable encoding techniques for QUBOs and their comparison with QUIOs. We also extend these experiments to several instances regarding solutions of a qudit-based quantum device, aiming for the global optimal solution of an exact solver.
3. Assess how different encoding techniques influence the performance of the QUIO and QUBO formulations.

The remainder of this paper is structured as follows. Section 2 provides an overview of the literature on problem reformulations for QUBOs and QUIOs. In Sect. 3, we review the mathematical forms of QIPs and their formulations into QUIOs and QUBOs. Section 4 explains the implementation details of the QUBO and QUIO reformulations. Next, Sect. 5 details the optimization problems and their corresponding formulations. Section 6 reports the experimental results on the tested encoding strategies, both on the resulting reformulated problem and when running them on a qudit-based Entropy Quantum Computer. Finally, Sect. 7 presents final remarks and outlines research gaps that could be addressed in future work.

2 Literature Review

Quantum optimization has largely been centered around *Quadratic Unconstrained Binary Optimization (QUBO)* formulations, which map discrete optimization problems into forms compatible with the transverse field Ising spin model [5]. This mapping is well aligned with hardware based on qubits, where variables assume binary values, and has been widely used in quantum gate-based and annealing systems.

To reformulate more general discrete problems, such as *Quadratic Integer Programs (QIPs)* into QUBOs, classical pre-processing is required. Typical steps include binarizing integer variables, introducing slack variables to encode inequalities, and penalizing constraint violations in the objective function [14]. These steps often lead to significant growth in problem size, increased density of quadratic terms, and degradation in sparsity. Such transformations can impose burdens on quantum devices, particularly when embedding and calibration budgets are limited.

Recently, attention has shifted to quantum hardware built from *qudits*, or d-level quantum systems, which generalize the qubit model [13]. Qudits can represent larger alphabets natively, reducing the need for binary encodings and potentially enabling more compact problem representations. In gate-based settings, qudit circuits have been shown to reduce circuit depth and offer improved connectivity and expressivity [1,13]. Such hardware capabilities enable a more natural and efficient representation of integer variables, motivating the development of integer-native reformulations.

The idea of exploiting qudit capabilities has been explored in algorithm design for certain structured problems [2,11]. A broader overview of high-dimensional quantum computing also highlights scenarios where qudits offer both algorithmic and architectural benefits [13].

However, these efforts have not extended to general-purpose reformulations of QIPs into *Quadratic Unconstrained Integer Optimization (QUIO)* problems. To our knowledge, there are currently no systematic studies or software infrastructures for directly reformulating QIPs into QUIOs, which are designed for qudit-compatible quantum optimization. This is a notable gap, especially given the growing availability of early-stage qudit hardware, and alternatives are limited in terms of features (e.g., PyQUBO [15]) or efficiency.

Libraries such as QUBO.jl [14] extensively support QUBO reformulations, but do not provide analogous support for QUIO targets. This work aims to fill that gap by introducing structured reformulations for QUIO, comparing them to standard QUBO approaches, and empirically evaluating them across representative QIP classes. By leveraging direct integer encodings, QUIO reformulations promise to improve scalability and maintain sparsity, aligning well with current and future qudit-based computing capabilities.

3 Mathematical Formulations for QIPs, QUIOs, and QUBOs

In this section, we reformulate optimization problems with quadratic objectives, linear constraints, and integer variables into quadratic integer programs (QIPs), which are more tractable problems and align well with the scope defined for this work. We then convert these QIPs into quadratic unconstrained integer optimization (QUIO) problems. QUIOs extend QUBOs (Quadratic Unconstrained Binary Optimization), which require binary encodings and a significantly larger number of variables.

QUIOs directly exploit quantum hardware with *qudits* (multi-level systems), unlike QUBOs, which rely on *qubits* (two-level systems). The availability of prototype qudit-based machines motivates the use of direct integer formulations, as QUIOs preserve sparsity and circumvent the exponential blowup of binary encodings in QUBOs. One such machine is Quantum Computing Inc. (QCi)'s `Dirac-3` machine [8] (see Sect. 4.1 for more information).

We consider problems of the form:

$$\min_{x \in \{0,\dots,U\}^n} x^\top Q x + c^\top x \quad \text{s.t.} \quad Ax \leq b, \tag{1}$$

with $Q \in \mathbb{R}^{n\times n}$, $A \in \mathbb{R}^{m\times n}$, $b \in \mathbb{R}^m$. Since QIPs are discrete, we assume without loss of generality that their data can be scaled to integer values: $Q \in \mathbb{Z}^{n\times n}$, $A \in \mathbb{Z}^{m\times n}$, $b \in \mathbb{Z}^m$. For all notations used in this section, we refer the reader to Table 1. Therefore, we consider QIPs in the following form:

$$\min_{x\in\mathbb{Z}^n} x^\top Qx + c^\top x \quad \text{s.t.} \quad Ax \leq b, \qquad Q \in \mathbb{Z}^{n\times n},\ A \in \mathbb{Z}^{m\times n},\ b \in \mathbb{Z}^m. \tag{2}$$

Table 1. Notation summary for QIP, QUIO, and QUBO reformulations.

Symbol	Description
n	Number of original decision variables
m	Number of linear inequality constraints
U	Upper bound on integer variables $x_i \in \{0,\ldots,U_i\}$
$Q \in \mathbb{Z}^{n\times n}$	Quadratic coefficient matrix
$c \in \mathbb{Z}^n$	Linear coefficient vector
$A \in \mathbb{Z}^{m\times n}$	Constraint coefficient matrix
$b \in \mathbb{Z}^m$	Constraint right-hand side
$x \in \mathbb{Z}^n$	Original integer decision variables
$s \in \mathbb{Z}^m$	Integer slack variables for relaxed constraints
$y = [x; s] \in \mathbb{Z}^{n+m}$	QUIO decision vector
$D = \text{diag}(\mu)$	Penalty weight matrix for QUIO constraints
$G \in \mathbb{Z}^{(n+m)\times(n+m)}$	Block-quadratic matrix in QUIO form
N	Number of binary variables encoding x in QUBO
K	Number of binary slack variables in QUBO
$E \in \mathbb{Z}^{n\times N}$	Binary encoding matrix for variables
$T \in \mathbb{Z}^{m\times K}$	Binary slack encoding matrix
$z \in \{0,1\}^N$	Binary vector encoding original variables in QUBO
$t \in \{0,1\}^K$	Binary vector encoding slack variables in QUBO
$w = [z; t] \in \{0,1\}^{N+K}$	Full QUBO decision vector
$\Lambda = \text{diag}(\lambda)$	Penalty weight matrix in QUBO
$H \in \mathbb{Z}^{(N+K)\times(N+K)}$	QUBO quadratic cost matrix

We demonstrate, in Table 2, a summary of the main differences among the QIP, QUBO, and QUIO. In Subsects. 3.1 and 3.2, we present the formal details of QIPs reformulated to QUIO and QUBO.

3.1 QUIO Reformulation (with Slack)

A reformulation of QIP (Eq. 2) to QUIO requires introducing slack variables $s \in \mathbb{Z}^m$ and penalizing infeasibility in the objective function instead of enforcing

Table 2. Comparison of QIP, QUBO, and QUIO formulations and their connection to quantum hardware.

Acronym	Decision variables	Constraints	Link to quantum hardware
QIP	integers $x \in \mathbb{Z}^n$	linear $Ax \le b$	natural mixed-integer model
QUBO	binaries $z \in \{0,1\}^N$	none	matches qubits (2-level)
QUIO	integers $w \in \mathbb{Z}^\ell$	none	matches *qudits* (multi-level)

$Ax \le b$ directly. We define residual as $r := Ax + s - b$, and weight violations using $\mu \in \mathbb{R}_+^m$, with $M = \mathrm{diag}(\mu)$, resulting in the following form:

$$\min_{x \in \mathbb{Z}^n,\ s \in \mathbb{Z}_+^m} x^\top Q x + c^\top x + r^\top M r, \quad \text{where } r := Ax + s - b. \tag{3}$$

Block-Quadratic Form in $y = [x;\ s]$ Define:

$$y = \begin{bmatrix} x \\ s \end{bmatrix} \in \mathbb{Z}^{n+m}, \quad r := Ax + s - b.$$

Expanding $r^\top M r$ yields:

$$\boxed{\min_{y \in \mathbb{Z}^{n+m}} y^\top G y + g^\top y + \text{const}} \tag{4}$$

where

$$G = \begin{bmatrix} Q + A^\top M A & A^\top M \\ MA & M \end{bmatrix}, \quad g = \begin{bmatrix} c - 2A^\top M b \\ -2Mb \end{bmatrix}, \quad \text{const} = b^\top M b.$$

This reformulation removes constraints from the QIP and yields a QUIO compatible with qudit hardware. In the next section, we compare this to the traditional QUBO encoding.

3.2 QUBO Reformulation for Comparison

Binary Encoding of x. To encode $x_i \in \{0, \ldots, U_i\}$, represent each component using $\lceil \log_2 U_i \rceil$ binary variables in z. Let:

$$x = Ez, \quad z \in \{0,1\}^N, \quad N \sim \sum_i \lceil \log_2 U_i \rceil,$$

where $E \in \mathbb{Z}^{n \times N}$ is the binary encoding matrix. This quantity represents a lower bound on the number of binary variables needed to encode x.

Constraint Encoding. Transform $Ax \le b$ into:

$$Bz \le d, \quad B := AE, \quad d := b.$$

Slack and Penalty. Introduce binary slack $t \in \{0,1\}^K$ such that:

$$Bz + Tt = d, \quad T \in \mathbb{Z}^{m \times K}, \quad K \sim m\lceil \log_2 V \rceil$$

for some slack bound V, these bounds can be calculated based on extreme values of z that maximizes Tt. Use penalty weights $\lambda \in \mathbb{R}^m_+$.

$$\min_{z \in \{0,1\}^N,\, t \in \{0,1\}^K} z^\top E^\top Q E z + (Bz + Tt - d)^\top \operatorname{diag}(\lambda)(Bz + Tt - d). \tag{5}$$

Here, $E^\top Q E$ is interpreted as the lifted quadratic term acting on the binary expansion z of the original variables. Constraint violation is then penalized via binary slack encoding. Note that a sufficiently large penalty factor must be chosen to enforce the constraints.

Quadratic Form in $w = [z;\, t]$

Define $w = \begin{bmatrix} z \\ t \end{bmatrix} \in \{0,1\}^{N+K}$, where $z \in \{0,1\}^N$ encodes the original integer variables and $t \in \{0,1\}^K$ encodes the binary slack variables. The expanded form becomes:

$$\boxed{\min_{w \in \{0,1\}^{N+K}} w^\top H w + h^\top w + \text{const}} \tag{6}$$

where the matrix $H \in \mathbb{Z}^{(N+K)\times(N+K)}$ and vector $h \in \mathbb{Z}^{N+K}$ are defined as:

$$H = \begin{bmatrix} E^\top Q E + B^\top \Lambda B & B^\top \Lambda T \\ T^\top \Lambda B & T^\top \Lambda T \end{bmatrix}, \quad h = \begin{bmatrix} E^\top c - 2B^\top \Lambda d \\ -2T^\top \Lambda d \end{bmatrix}, \quad \text{const} = d^\top \Lambda d,$$

with $\Lambda = \operatorname{diag}(\lambda) \in \mathbb{Z}^{m \times m}$.

4 Implementation Details

QUBO reformulation of QIP was implemented using an open-source package, `ToQUBO.jl`[1]. This package allows for the conversion of JuMP models of QIP into QUBOs and provides an interface with various classical and quantum solvers. Additional details of implementation can be found in reference [14].

A similar approach was taken for QUIO reformulation. Another open-source package, `ToQUIO.jl`[2] is an extension developed for conversion to QUIO models, also developed by the authors of `ToQUBO.jl`, and constitutes one of the contributions of this work. This tool also builds on the interface provided by the JuMP [4] algebraic modeling language and its low-level backend MathOptInterface (MOI) [3]. Users define a QIP in JuMP as usual, but are able to wrap their QUIO-capable solver constructor within a `ToQUIO.Optimizer` instance, which will automatically construct the QUIO form and handle the necessary transformations regarding model data and results.

[1] https://github.com/JuliaQUBO/ToQUBO.jl.

[2] https://github.com/SECQUOIA/ToQUIO.jl.

More specifically, the translation procedure reads the constraint structure from MOI, adds slack variables, and augments the objective with a penalty term. Penalty weights $\mu \in \mathbb{R}_{+}^{m}$ are introduced in a diagonal matrix $M = \text{diag}(\mu)$, which penalizes constraint violations with the quadratic form rMr that follows from the residual $r = Ax + s - b$. Users may supply μ manually or rely on a default heuristic that sets all $\mu_i = \delta_+ - \delta_-$, where

$$\delta_+ = \sum_i \max(g_i u_i, g_i \ell_i) + \sum_{i,j} \max(G_{i,j}\ell_i\ell_j, G_{i,j}\ell_i u_j, G_{i,j}u_i\ell_j, G_{i,j}u_i u_j)$$
$$\delta_- = \sum_i \min(g_i u_i, g_i \ell_i) + \sum_{i,j} \min(G_{i,j}\ell_i\ell_j, G_{i,j}\ell_i u_j, G_{i,j}u_i\ell_j, G_{i,j}u_i u_j)$$

This guarantees that any variation in the objective function that happens within variable bounds $[\ell_i, u_i] \ni x_i$ will not be enough to compensate for the penalty incurred upon violation.

Once penalty terms are applied, the package computes the complete augmented objective in matrix-vector form $z^\top Gz + g^\top z + \text{const}$ as specified in Sect. 3. Intermediate structures—including variable maps and reformulation metadata—are exposed via standard JuMP interfaces. The package supports solver-agnostic workflows and integrates with `ToQUBO.jl` for side-by-side benchmarking.

4.1 Hardware Implementation

`Dirac-3` is an advanced piece of hardware designed and manufactured by QCi to solve polynomial optimization problems over integer or continuous variables [8]. Building on a recent research trend within the quantum and physics-inspired hardware community, QCi's machine utilizes quantum noise and decoherence as a fundamental component of its approach, rather than attempting to suppress these phenomena. More importantly, `Dirac-3` implements the *qudit* abstraction [13], a representation of quantum data that supports direct operations over integer numbers instead of binary values as in the usual *qubit* framework.

The vast majority of emergent devices developed for optimization purposes, including `Dirac-3`'s predecessor, `Dirac-1`, were built around the formulations QUBO and Ising [7]. `Dirac-3` stands out for its native support of integer variables and higher-order polynomials without the need for additional *encoding* and *quadratization* steps. Moreover, besides enjoying full connectivity between its variables with respect to their nonlinear interaction terms, the machine's capacity is not directly determined by the number of variables it has. Instead, the main bottleneck for representing larger problems lies in its nominal limit on the total number of *levels*, i.e., distinct integer values it can represent across all variables. For example, each binary variable incurs a cost of two levels for the device.

A similar limitation is present when solving problems with continuous variables. In this case, the input model must contain a constraint of the form

$$\sum_{i=1}^{n} x_i = R$$

over every one of its n variables $x_1, \ldots, x_n$ given some value R up to 10000. The resolution for each variable is approximately R divided by the device's *dynamic range*, currently estimated to be approximately 200 [10].

As part of the hardware implementation of this study, we also provide integration with quantum-backend submission via `QCIOpt.jl`[3], a Julia wrapper for the QCi REST API and `QUBODrivers`. This package allows JuMP-based model definitions to interface directly with QCi's devices. Users can specify attributes such as solver type, device, and number of samples via `MOI.set` or the high-level call `JuMP.set_attribute`. Model results—including objective values and samples—are returned in JuMP-compatible structures for analysis and benchmarking. This integration supports both QUBO and QUIO workflows and is used in hybrid quantum-classical experiments.

5 Methodology

5.1 Test Problems

To evaluate the performance of our QUIO and QUBO encodings, we have defined four optimization problems that use integer decision variables in their QIP form. In the following, we will introduce the problems that will be used in our evaluation reported in Sect. 6:

- Quadratic Facility Location Problem (QFLP)
- Quadratic Inventory Management Problem (QIMP)
- Quadratic Vehicle Routing Problem (QVRP)
- Quadratic Knapsack Problem (QKP)

The experiments were carried out on a system running Linux Ubuntu 22.04 with Python 3.10.12 and Julia 1.11. The machine was a Dell Latitude 5540, equipped with an Intel® i7-1365U processor (base frequency 1.80 GHz) and 32.0 GB of memory. The software stack included `JuMP` v1.26.0, `ToQUBO.jl` v0.1.10, `ToQUTO.jl` v0.1.0, and `QCIOpt` v0.1.0. For solving, we used `Gurobi` v12.0.2 as the branch-and-bound solver and the `Dirac-3` device as the quantum backend.

Quadratic Facility Location Problem. The Quadratic Facility Location Problem (QFLP) is a complex extension of the classical Facility Location Problem (FLP), where the objective of the latter is to determine optimal locations for facilities at minimal cost of opening them and serving a set of clients. Unlike the

[3] https://github.com/SECQUOIA/QCIOpt.jl.

linear version, QFLP incorporates quadratic cost terms that account for interactions between pairs of facilities. These interactions might represent synergies or conflicts, such as economies of scale, competition, or shared infrastructure costs. The inclusion of these quadratic terms makes the problem nonlinear and non-convex, significantly increasing its computational complexity.

QFLP is particularly useful in scenarios where the placement of one facility affects the performance or cost of another. For example, in telecommunications, placing two data centers too close might lead to redundancy, while putting them strategically apart could improve network resilience. Similarly, in retail, opening stores too close to each other might cannibalize sales. Its applications span logistics, urban planning, and network design, making it a vital problem in operations research and strategic decision-making.

Our version of this problem includes the penalization of connectivity between two routes x from i to j and from k to l. Additionally, a cost f of opening a number of units y at a facility j is included in the objective function.

$$\text{minimize} \quad \sum_{j=1}^{m} f_j y_j + \sum_{i=1}^{n}\sum_{j=1}^{m}\sum_{k=1}^{n}\sum_{l=1}^{m} c_{ij,kl} x_{ij} x_{kl} \tag{7}$$

$$\text{subject to:} \quad \sum_{j=1}^{m} x_{ij} = d_i \quad \text{for } i = 1,\dots,n \tag{8}$$

$$\sum_{i=1}^{n} x_{ij} \leq M_j y_j \quad \text{for } j = 1,\dots,m \tag{9}$$

$$x_{ij} \leq d_i y_j \quad \text{for } i = 1,\dots,n; j = 1,\dots,m \tag{10}$$

$$x_{ij} \geq 0 \quad \text{for } i = 1,\dots,n; j = 1,\dots,m \tag{11}$$

$$y_j \leq y_{max} \quad \text{for } j = 1,\dots,m \tag{12}$$

$$y_j \in \mathbb{N} \quad \text{for } j = 1,\dots,m \tag{13}$$

A demand constraint ensures that all the transactions to one location must match the demand d of this location i as shown in Constraints (8). Constraints (9) limit the sum of transactions that go to one location, restricted by the capacity M of this facility j and the number of facilities opened at the same location y. The remaining constraints serve to bound each variable domain.

Quadratic Inventory Management Problem. The Quadratic Inventory Management Problem (QIMP) is a specialized form of inventory optimization that incorporates a cost function with quadratic terms, reflecting more realistic and complex cost behaviors. Unlike traditional linear models, which assume constant marginal costs, quadratic models account for nonlinear cost structures such as increasing holding costs, penalties for overstocking, or economies of scale in ordering. In our case, we are modeling a setting where we have linear costs O

associated with x goods of type i that were stocked and y goods of type i that were sold. Our quadratic costs in the objective function represent a mixed cost coming from the interaction of two items of different types (i and j). The revenue from the sale of y has a unit cost of R. We used the following mathematical formulation for this problem.

$$\text{minimize} \quad \sum_{i=1}^{n}\sum_{j=1}^{n} H_{i,j}x_i x_j + \sum_{i=1}^{n} O_i x_i - \sum_{i=1}^{n} R_i y_i \tag{14}$$

$$\text{subject to:} \quad \sum_{i=1}^{n} x_i \leq s \tag{15}$$

$$\sum_{i=1}^{n} y_i \leq D \tag{16}$$

$$y_i - x_i \leq 0 \quad \text{for } i = 1, \ldots, n \tag{17}$$

$$x_i \in \{0, \ldots, s\} \quad \text{for } i = 1, \ldots, n \tag{18}$$

$$y_i \in \{0, \ldots, \min(D, s)\} \quad \text{for } i = 1, \ldots, n \tag{19}$$

where constraint (15) limits the total supply capacity, while Constraint (16) establishes a maximum demand. Constraint (17) states that for each item type i, the number of sold items cannot exceed the number of stored items. Finally, the remaining constraints define the integer domains of the decision variables.

Quadratic Vehicle Routing Problem. The Quadratic Vehicle Routing Problem (QVRP) is an extension of the canonical Vehicle Routing Problem (VRP), where the objective of the former is not only to minimize the total distance or cost of routes but also to account for interactions between pairs of customers or deliveries. In QVRP, the cost function includes quadratic terms that represent these pairwise interactions, which can model real-world complexities such as traffic congestion, delivery synchronization, or environmental impact. This makes the problem significantly more challenging to solve, as the solution space becomes nonlinear and more computationally intensive to explore. The problem's complexity and practical relevance make it a rich area for research and application in modern supply chain and urban mobility planning. The QIP formulation we chose for this problem is shown as follows:

$$\text{minimize} \quad \sum_{i=1}^{n}\sum_{j=1}^{n} D_{i,j}x_i x_j + \sum_{i=1}^{n} T_i x_i \tag{20}$$

$$\text{subject to:} \quad \sum_{j=1}^{m} y_j = \sum_{i=1}^{n} x_i \tag{21}$$

$$x_i \in \{1, \ldots, m\} \quad \text{for } i = 1, \ldots, n \tag{22}$$

$$y_j \in \{1, \ldots, n\} \quad \text{for } j = 1, \ldots, m \tag{23}$$

where the objective function (20) penalizes interactions between pairs of routes (i,j) in relation to the number of vehicles assigned to each of them x, considering a cost $D_{i,j}$. Additionally, the base cost of assigning vehicles to route i under a route cost T_i per path i is also computed in the objective function. Constraint (21) was created to affect the number of decision variables in the resulting reformulation, and it can be interpreted as a way of reducing the feasible search space by equalizing the number of vehicles assigned to every route (left-hand side) and the number of vehicles assigned to the routes (right-hand side). Finally, constraints (22) and (23) refer to the decision variables defined on an integer domain.

Quadratic Knapsack Problem. The Quadratic Knapsack Problem (QKP) consists of selecting x_i items of each kind i so as to maximize the overall value $x^\top Qx$, given that the overall weight of the selected items, i.e. $\sum_i w_i x_i$, does not surpass the maximum capacity C. For simplicity, in this example, we analyze a variant commonly known as the *Constrained Quadratic Knapsack Problem*, in which all variables have a predetermined upper bound $\leq m$. The QIP formulation we chose for this problem is shown as follows:

$$\text{maximize} \qquad \sum_{i=1}^{n}\sum_{j=1}^{n} Q_{i,j} x_i x_j \tag{24}$$

$$\text{subject to:} \qquad \sum_{i=1}^{n} w_i x_i \leq C \quad \text{for } i = 1, \cdots, n \tag{25}$$

$$x_i \in \mathbb{Z}, \quad 0 \leq x_i \leq m \quad \forall i = 1, \ldots, n \tag{26}$$

where the objective function (24) maximizes the total quadratic value between pairs of product types (i,j) with respect to the number of products of each type x, considering a cost $Q_{i,j}$. Constraint (25) limits the total weight capacity, and Constraint (26) specifies the domain and bounds of the decision variables, including the range for the number of products of each type x.

Moreover, besides enjoying full connectivity between its variables with respect to their nonlinear interactions, the machine's capacity is not directly determined by the number of variables it has.

6 Computational Experiments

This section includes two kinds of computational experiments. We computed the QUIO and QUBO reformulations of the quadratic integer programs presented above and compared these reformulations. The main object of comparison is the quadratic matrix arising from each reformulation, where we computed its size, equivalent to the number of individual variables to be included in the problem, and its density, accounting for the number of pairwise interaction terms in the quadratic unconstrained program. The second comparison involves solving several instances of a single problem using both reformulations with the QCi `Dirac-3` device and comparing the solutions to the global optimal solution found using the state-of-the-art branch-and-bound solver `Gurobi`.

6.1 Formulation Comparisons

We evaluate the QUIO and QUBO formulations on four representative QIP classes: the quadratic facility location problem (QFLP), the quadratic inventory management problem (QIMP), the quadratic vehicle routing problem (QVRP), and the quadratic knapsack problem (QKP). Each case is parameterized by problem size, and we track two key metrics: the number of variables and the density of quadratic terms. In this experimental setting, problem sizes are considered beyond the current availability of qudit-based machines, allowing for verification of the reformulations' performance alone. All experiments use identical input data, and constraint bounds are scaled to ensure feasibility.

In this evaluation, each case is modeled using a recent version of QUBO.jl [14], which extends the standard implementation that reformulates QIPs to QUBOs (`ToQUBO.jl`) by also supporting reformulation to QUIOs (`ToQUIO.jl`). For all QIPs, their results are reported as follows.

Quadratic Facility Location Problem (Figure 1, left). As the number of clients n increases from 10 to 50, the number of QUBO variables grows rapidly due to binary encoding, exceeding 10,000 in the largest case. In contrast, QUIO exhibits slower scaling and stays under 1,000 variables. The QUBO density increases beyond 15% due to the penalty structure and binary slack transformations. QUIO maintains densities below 15% across all sizes, reflecting the sparsity preserved from the original constraint matrix A.

Quadratic Inventory Management Problem (Figure 1, right). The same trend persists for the knapsack-like inventory management problem. For item counts up to $n = 200$, the QUBO formulation reaches nearly 6,000 variables, whereas the QUIO formulation remains compact at around 600 variables. For this problem, QUBO and QUIO have comparable densities at each scale. While they have comparable sparsity due to the nature of the problem, the drastic difference in the growth rate of the number of variables with problem scale indicates that the QUIO formulation provides a more compact and scalable representation.

Quadratic Vehicle Routing Problem (Figure 2, left). This class shows a notable saturation behavior. Both QUBO and QUIO densities approach 100% as the number of routes grows from 50 to 200. This is because all possible pairwise interactions are activated in the quadratic model, leading to a fully dense structure. However, QUIO still preserves a significant advantage in variable count, remaining under 500 variables, compared to nearly 3,000 variables in QUBO.

Quadratic Knapsack Problem (Figure 2, right). Similar to QVRP, both QUBO and QUIO densities remain at 100% across all experiments, regardless of

the number of products n or upper bound m. In variable count, QUIO remains under 30 variables, even as n and m increase, reflecting its compact encoding. In contrast, QUBO grows substantially, reaching over 110 variables for the largest problem instances. The gap between QUBO and QUIO becomes more pronounced as either the number of product types or their bounds increases, highlighting the scalability advantage of QUIO.
Across all cases, QUIO consistently yields smaller and sparser formulations. This structural simplicity leads to better numerical conditioning and potentially better scalability with solvers. Empirical trends validate the theoretical motivations introduced in Sect. 3.1.

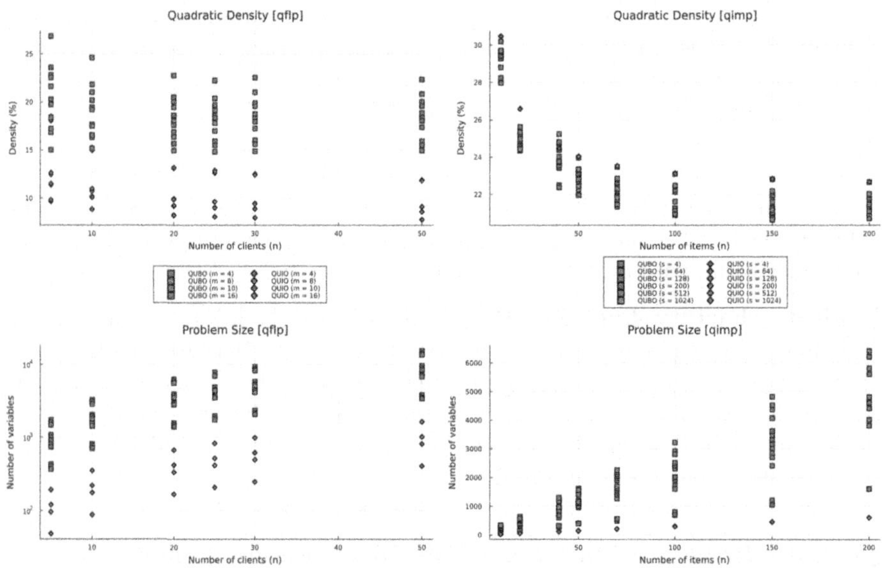

Fig. 1. Variable count and quadratic density vs. number of clients for Quadratic Unconstrained Integer Optimization (QUIO) and Quadratic Unconstrained Binary Optimization (QUBO) reformulations. Quadratic Facility Location Problem (QFLP) (left): QUIO maintains a compact and sparse structure. Quadratic Inventory Management Problem (QIMP) (right): Scaling trends for variable count and density with number of items. QUBO introduces high overhead.

In comparing the QUIO formulation (4) and the QUBO formulation (6), we observe the following empirical implications. QUIO retains $n + m$ integer variables, while QUBO requires $N + K$ binary variables, where $N \gg n$ and $K \gg m$ when U and V are moderate to large. QUIO preserves sparsity from the original constraint matrix A, while QUBO introduces denser matrices due to binary encoding and slack transformations. This results in a denser H compared to G.

The binary encoding in QUBO often results in poorly conditioned matrices H, particularly when the encoding matrix E or slack matrix T contains large

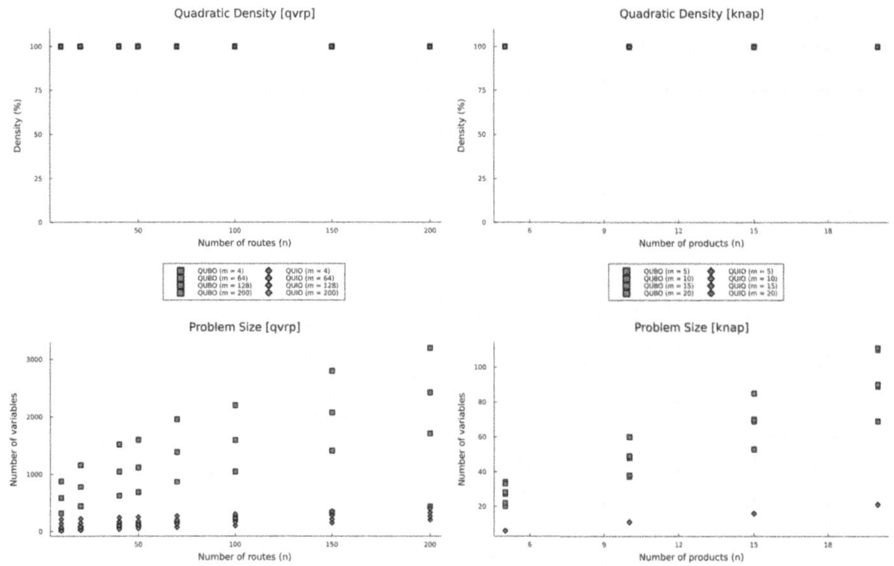

Fig. 2. Variable count and quadratic density vs. number of clients for Quadratic Unconstrained Integer Optimization (QUIO) and Quadratic Unconstrained Binary Optimization (QUBO) reformulations. Quadratic Vehicle Routing Problem (QVRP) (left): Both encodings yield dense matrices, but QUIO remains significantly more compact. Quadratic Knapsack Problem (QKP) (right): Similar to QVRP, the quadratic matrices remain dense across the experiments, and QUBO has faster growth in the number of variables with n than QUIO.

coefficients. This can degrade optimization performance. Due to fewer variables and better sparsity, QUIO could be more amenable to classical and quantum heuristics tailored for structured integer spaces, while QUBO suffers from a variable increase.

6.2 Hardware Implementation Results

The QKP example was run on QCi `Dirac-3` to demonstrate the hardware implementation for QUBO and QUIO reformulations. The problem was also modeled using `JuMP` and solved with `Gurobi` for comparison with the QUBO and QUIO reformulations that were automatically generated by `ToQUBO.jl` and `ToQUIO.jl`. For computational time comparison, the time-to-feasibility (TTFeas) was calculated for the `Dirac-3` results to account for their probabilistic nature.

The TTFeas metric is calculated as follows: $\text{TTFeas}_s = \tau \cdot \frac{\log(1-s)}{\log(1-p_{\text{feas}})}$ where TTfeas_s is the time required to achieve success with probability s (typically 0.99) in reaching feasibility. τ is the execution time of the algorithm, and p_{feas} is the probability of reaching feasible solutions. If $p_{\text{target}} = 1$, which applies

Table 3. Raw Execution Times (τ) and Time to Feasibility ($TTFeas_{99}$) [s] Across 3 Runs for Each (n, m) Instance and Solver

n	m	Gurobi QIP	Dirac-3 QUBO		Dirac-3 QUIO	
		τ [s]	τ [s]	TTFeas$_{99}$ [s]	τ [s]	TTFeas$_{99}$ [s]
5	5	{0.02, 0.02, 0.02}	{81.94, 79.58, 56.13}	{544.37, 79.58, 214.72}	{77.31, 69.78, 266.42}	{295.71, 69.78, 266.42}
	10	{0.03, 0.01, 0.06}	{90.21, 93.47, 80.03}	{345.06, 469.79, 306.12}	{73.62, 50.78, 30.17}	{370.02, 337.36, 200.46}
	15	{0.02, 0.01, 0.09}	{91.56, 88.35, 84.88}	{350.23, 252.79, 426.58}	{82.62, 86.07, 76.80}	{415.22, 86.07, 385.96}
	20	{0.19, 0.18, 0.07}	{87.87, 95.87, 91.38}	{792.20, 274.32, 261.46}	{75.81, 75.92, 81.81}	{381.00, 290.39, 234.10}
10	5	{0.06, 0.09, 0.02}	{100.36, 96.85, 99.11}	{383.88, 486.76, 379.09}	{75.84, 43.20, 30.11}	{151.67, 86.40, 60.23}
	10	{0.08, 0.08, 0.46}	{67.03, 68.17, 74.66}	{67.03, 136.34, 285.58}	{35.55, 38.82, 27.97}	{35.55, 38.82, 106.98}
	15	{0.12, 0.23, 0.36}	{65.93, 116.92, 114.51}	{188.64, 587.61, 229.02}	N/A	N/A
	20	{0.38, 0.58, 0.41}	{122.12, 133.62, 135.16}	{244.25, 887.76, 516.98}	N/A	N/A
15	5	{0.14, 0.08, 0.07}	{124.00, 122.95, 125.45}	{248.00, 245.91, 479.86}	{55.90, 64.71, 66.17}	{111.81, 129.41, 66.17}
	10	{0.12, 3.20, 1.72}	{88.70, 144.51, 83.40}	{177.40, 726.30, 238.64}	N/A	N/A
	15	{0.60, 300+, 0.78}	{153.94, 132.72, 130.85}	{440.49, 265.44, 130.85}	N/A	N/A
	20	{8.12, 3.05, 300+}	{139.14, 142.68, 131.20}	{398.13, 142.68, 375.40}	N/A	N/A
20	5	{1.27, 3.15, 1.90}	{117.72, 118.12, 131.95}	{235.45, 236.23, 504.71}	{396.15, 45.33, 58.37}	{396.15, 45.33, 58.37}
	10	{0.90, 0.23, 26.13}	{138.25, 111.75, 130.18}	{276.49, 111.75, 654.29}	N/A	N/A
	15	{300+, 300+, 105.61}	{148.31, 149.08, 150.28}	{424.37, 570.24, 574.83}	N/A	N/A
	20	300+	{167.05, 189.96, 189.43}	{334.10, 543.54, 378.86}	N/A	N/A

for deterministic solvers like `Gurobi`, then $\text{TTFeas}_s = \tau$ [6]. The computation execution and calculated TTFeas_{99} times for QKP instances are summarized in Table 3 (FIg. 3).

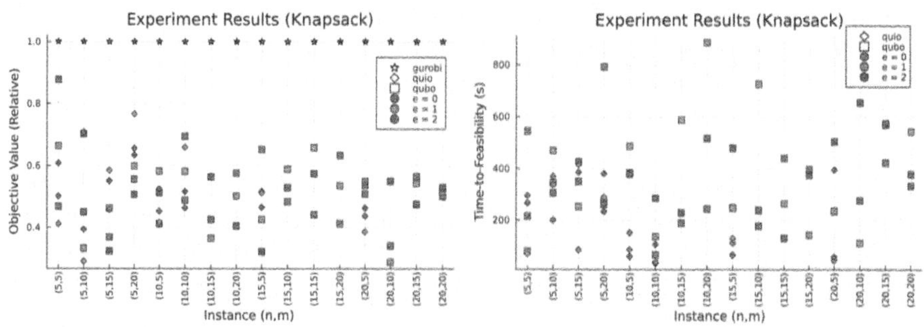

Fig. 3. QKP: (left) hardware run summary-relative objective function value. (right) hardware run summary-ttfeas.

7 Discussion and Conclusions

Reformulating and running QIP problems as both QUBO and QUIO reformulated models demonstrated that some of the metrics desirable for optimizing QUBO reformulation are not the same for creating QUIO models. Moreover, the development of (quantum) solvers that natively address QUIO problems, such as `Dirac-3`, motivates the work presented herein. At the same time, it also presents new challenges to overcome when addressing practical problems in novel hardware. Folklore rules, such as the preference for reformulations that use fewer variables, might be replaced by more elaborate criteria, such as variable levels, whose sum is constrained in `Dirac-3`. This means that instead of trying to use fewer integer variables directly, one should consider using an encoding scheme, as in QUBO, and limit the bounds of those variables accordingly.

On the other hand, using integer variables with more levels brings in some desirable properties, as the produced QUIO (Eq. (4)) models maintain their sparsity in contrast to purely QUBO (Eq. (6)) models, which tend to expand in dimensionality and typically result in a denser matrix of coefficients. Unlike QUBO, the QUIO literature did not evolve significantly towards the development of variable encoding schemes, in part because it was assumed that variables would require no more than translation and scaling transformations to represent decision values. In practice, however, the limitation imposed by the number of levels passed on to the hardware makes the adoption of encoding schemes essential if one wants to further explore the hardware's capabilities. The impact of variable encoding is highlighted by the fact that, despite requiring significantly more variables, QUBO reformulations can fit into the machine more efficiently than their respective QUIO counterparts.

Naturally, these results suggest the need for a deeper understanding of integer variable encoding methods for QUIO. Departing from the binary case, where only the encoding expression is to be decided, one must also choose the appropriate intervals for integer variables. This is undoubtedly a research track in its own right, given the numerous possibilities that can be evaluated as a combination of these two factors. Still within the realm of machine precision, another step forward is to investigate the role that the magnitude of the penalty factors and the overall coefficients of the reformulated problem play in the solver's performance. It also remains unclear whether better heuristics could be adopted to automatically select the penalization factors for QUIO reformulation.

As a final note, we highlight that this is the largest published benchmark of QUIO and QUBO reformulations of quadratic integer programs, only possible through the software development of reformulation code, `ToQUBO.jl` and `ToQUIO.jl`, and integrated solvers with QCi devices via `QCIOpt.jl`. We anticipate that this innovative hardware will enable the development of further algorithms to tackle practical optimization problems using novel computational platforms and problem reformulations.

Acknowledgments. The Quantum Collaborative, led by Arizona State University, provided valuable expertise and resources for this project. The Quantum Collaborative connects top scientific programs, initiatives, and facilities with prominent industry partners to advance the science and engineering of quantum information science. This material is based upon work supported by the Center for Quantum Technologies under the Industry-University Cooperative Research Center Program at the US National Science Foundation under Grant No. 2224960. We acknowledge the support of QCi in providing access to the `Dirac-3` Entropy Quantum Computer and to Wesley Dyk for his invaluable support throughout the development of this paper.

References

1. Deller, Y., Weimer, H., Glaetzle, A.W.: Quantum approximate optimization algorithm for qudit systems. Phys. Rev. A **107**(6), 062410 (2023)
2. Jansen, D., Heightman, T., Mortimer, L., Perito, I., Acín, A.: Qudit-inspired optimization for graph coloring. Phys. Rev. Appl. **22**, 064002 (2024)
3. Legat, B., Dowson, O., García, J.D., Lubin, M.: MathOptInterface: a data structure for mathematical optimization problems. INFORMS J. Comput. (2021)
4. Lubin, M., Dowson, O., García, J.D., Huchette, J., Legat, B., Vielma, J.P.: JuMP 1.0: recent improvements to a modeling language for mathematical optimization. Math. Program. Comput. (2023)
5. Lucas, A.: Ising formulations of many NP problems. Front. Phys. **2**, 5 (2014)
6. McGeoch, C.C., Wang, C.: Experimental evaluation of an adiabiatic quantum system for combinatorial optimization. In: Proceedings of the ACM International Conference on Computing Frontiers, CF 2013. Association for Computing Machinery, New York (2013)
7. Mohseni, N., McMahon, P.L., Byrnes, T.: Ising machines as hardware solvers of combinatorial optimization problems. Nat. Rev. Phys. **4**(6), 363–379 (2022)
8. Nguyen, L., et al.: Entropy computing: a paradigm for optimization in an open quantum system (2025)
9. Del Pia, A., Dey, S.S., Molinaro, M.: Mixed-integer quadratic programming is in NP. Math. Program. **162**(1), 225–240 (2017)
10. Quantum Computing Inc. Dirac-3 User Guide (2025)
11. Smith, A., Lee, J., Zhou, S., Eldredge, Z., Monroe, C., Gorshkov, A.V.: Graph coloring via quantum optimization on a Rydberg-qudit atom array. arXiv preprint arXiv:2504.08598 (2025)
12. Volkovich, O.V., Roshchin, V.A., Sergienko, I.V.: Models and methods of solution of quadratic integer programming problems. Cybernetics **23**(3), 289–305 (1987)
13. Wang, Y., Hu, X., Li, L., Zeng, B., Wang, X.: Qudits and high-dimensional quantum computing. arXiv preprint arXiv:2008.00959 (2020)
14. Xavier, P.M., Ripper, P., Andrade, T., Garcia, J.D., Maculan, N., Bernal Neira, D.E.: QUBO.jl: a Julia ecosystem for quadratic unconstrained binary optimization (2023)
15. Zaman, M., Tanahashi, K., Tanaka, S.: PyQUBO: python library for mapping combinatorial optimization problems to QUBO form. IEEE Trans. Comput. **71**(4), 838–850 (2022)

Quantum Interior Point Methods: A Review of Developments and an Optimally Scaling Framework

Mohammadhossein Mohammadisiahroudi[1,2(✉)], Zeguan Wu[3,4], Pouya Sampourmahani[3], Adrian Harkness[3,5], and Tamás Terlaky[3]

[1] Mathematics and Statistics, University of Maryland, Baltimore County, Baltimore, MD 21250, USA

[2] Quantum Science Institute, University of Maryland, Baltimore County, Baltimore, MD 21250, USA

mhms379@umbc.edu

[3] Industrial and Systems Engineering, Lehigh University, Bethlehem, PA 18015, USA

[4] Department of Computer Science, University of Pittsburgh, Pittsburgh, PA 15260, USA

[5] Physical and Computational Sciences, Pacific Northwest National Laboratory, Richland, WA 99354, USA

Abstract. The growing demand for solving large-scale, data-intensive linear and conic optimization problems, particularly in applications such as artificial intelligence and machine learning, has highlighted the limitations of classical interior point methods (IPMs). Despite their favorable polynomial-time convergence, conventional IPMs often suffer from high per-iteration computational costs, especially for dense problem instances. Recent advances in quantum computing, particularly quantum linear system solvers, offer promising avenues to accelerate the most computationally intensive steps of IPMs. However, practical challenges such as quantum error, hardware noise, and sensitivity to poorly conditioned systems remain significant obstacles. In response, a series of Quantum IPMs (QIPMs) has been developed to address these challenges, incorporating techniques such as feasibility maintenance, iterative refinement, and preconditioning. In this work, we review this line of research with a focus on our recent contributions, including an almost-exact QIPM framework. This hybrid quantum-classical approach constructs and solves the Newton system entirely on a quantum computer, while performing solution updates classically. Crucially, all matrix-vector operations are executed on quantum hardware, enabling the method to achieve an optimal worst-case scalability w.r.t dimension, surpassing the scalability of existing classical and quantum IPMs.

Keywords: Quantum Interior Point Method · Quantum Linear System Algorithm · Iterative Refinement · Preconditioning · Linear Optimization · Conic Optimization

V. Aggarwal et al. (Eds.): PQAI 2025, CCIS 2724, pp. 123–139, 2026.
https://doi.org/10.1007/978-981-95-7829-0_11

1 Introduction

In this paper, we review recent advances in Quantum Interior Point Methods (QIPMs) for linear optimization (LO) problems. The standard form LO problem is minimizing a linear objective function over a polyhedron, formally defined as

$$\begin{aligned} \min_{x\in\mathbb{R}^n} \ & c^T x \\ \text{s.t. } & Ax = b, \\ & x \geq 0, \end{aligned} \tag{P}$$

where $A \in \mathbb{R}^{m\times n}$ with $\operatorname{rank}(A) = m$, $b \in \mathbb{R}^m$, and $c \in \mathbb{R}^n$. It is well-known that there is a dual problem associated with the primal problem, as

$$\begin{aligned} \max_{(y,s)\in\mathbb{R}^m\times\mathbb{R}^n} \ & b^T y \\ \text{s.t. } & A^T y + s = c, \\ & s \geq 0. \end{aligned} \tag{D}$$

By the strong duality theorem [34], all optimal solutions, if they exist, belong to the set $\mathcal{PD}^*$, which is defined as

$$\begin{aligned} \mathcal{PD}^* = & \{(x, y, s) \in \mathbb{R}^{n+m+n} : \ Ax = b, \ A^T y + s = c, \\ & x^T s = 0, \ (x, s) \geq 0\}. \end{aligned}$$

Linear optimization plays a foundational role in a broad range of fields, including machine learning, operations research, logistics, and finance. Historically, the Simplex algorithm [7] was among the first prominent methods to solve LO problems. While highly effective in many practical instances, Simplex methods can exhibit exponential-time behavior in the worst case [21]. In contrast, the introduction of Interior Point Methods (IPMs) marked a major breakthrough in optimization. Starting with Karmarkar's projective algorithm [18], IPMs have evolved into the most theoretically efficient class of algorithms for solving LO problems, offering polynomial-time complexity with robust convergence guarantees. Using the standard form of LO problems, the set of feasible primal-dual solutions is defined as

$$\mathcal{PD} = \left\{(x, y, s) \in \mathbb{R}^n \times \mathbb{R}^m \times \mathbb{R}^n \mid Ax = b, \ A^T y + s = c, \ (x, s) \geq 0\right\}.$$

Then, the set of all feasible interior solutions is

$$\mathcal{PD}^0 = \{(x, y, s) \in \mathcal{PD} \mid (x, s) > 0\}.$$

By the Strong Duality theorem, all optimal solutions, if there exist any, belong to the set $\mathcal{PD}^*$ defined as

$$\mathcal{PD}^* = \left\{(x, y, s) \in \mathcal{PD} \mid x^T s = 0\right\}.$$

Let $\zeta \geq 0$, then the set of ζ-optimal solutions can be defined as

$$\mathcal{PD}_\zeta = \left\{ (x, y, s) \in \mathcal{PD} \;\middle|\; \frac{x^T s}{n} \leq \zeta \right\}.$$

Modern IPMs exploit the geometry of the central path, an analytic trajectory defined by a set of perturbed optimality conditions, which guides iterates toward the optimal solution [32,34]. When initialized appropriately, IPMs follow this path using Newton's method, requiring at most $\mathcal{O}(\sqrt{n}\log(1/\epsilon))$ iterations to obtain an ϵ-approximate solution [34]. However, a significant computational bottleneck in IPMs lies in solving the Newton linear system at each iteration. Classical direct methods such as Cholesky factorization incur $\mathcal{O}(n^3)$ complexity, which becomes intractable for large-scale, dense problems. Iterative methods, including conjugate gradient (CG) solvers [1,30], mitigate this challenge with lower per-iteration costs but at the expense of increased sensitivity to matrix conditioning and convergence accuracy.

To further enhance the scalability of IPMs, several improvements have been introduced. These include partial update techniques and low-rank updates, which reduce the cost of computing Newton directions and yield the best-known classical total complexity of $\mathcal{O}(n^3 L)$ for LO problems [34]. More recently, the incorporation of advanced tools such as fast matrix multiplication, spectral sparsification, and stochastic methods have pushed the complexity to $\mathcal{O}(n^\omega \log(n/\epsilon))$, where $\omega < 2.3729$ is the matrix multiplication exponent [13,22,39]. Alternatively, first-order methods like the primal-dual hybrid gradient (PDHG) algorithm have demonstrated empirical success in solving large-scale LO problems, although they lack rigorous theoretical complexity bounds [4,11].

Alongside these classical advances, quantum computing has emerged as a powerful paradigm capable of accelerating various computational tasks. Quantum algorithms such as Shor's factoring algorithm [35] and Grover's search algorithm [16] have showcased the potential of quantum computers to achieve polynomial or even exponential speedups. Of particular interest for optimization is the class of quantum linear system algorithms (QLSAs), pioneered by the Harrow-Hassidim-Lloyd (HHL) algorithm [17]. HHL and its successors [2,12,40] solve sparse quantum linear systems with exponential speedups under certain assumptions, although they exhibit unfavorable dependence on condition number, sparsity, and required precision.

Motivated by the capabilities of quantum computing, researchers have sought to integrate quantum solvers into classical optimization frameworks. This effort has led to the development of Quantum Interior Point Methods (QIPMs), which aim to exploit quantum acceleration in solving the Newton systems arising in IPMs. Notable contributions include quantum subroutines for the Simplex method [31], QAOA for binary optimization [14], and quantum multiplicative weight update methods for semidefinite optimization [8,38]. For linear and semidefinite programming, QIPMs have demonstrated potential polynomial speedups in terms of problem dimension [6,19]. However, these early QIPMs faced substantial challenges. The hybrid nature of QIPMs necessitates the extraction of classical information from quantum states at each iteration, typically via quantum tomography algorithms (QTAs). These steps often introduce significant errors and computational overhead, diminishing the overall efficiency of the method.

To overcome these limitations, a series of research efforts has led to the development of improved QIPMs. By incorporating iterative refinement and preconditioning techniques, recent frameworks reduce the impact of quantum errors and ill-conditioning, achieving exponential improvements with respect to precision and condition number compared to earlier quantum methods [24,25,27,42]. For instance, Wu et al. [43] introduced a dual logarithmic barrier-based QIPM with improved iteration complexity and memory access efficiency via QRAM.

In this work, we review some of these advancements with a focus on the novel, almost-exact QIPM framework that achieves provable quantum advantage. In our proposed approach, the Newton system is both constructed and solved entirely on a quantum computer, while classical computation is reserved only for solution updates. All matrix-vector products, the most expensive components in classical QIPMs, are offloaded to quantum hardware, reducing total runtime. Our hybrid quantum-classical framework achieves optimal worst-case scaling of $\mathcal{O}(n^2)$ for fully dense linear optimization problems, outperforming both classical IPMs and existing QIPMs in terms of dimensional complexity.

This framework supports inexact quantum operations, such as quantum matrix inversion and matrix-vector/matrix-matrix multiplication, through the use of iterative refinement. Crucially, unlike prior QIPMs, our method eliminates all classical matrix operations, resulting in a total classical arithmetic cost of $\mathcal{O}(n^2 \log(1/\epsilon))$. This asymptotically improves upon previous QIPMs by a factor of $\mathcal{O}(\sqrt{n})$ and offers a provable quantum speedup, as any classical analog would require at least $\mathcal{O}(n^{2.5})$ operations.

The structure of the paper is as follows. In Sect. 2, we discuss how novel reformulations aid in maintaining feasibility and achieving the best-known iteration complexity for QIPMs. Section 3 reviews how iterative refinement and preconditioning techniques can mitigate the effects of ill-conditioning and enhance the precision of QIPMs. In Sect. 4, we review recent advancements in Quantum Linear System Algorithms (QLSAs) and Quantum Tomography. Section 5 presents the state-of-the-art QIPM based on a novel Almost-Exact IPM framework that achieves optimal scaling. In Sect. 6, we explore the applications and implications of recent QIPM advancements in artificial intelligence and machine learning. Finally, Sect. 7 concludes the paper and outlines directions for future work.

2 Inexact Feasible QIPMs

In the general scheme of IPMs, we apply the Newton method to the perturbed optimality conditions iteratively to approach an optimal solution by tracing the so-called central path. There are three reformulations of Newton systems to calculate the Newton direction at each step of IPMs in the classical IPM literature. The prevailing system is the Normal Equation System (NES) defined as

$$AD^2 A^T \Delta y = Ax - \beta\mu A S^{-1} e,$$

where $\beta < 1$, $A \in \mathbb{R}^{m \times n}$ is the constraint matrix, $S = \text{diag}(s)$,

$$D = \text{diag}(x)^{1/2}\text{diag}(s)^{-1/2}$$

is the diagonal scaling matrix, and $\mu = \frac{x^T s}{n}$ is the complementarity measure.

One major issue is that an inexact solution to any traditional Newton systems calculated by a QLSA+QTA subroutine may lead to infeasibility. To properly address this infeasibility, inexact infeasible QIPM (II-QIPM) [25] has been developed which has $\mathcal{O}(n^2 \log(\frac{1}{\epsilon}))$ iteration complexity, where n is the number of variables and ϵ is the target optimality gap.

To improve this complexity, we propose two inexact feasible QIPMs (IF-QIPMs) using two novel reformulations of Newton systems. First, we use a basis for null-space of A, stored in columns of the matrix V to reformulate the Newton system in the Orthogonal Subspaces system (OSS) [26] as

$$\begin{bmatrix} -XA^T & SV \end{bmatrix} \begin{bmatrix} \Delta y \\ \lambda \end{bmatrix} = \beta\mu e - Xs. \tag{OSS}$$

We prove that the inexact solution for OSS solution provides a feasible Newton direction.

In another paper, we propose another system that is a modified version of the NES and it is more adaptable for quantum singular value transformation [27]. We prove the iteration complexity for both IF-QIPMs is $\mathcal{O}(\sqrt{n} \log(\frac{1}{\epsilon}))$, which leads to considerable polynomial speed-up in the complexity of QIPMs.

3 Iterative Refinement and Preconditioning

Another challenge in QIPMs is that their complexity has polynomial dependence on $\frac{1}{\epsilon}$ because of the QTA's overhead. This means previous QIPMs are not polynomial time algorithms as one needs to reach $\frac{1}{\epsilon} = \mathcal{O}(2^L)$ to find an exact optimal solution for an LO problem, where L is the binary length of input data. We use an iterative refinement technique to address this issue [26,27].

Iterative refinement has been widely used in classical numerical algorithms to improve the accuracy of solutions to linear systems. We adapt this technique to iteratively use limited-precision IF-QIPM to obtain a higher-precision solution. We prove that iteratively refined IF-QIPMs (IR-IF-QIPMs) have exponentially improved complexity w.r.t precision compared to previous QIPMs.

The last challenge in QIPMs is that QLSAs are sensitive to the condition number of linear systems arising in QIPMs and Newton systems are usually ill-conditioned. There are two major sources of ill-conditioning in QIPMs. First, for degenerate LO problems, the sequence of coefficient matrices of Newton systems converge to a singular matrix. i.e., their condition number grows to infinity. We show that a properly adapted iterative refinement technique helps with issues, as we stop QIPMs early when the condition number is comparatively small enough. Another source of ill-conditioning is the ill-conditioned input matrix A. We address this issue by preconditioning the Newton system. In addition, we show how this particular preconditioner can be applied on a quantum machine without excessive cost [27].

4 Improved QLSA+QTA Subroutine for QIPMs

The idea of using iterative refinement for quantum algorithms is further used to develop improved QLSA+QTA subroutine for QIPMs [23]. The most efficient QLSA to solve a linear system of the form $Mz = \sigma$ with $\widetilde{\mathcal{O}}_{\frac{p}{\epsilon}}(\kappa\|M\|_F)$ queries[1] to QRAM [9], where p is the system dimension, representing an exponential speedup over classical algorithms. A major hurdle lies in the fact that QLSAs solve Quantum Linear System Problems (QLSPs) and so the result is a quantum state, which deviates from the classical definition of the solution of LSPs. Consequently, a Quantum Tomography Algorithm (QTA) is essential to extract a classical solution. The best time complexity of QTA is $\mathcal{O}(\frac{p\varrho}{\epsilon})$, where ϱ represents the upper bound on the norm of the solution.

The overall complexity of QLSA and QTA combined is $\widetilde{\mathcal{O}}_{\frac{p}{\epsilon}}(\frac{p\kappa^2\|\sigma\|}{\epsilon})$. In comparison to the conjugate gradient method (CGM), its query complexity exhibits a better dependence on sparsity with unfavorable dependence on precision and condition number. An iterative classical-quantum linear system algorithm (ICQLSA) has been proposed which exponentially improves the time complexity of Quantum Linear Solvers, providing a classical solution with high precision up to $\widetilde{\mathcal{O}}_{\frac{p\|\sigma\|}{\epsilon}}(p\kappa^2)$ queries to QRAM [23].

This new advancement enables us to do the calculations in high precision settings where $\epsilon = 2^{-2L}$, which is almost exact for the solution of the LO problems. Thus we can also do matrix-vector multiplications on the quantum machine. Using ICQLSA and Quantum mat-vec product within the IR-IF-QIPM leads to optimal scaling $\mathcal{O}(n^2\kappa_A L)$ as the worst case complexity for solving LO can not have better dimension dependence than quadratic as storing and reading dense matrix A need $\mathcal{O}(n^2)$ arithmetic operations.

5 The State-of-the-Art QIPM

In this section, we develop an almost exact quantum interior point method for solving linear optimization problems. Assuming that the input data is all integers, we denote the binary length of the input data by

$$\begin{aligned} L = mn + m + n + \sum_{i,j} \lceil \log_2(|a_{ij}| + 1) \rceil \\ + \sum_i \lceil \log_2(|c_i| + 1) \rceil + \sum_j \lceil \log_2(|b_j| + 1) \rceil, \end{aligned}$$

where a_{ij} represents the ij-element of matrix A. The optimal partition is also defined as

$$\begin{aligned} \mathcal{B} &= \{j \in \{1, \ldots, n\} : x_j^* > 0 \text{ for some } (x^*, y^*, s^*) \in \mathcal{PD}^*\}, \\ \mathcal{N} &= \{j \in \{1, \ldots, n\} : s_j^* > 0 \text{ for some } (x^*, y^*, s^*) \in \mathcal{PD}^*\}. \end{aligned}$$

[1] The $\widetilde{\mathcal{O}}_{\alpha,\beta}(g(x))$ notation indicates that quantities polylogarithmic in α, β and $g(x)$ are suppressed.

The following lemma is a classical result first proved by [20].

Lemma 1. *Let* $(x^*, y^*, s^*) \in \mathcal{PD}^*$ *be a basic solution. If* $x_i^* > 0$*, then we have* $x_i^* \geq 2^{-L}$*. If* $s_i^* > 0$*, then we have* $s_i^* \geq 2^{-L}$*.*

Lemma 1 is a fundamental result in the complexity analysis of IPMs. It means that after enough number of iterations of IPMs, a decision variable can be rounded to zero if it is smaller than 2^{-L}. Then, by a rounding procedure, one can find an exact optimal solution for linear optimization [34,41]. In the proposed algorithm, all calculations happen on a quantum machine with precision $\epsilon = 2^{-tL}$ where t is a small constant, less than 10. This high level of accuracy justifies describing the algorithm as *almost-exact*. The only calculation that happens on a classical computer is updating the solution and vector-vector summation. In this paper, we use the dual logarithmic barrier method, which has a simple framework. At each step of the dual log barrier IPM, we need to solve the following Newton system

$$\begin{bmatrix} I & A^T \\ AS^{-2} & 0 \end{bmatrix} \begin{bmatrix} \Delta s \\ \Delta y \end{bmatrix} = \begin{bmatrix} 0 \\ \frac{1}{\mu}(b - AS^{-1}e) \end{bmatrix}, \tag{1}$$

where $S = \text{diag}(s)$. Let $\hat{\Delta s} = S^{-2}\Delta s$, we can have the system

$$\begin{bmatrix} S^2 & A^T \\ A & 0 \end{bmatrix} \begin{bmatrix} \hat{\Delta s} \\ \Delta y \end{bmatrix} = \begin{bmatrix} 0 \\ \frac{1}{\mu}(b - AS^{-1}e) \end{bmatrix}. \tag{2}$$

One can easily verify that $M = \begin{bmatrix} S^2 & A^T \\ A & 0 \end{bmatrix}$ is a symmetric positive definite matrix, and so system (2) has a unique solution [34]. Given s, one can build block-encodings of implementing matrix M and preparing state $\sigma = \begin{bmatrix} 0 \\ \frac{1}{\mu}(b - AS^{-1}e) \end{bmatrix}$ efficiently, assuming that matrix A stored in QRAM in advance. The general steps of the proposed almost exact QIPM using a short-step framework are described in Algorithm 1.

The dual logarithmic barrier method starts with a strictly feasible dual solution $(y^0,\ s^0) \in \mathcal{D}^\circ$ and a $\mu^0 > 0$ such that $(y^0,\ s^0)$ is close to the μ^0-center in the sense of the proximity measure $\delta(s^0, \mu^0)$, which, for given s and μ, is defined as,

$$\delta(s, \mu) := \left\|s^{-1}\Delta s\right\|_2.$$

As we analyze the worst-case complexity, we assume $m = \mathcal{O}(n)$ and matrices are fully dense.

Theorem 1. *The number of iterations in Algorithm 1 is at most*

$$\mathcal{O}(\sqrt{n}L).$$

Algorithm 1. Almost Exact QIPM (AE-QIPM)

INPUT Dual feasible solution (y^0, s^0), $\mu^0 > 0$, $0 < \theta < 1$, and $\delta\left((y^0, s^0), \mu^0\right) < \frac{1}{2}$, where δ is the proximity measure from [34,43]
Store A, b, c on QRAM
$k \leftarrow 1$
while $\mu > 2^{-2L}$ **do**
  $(\Delta y^k, \hat{\Delta s}^k) \leftarrow$ Solve system 2 with precision $\epsilon = 2^{-tL}$
  $y^{k+1} \leftarrow y^k + \Delta y^k$
  $s^{k+1} \leftarrow s^k + (S^k)^2 \hat{\Delta s}^k$
  $\mu^{k+1} = (1-\theta)\mu^k$
  $k \leftarrow k+1$
end while

Proof of Theorem 1. We start with a strictly feasible solution (x^0, y^0, s^0). In the dual logarithmic barrier IPM, we do not compute the value of x and y, but they exist. We have

$$Ax^0 = b,\ A^T y^0 + s^0 = c,\ s^0 > 0.$$

Then we use a quantum subroutine to compute an inexact Δs^0 with associated error ξ^1. After a full Newton step, we have

$$Ax^1 = b,\ A^T y^1 + s^1 = c + \xi^1,\ s^1 > 0.$$

Now we get a feasible solution for problem $(A, b, c + \xi_1)$. We do another full Newton step, then we have

$$Ax^2 = b,\ A^T y^2 + s^2 = c + \xi^1 + \xi^2,\ s^2 > 0.$$

We can keep doing this until we have

$$Ax^k = b,\ A^T y^k + s^k = c + \sum_{i=1}^{k} \xi^i,\ s^k > 0.$$

Then, we can rewrite all of them into

$$Ax^k = b,\ A^T y^k + s^k + \sum_{j=1}^{k} \xi^j = c + r^k,\ s^k > 0,$$

where $r^k = \sum_{j=1}^{k} \xi_j$. This implies we obtained a series of feasible iterates for problem $(A, b, c + r^k)$. The Newton steps are inexact but satisfy the conditions in [43], the $\mathcal{O}(\sqrt{n}L)$ complexity still holds. However, these Newton steps are artificial steps because we do not know exactly the errors ξ^i. We need to show that the actual Newton steps we inexactly compute are close enough to these artificial Newton steps, and the inexactness is acceptable for the convergence conditions.

In the first iteration, the actual and artificial Newton steps are computed as

$$\begin{aligned}\Delta s^0 &= -A^T \left(A(S^0)^{-2}A^T\right)^{-1} \frac{1}{\mu^0} \left(b - \mu^0 A(S^0)^{-1}e\right) + \xi^1 \\ \Delta \tilde{s}^0 &= -A^T \left(A(\tilde{S}^0)^{-2}A^T\right)^{-1} \frac{1}{\mu^0} \left(b - \mu^0 A(\tilde{S}^0)^{-1}e\right),\end{aligned}$$

where

$$\tilde{S}^0 = s^0 + r^k.$$

According to [43], we need

$$\left\|(\tilde{S}^0)^{-1}(\Delta s^0 - \Delta \tilde{S}^0)\right\|_2 \leq 0.1\delta_{\tilde{c}}(\tilde{s}^0, \mu^0),$$

where $\delta_{\tilde{c}}$ is the proximity measure for the perturbed problem $(A, b, \tilde{c})$ with $\tilde{c} = c + r^k$. This condition can be guaranteed when

$$\begin{aligned}\left\|(S^0(\tilde{S}^0)^{-1}(I - S^0(\tilde{S}^0)^{-1}))\right\|_2 &\leq 0.033\delta_{\tilde{c}}(\tilde{s}^0, \mu^0), \\ \left\|I - (S^0(\tilde{S}^0)^{-1})^2\right\|_2 &\leq 0.033, \\ \left\|(\tilde{S}^0)^{-1}\xi^1\right\|_2 &\leq 0.033\delta_{\tilde{c}}(\tilde{s}^0, \mu^0).\end{aligned} \tag{3}$$

Notice that all three conditions can be satisfied by pushing $\xi^i \leq 2^{-tL}$. This proves that our inexact Newton step is a feasible inexact Newton step for the perturbed problem. Then, according to Theorem 3.3 of [43], we have the $\mathcal{O}(\sqrt{n}L)$ complexity. Additionally, it is easy to verify that $t = 4$ satisfies all requirements.

After $\mathcal{O}(\sqrt{n}L)$ iterations, we have an $\tilde{x} > 0$ such that

$$\begin{aligned}A\tilde{x} &= b, \\ A^T y^k + s^k &= c + r^k, \\ (\tilde{x})^T s^k &\leq 2^{-2L},\end{aligned}$$

where $r^k = \sum_{i=1}^k \xi^i$. It is easy to verify that $(\tilde{x}, y^k, s^k)$ is a 2^{-2L}-optimal solution for the perturbed problem $(A, b, c+r^k)$, and one can calculate the exact optimal solution by a rounding procedure. It is easy to verify that $\|r^k\| \leq 2^{(1-t)L}$. In the remaining, we show how we can retrieve an optimal solution of the original problem with a rounding procedure from the optimal solution for the perturbed problem.

It is straightforward to see that $(\tilde{x}, y^k, s^k)$ is in a $2^{(1-t)L}$-neighborhood of the optimal set for the original problem (A, b, c). As the smallest nonzero element of s^* and x^* is greater than 2^{-L}, using partitions B and N of this solution, by solving a constrained least squares problem, an optimal solution for the original problem can be obtained. For the details of the rounding procedures, refer to Chap. 7 of [41].

It is worth noting that the rounding procedures are strongly polynomial-time methods. They can also be quantized using quantum linear system solvers; however, we do not explore the cost and implementation details of rounding procedures in this paper, as it is beyond the scope of this paper.

5.1 Quantum Subroutine

In this section, we analyze the complexity of building and solving system (2). We use the general scheme of the Quantum Tomography framework of [23,29]. We assume that we have access to a large enough QRAM, and we store data A, b, c initially on QRAM with worst-case $\mathcal{O}(n^2)$ complexity. At each state we need to store s on QRAM and build and solve System (2) using the iterative quantum linear solver of [29].

Algorithm 2. Quantum Linear Solver

INPUT (A, b, c) stored on QRAM,
Store s on QRAM
$k \leftarrow 1$
$z^k \leftarrow 0$
while $\|\Delta y^k - \Delta y^{k-1}\| > 2^{-4L}$ **do**
  Prepare State $|r^k\rangle = |\sigma - Mz^k\rangle$
  Apply inverse of block encoding of M using QSVT [9]
  Extract classical solution $\frac{p^k}{\|p^k\|} = \frac{M^{-1}r^k}{\|M^{-1}r^k\|}$ via Tomography [37] with precision $\epsilon = 10^{-2}$
  Estimate norm of $\|p^k\|$ and $\|r^k\|$
  $z^{k+1} \leftarrow z^k + \frac{p^k}{\|r^k\|}$
  $k \leftarrow k + 1$
end while

At each iteration of Algorithm 2, the only classical operation is updating the solution by a vector summation with $\mathcal{O}(n)$ arithmetic operations. In the following, we calculate the cost of quantum operations.

Lemma 2. *Given A and S stored on QRAM, the following statements are true:*

- *We can construct a block-encoding of M using $\mathcal{O}(polylog(\frac{n}{\epsilon}))$ queries to QRAM.*
- *We can prepare the the state $|r\rangle$ using $\mathcal{O}(polylog(\frac{n}{\epsilon}))$ queries to QRAM.*
- *We can apply M^{-1} using $\tilde{\mathcal{O}}_{n,\kappa,\frac{1}{\epsilon}}(\kappa\|A\|_F)$ queries to QRAM.*
- *Norm estimation of p^k and r^k costs $\tilde{\mathcal{O}}_{n,\kappa,\frac{1}{\epsilon}}(\kappa\|A\|_F)$ queries to QRAM.*

The proof of Lemma 2 is the direct result of Prepositions 1 to 6 of [6].

Lemma 3. *The number of iterations of Algorithm 2 is at most $\mathcal{O}(L)$.*

The proof of Lemma 3 is based on [29]. Additionally, the total complexity of Algorithm 2 is based on the analysis provided by [29].

Theorem 2. *Assuming (A, b, c) is stored on QRAM, Algorithm 2 can find a 2^{-tL}-precise solution for System (2) with*

$$\tilde{\mathcal{O}}_{n\kappa L}(n\kappa\|A\|_F)$$

iterations.

5.2 Proposed IR-AE-QIPM

In this section, we discuss how to use the iterative refinement method (IR) for LO problems to improve complexity as in [43], and provide the full description of our proposed algorithm. The first IR method for LO has been proposed by [15]. Mohammadisiahroudi et al. [25] first showed that using iterative refinement can improve the complexity of QIPMs w.r.t precision and condition number. Further, in [26], a quadratically convergent iterative refinement scheme was proposed for feasible IPMs. An IR method for dual log-barrier QIPM has been developed in [43].

In [43], the iterative refinement method for the LO problem works as follows:

1. Start with the original problem and solve it to a low accuracy;
2. If the accuracy of the original problem is not enough, construct a refining problem using the current iteration values; otherwise, the algorithm halts;
3. Solve the refining problem to a low accuracy and update the solution to the original problem; then, go to step 2.

In our proposed algorithm, after each solve, we have a feasible solution to a perturbed problem. To use the iterative refinement method, we need to construct a solution to the original problem from the solution to a perturbed problem. To do so, we need a projection procedure. We use Algorithm 2 to solve the following problem

$$\min_y \|A^T y + s_k - c\|_2,$$

which is equivalent to

$$AA^T y = A(c - s_k).$$

Then we have

$$s = c - A^T y.$$

According to the argument in the previous section, this (y, s) is feasible for the original problem with a duality gap bounded by twice of the low accuracy. Then, we can use the IR scheme to refine the solution to high accuracy as in [43].

To get the full complexity of the proposed algorithm, we discuss the accuracy needed for ξ^i. In the first iteration, we need conditions (3). Theoretically, $\Delta_{\tilde{c}}$ might be zero, which implies the corresponding Newton system right-hand side is zero. We do not need to solve such Newton systems. Instead, we check the norm of the right-hand side vector. If the norm is too small ($\leq 2^{-4L}$), we update μ without computing the Newton step. Then, conditions (3) can be guaranteed when

$$\|\xi^i\|_2 \leq \text{poly}\left(\frac{2^{-4L}}{n\kappa_{A(s^0)^{-1}}}\right) \approx \text{poly}(2^{-4L}), \ \forall i \in [k].$$

This bound also works for the remaining iterations. Now, we present the pseudocode of our proposed algorithm and the main theorem.

Theorem 3 (Lemma 13 of [43]). *Algorithm 3 terminates after* $\mathcal{O}\left(\frac{\log(\zeta)}{\log(\hat{\zeta})}\right)$ *iterations.*

Algorithm 3. Iteratively Refined Almost Exact QIPM (IR-AE-QIPM)

INPUT Dual feasible solution (y^0, s^0), $\mu^0 > 0$, $0 < \theta < 1$, $\delta\left((y^0, s^0), \mu^0\right) < \frac{1}{2}$, $\nabla^{(0)} = 1$, $0 < \zeta \ll \tilde{\zeta}$
Store A, b, c on QRAM
$k \leftarrow 1$
$(y_1, s_1) \leftarrow$ Solve dual problem with accuracy $\tilde{\zeta}$
while $\nabla^{(k-1)} < \frac{1}{\zeta}$ **do**
  $\nabla^{(k)} \leftarrow \nabla^{(k-1)} \times \frac{1}{\tilde{\zeta}}$
  Construct the IR problem as in [43]
  $(\hat{y}, \hat{s}) \leftarrow$ Solve IR problem with accuracy $\tilde{\zeta}$ and project into proper subspace
  $y^{k+1} \leftarrow y^k + \frac{1}{\nabla^{(k)}}\hat{y}$
  $s^{k+1} \leftarrow c - A^T y^{(k)}$
  $k \leftarrow k + 1$
end while

For our purpose, we use $\zeta = 2^{-tL}$ and $\hat{\zeta}$ is constant. Thus the outer loop has $\mathcal{O}(L)$ iteration bound. We also have a matrix-vector product at each step of this IR scheme with cost $\mathcal{O}(n^2)$ classic arithmetic operations.

The major challenge in AE-QIPM Algorithm 1 is that the complexity of the quantum solver depends on the condition number, and the condition number grows in each iteration of AE-QIPM. As in IR Algorithm 3, we stop AE-QIPM early at fixed precision. It has been shown that with early termination $\kappa^{(k)} = \mathcal{O}(\kappa_0)$ where κ_0 is the condition number of the coefficient matrix for (y^0, s^0) and it is constant [43].

5.3 Total Complexity

In this section, we put together all the elements discussed in the previous sections to calculate the total worst-case complexity of IR-AE-QIPM.

Theorem 4. *Algorithm 3 produces a $2^{(1-t)L}$ precise optimal solution of the LO problem using at most*

$$\tilde{\mathcal{O}}_{\kappa_0, n, \|A\|_F}(n^{1.5} L \kappa_0))$$

queries to QRAM and $\mathcal{O}(n^2 L)$ classical arithmetic operations.

Proof. The number of IR iterations is bounded by $\mathcal{O}(L)$ based on Theorem 3. At each iteration, we have $\mathcal{O}(n^2)$ classic arithmatic operations due to a classical matrix-vector product and the cost of AE-QIPM to solve the refining problem. Additionally, to address $\|A\|_F$ in the complexity, one can initially normalize data by $\|A\|_F$, and inconsequence the final precision should be increased by $\|A\|_F$, which appears in polylog. The quantum complexity is $\tilde{\mathcal{O}}_{n, L, \|A\|_F}(n^{1.5} \kappa_0 L)$

queries to QRAM, and $\mathcal{O}(nL)$ arithmetic operations at each step of AE-QIPM. Thus, the total number of queries to QRAM is

$$\tilde{\mathcal{O}}_{\kappa_0,n}(n^{1.5}L\kappa_0)),$$

and the total number of classical arithmetic operations is bounded by $\mathcal{O}(n^2L)$.

Table 1 compares the complexity of the proposed IR-AE-QIPM with other classical and quantum IPMs. As we can see, the total complexity of our approaches outperforms previous complexities. In the last line of the table, we show the complexity of the classical counterpart of the IR-AE-IPM using CG to solve the system. As we can see, the total complexity can not be better than $n^{2.5}$ in the classical version. This exhibits a clear quantum advantage compared to other algorithms in the literature. It should be mentioned that the quantum complexity of all QIPMs is the number of queries to QRAM. Without QRAM assumptions, some overhead cost may appear in complexities, although the quantum central path method of [5] is QRAM-free.

Table 1. Worst-case Complexity of different IPMs for LO ($m = n$)

Algorithm	Linear System Solver	Quantum Complexity	Classical Complexity
IPM with Partial Updates [34]	Low rank updates		$\mathcal{O}(n^3L)$
Feasible IPM [34]	Cholesky		$\mathcal{O}(n^{3.5}L)$
II-IPM [30]	PCGM		$\mathcal{O}(n^5L\bar{\chi}^2)$
Robust IPM [39]	Fast Mat-Mul and Partial Update		$\mathcal{O}(n^wL)$
Quantum Central Path [5]	Hamiltonian Evolution	$\tilde{\mathcal{O}}(n^{3.5}\frac{\omega}{\epsilon})$	
IR-IF-IPM [27]	PCGM		$\tilde{\mathcal{O}}_{\mu^0}(n^{3.5}L\bar{\chi}^2)$
IR-IF-QIPM [26]	QLSA+QTA	$\tilde{\mathcal{O}}_{n,\kappa_A,\Vert A\Vert,\Vert b\Vert,\mu^0}(n^{1.5}L\kappa_A^2\omega^5)$	$\tilde{\mathcal{O}}_{\mu^0}(n^{2.5}L)$
IR-IF-QIPM [27]	Precond+QLSA+QTA	$\tilde{\mathcal{O}}_{n,\Vert A\Vert_F,\frac{1}{\epsilon}}(n^{1.5}L\bar{\chi}^2)$	$\tilde{\mathcal{O}}_{\mu^0}(n^{2.5}L)$
IPM with approximate Newton steps [3]	Q-spectral Approx.	$\tilde{\mathcal{O}}_{n,\frac{1}{\zeta}}(n^{5.5})$	$\tilde{\mathcal{O}}_{\frac{1}{\zeta}}(n^{1.5})$
Quantum Dual-log Barrier [43]	QLSA+QTA	$\tilde{\mathcal{O}}_{n,\kappa_0,\mu^0,\Vert A\Vert_F}(n^{1.5}\kappa_0L)$	$\mathcal{O}(n^{2.5}L)$
Proposed IR-AE-QIPM	IQLSA+Quant Mat-Vec	$\tilde{\mathcal{O}}_{n,\kappa_0,\Vert A\Vert_F}(n^{1.5}L\kappa_0)$	$\mathcal{O}(n^2L)$
Classical IR-AE-IPM	CGM		$\mathcal{O}(n^{2.5}L\kappa_0)$

Note. Quantum complexity is expressed in the form of query complexity. Here, ω is the upper bound on the norm of the optimal solution, and $\bar{\chi}^2$ is an upper bound on the condition number of the preconditioned NES). w is the matrix multiplication parameter. PCGM stands for Preconditioned Conjugate Gradient Method.

6 Applications in AI and Machine Learning

The integration of QLSAs and QIPMs has shown the potential to accelerate core problems in machine learning. This section highlights some key applications that can benefit from these quantum techniques, as demonstrated in recent studies including [28,42].

Quantum-enhanced regression is one of the most direct applications of QLSAs in machine learning. Ordinary Least Squares (OLS), Weighted Least Squares (WLS), and Generalized Least Squares (GLS) problems can all be reduced to solving linear systems of the form $(X^T X)\beta = X^T y$, which QLSAs can handle efficiently. When paired with quantum tomography algorithms (QTAs), these solvers can retrieve classical solutions for model training and inference. The incorporation of iterative refinement techniques further enables exponential speedups with respect to precision, overcoming the classical bottleneck caused by ill-conditioning. Specifically, QLSAs offer:

- Exponential speedup w.r.t. dimension in state preparation.
- Exponential speedup w.r.t. precision via iterative refinement.
- Milder dependence on condition number through adaptive regularization.

Many sophisticated machine learning models, such as Support Vector Machines (SVMs) and Lasso Regression, can be formulated as Linearly Constrained Quadratic Optimization (LCQO) problems. These problems are ideal candidates for QIPMs, which leverage QLSAs to solve Newton systems arising in Interior Point Methods. Wu et al. [42] proposed an Inexact Feasible QIPM (IF-QIPM) that preserves the feasibility of iterates using orthogonal subspace systems (OSS), enabling the solution of LCQO problems, including:

- Lasso Regression: Promotes sparse solutions using ℓ_1 regularization. Reformulated as an LCQO problem, it can be solved with improved complexity using IF-QIPMs.
- Soft-Margin Support Vector Machines: Reformulated as LCQO using variable splitting and slack variables. QIPMs achieve better complexity in high-dimensional regimes.

These quantum-enhanced formulations demonstrate:

- Polynomial speedup w.r.t. dimension n over classical IPMs.
- Exponential speedup w.r.t. precision and condition number over previous QIPMs.
- Improved feasibility guarantees through OSS-based feasibility maintenance.

Together, these applications showcase the growing relevance of QLSAs and QIPMs in machine learning, particularly as hardware capabilities advance. The combination of quantum speedups in dimension, precision, and matrix conditioning illustrates a compelling path forward for quantum-enhanced data science.

7 Conclusions

In this paper, we presented recent advances in the development of Quantum Interior Point Methods (QIPMs) for Linear Optimization. By integrating iterative refinement and preconditioning techniques, we tackled two major challenges inherent in QLSA-based QIPMs: the inexactness of quantum solvers and their sensitivity to the condition number of the Newton system. We further introduced a novel Almost-Exact Interior Point Method framework, in which all matrix-vector operations and Newton system computations are performed on a quantum computer. This approach delivers a provable quantum speedup over classical IPMs.

To achieve an exponentially small error in the computed Newton steps, we embed iterative refinement both internally within the quantum solver and externally across IPM iterations. As a result, the overall algorithm attains an optimal worst-case complexity of $\mathcal{O}(n^2 L)$ for fully dense linear optimization problems.

A key limitation of the proposed method is its dependence on Quantum Random Access Memory (QRAM), the physical implementation of which remains an open challenge. However, alternative approaches can be explored to mitigate this dependency. For example, circuit-based QRAM constructions [33], recent developments in Quantum Singular Value Transformation (QSVT) without block encoding [10], or sparse-access input models offer promising directions for developing QRAM-free variants of the algorithm. Moreover, a detailed resource estimation study, such as the framework in [36], is essential for evaluating the real-world feasibility and quantum advantage of the proposed method.

Future research could also focus on extending this framework to a primal-dual Almost-Exact QIPM applicable to both linear and semidefinite optimization problems. Primal-dual methods, particularly those based on self-dual embedding formulations, offer the advantage of not requiring an initial strictly feasible interior point, thus expanding the applicability of QIPMs in practice.

Acknowledgments. This work is supported by Defense Advanced Research Projects Agency as part of the project W911NF2010022: *The Quantum Computing Revolution and Optimization: Challenges and Opportunities.*

Disclosure of Interests. The authors have no competing interests to declare that are relevant to the content of this article.

References

1. Al-Jeiroudi, G., Gondzio, J.: Convergence analysis of the inexact infeasible interior-point method for linear optimization. J. Optim. Theory Appl. **141**, 231–247 (2009)
2. Ambainis, A.: Variable time amplitude amplification and quantum algorithms for linear algebra problems. In: 29th Symposium on Theoretical Aspects of Computer Science (STACS'12), vol. 14. LIPIcs, pp. 636–647, 2012
3. Apers, S., Gribling, S.: Quantum speedups for linear programming via interior point methods, arXiv preprint arXiv:2311.03215, 2023
4. Applegate, D. et al.: Practical large-scale linear programming using primal-dual hybrid gradient. Adv. Neural Inf. Process. Syst. **34**, 20 243–20 257 (2021)
5. Augustino, B., Leng, J., Nannicini, G., Terlaky, T., Wu, X.: A quantum central path algorithm for linear optimization, arXiv preprint arXiv:2311.03977, 2023
6. Augustino, B., Nannicini, G., Terlaky, T., Zuluaga, L.F.: Quantum interior point methods for semidefinite optimization. Quantum **7**, 1110 (2023)
7. Bertsimas, D., Tsitsiklis, J.N.: Introduction to Linear Optimization. Athena Scientific Belmont, MA, 1997, vol. 6

8. Brandão, F.G., Kalev, A., Li, C.T., Lin, Y.Y., Svore, K.M., Wu, X.: Quantum SDP solvers: large speed-ups, optimality, and applications to quantum learning. In: 46th International Colloquium on Automata, Languages, and Programming (ICALP 2019). Schloss-Dagstuhl-Leibniz Zentrum für Informatik, 2019
9. Chakraborty, S., Gilyén, A., Jeffery, S.: The power of block-encoded matrix powers: improved regression techniques via faster Hamiltonian simulation, arXiv preprint arXiv:1804.01973, 2018
10. Chakraborty, S., Hazra, S., Li, T., Shao, C., Wang, X., Zhang, Y.:"Quantum singular value transformation without block encodings: near-optimal complexity with minimal ancilla," arXiv preprint arXiv:2504.02385, 2025
11. Chambolle, A., Pock, T.: A first-order primal-dual algorithm for convex problems with applications to imaging. J. Math. Imaging Vis. **40**, 120–145 (2011)
12. Childs, A.M., Kothari, R., Somma, R.D.: Quantum algorithm for systems of linear equations with exponentially improved dependence on precision. SIAM J. Comput. **46**(6), 1920–1950 (2017)
13. Cohen, M.B., Lee, Y.T., Song, Z.: Solving linear programs in the current matrix multiplication time. J. ACM (JACM) **68**(1), 1–39 (2021)
14. Farhi, E., Goldstone, J., Gutmann, S.: "A quantum approximate optimization algorithm," arXiv preprint arXiv:1411.4028, 2014
15. Gleixner, A.M., Steffy, D.E., Wolter, K.: Iterative refinement for linear programming. INFORMS J. Comput. **28**(3), 449–464 (2016)
16. Grover, L.K.: A fast quantum mechanical algorithm for database search. In: Proceedings of the Twenty-Eighth Annual ACM Symposium on Theory of Computing, pp. 212–219, 1996
17. Harrow, A.W., Hassidim, A., Lloyd, S.: Quantum algorithm for linear systems of equations. Phys. Rev. Lett. **103**(15), 150502 (2009)
18. Karmarkar, N.: A new polynomial-time algorithm for linear programming. Combinatorica **4**(4), 373–395 (1984)
19. Kerenidis, I., Prakash, A., Szilágyi, D.: Quantum algorithms for second-order cone programming and support vector machines. Quantum **5**, 427 (2021)
20. Khachiyan, L.: Polynomial algorithms in linear programming. USSR Comput. Math. Math. Phys. **20**(1), 53–72 (1980). https://www.sciencedirect.com/science/article/pii/0041555380900610
21. Klee, V., Minty, G.J.: How good is the simplex algorithm. Inequalities **3**(3), 159–175 (1972)
22. Lee, Y.T., Sidford, A.: Efficient inverse maintenance and faster algorithms for linear programming. In: 2015 IEEE 56th Annual Symposium on Foundations of Computer Science. IEEE, pp. 230–249, 2015
23. Mohammadisiahroudi, M.: "Quantum computing and optimization methods," Ph.D. dissertation, Lehigh University, 2024
24. Mohammadisiahroudi, M., Augustino, B., Sampourmahani, P., Terlaky, T.: Quantum computing inspired iterative refinement for semidefinite optimization. Math. Program. 1–40 (2025)
25. Mohammadisiahroudi, M., Fakhimi, R., Terlaky, T.: Efficient use of quantum linear system algorithms in inexact infeasible IPMs for linear optimization. J. Optim. Theory Appl. 1–38 (2024)
26. Mohammadisiahroudi, M., Fakhimi, R., Wu, Z., Terlaky, T.: An inexact feasible interior point method for linear optimization with high adaptability to quantum computers. SIAM J. Optim. **35**(4), 2203–2233. SIAM (2025)

27. Mohammadisiahroudi, M., Wu, Z., Augustino, B., Carr, A., Terlaky, T.: Improvements to quantum interior point method for linear optimization. ACM Trans. Quantum Comput. **6**(1), 1–24 (2025)
28. Mohammadisiahroudi, M., Wu, Z., Augustino, B., Terlaky, T., Carr, A.: "Quantum-enhanced regression analysis using state-of-the-art QLSAs and QIPMs." In: 2022 IEEE/ACM 7th Symposium on Edge Computing (SEC), pp. 375–380, 2022
29. Mohammadisiaroudi, M., Augustino, B., Fakhimi, R., Nannicini, G., Terlaky, T.: "Exponentially more precise tomography for quantum linear system solutions via iterative refinement," Lehigh University, 23T-007, Tech. Rep., 2023
30. Monteiro, R.D., O'Neal, J.W.: Convergence analysis of a long-step primal-dual infeasible interior-point LP algorithm based on iterative linear solvers. Georgia Institute of Technology, 2003. https://bit.ly/3ss0kg1
31. Nannicini, G.: Fast quantum subroutines for the simplex method. Oper. Res. **72**(2), 763–780 (2024)
32. Nesterov, Y.E., Nemirovskii, A.: Interior-Point Polynomial Algorithms in Convex Programming. SIAM, Philadelphia, PA, USA, vol. 13 (1995)
33. Park, D.K., Petruccione, F., Rhee, J.-K.K.: Circuit-based quantum random access memory for classical data. Sci. Rep. **9**(1), 3949 (2019)
34. Roos, C., Terlaky, T., Vial, J.-P.: Interior Point Methods for Linear Optimization. Springer, Cham (2005). https://doi.org/10.1007/b100325
35. Shor, P.W.: Algorithms for quantum computation: discrete logarithms and factoring. In: Proceedings 35th Annual Symposium on Foundations of Computer Science. IEEE, pp. 124–134, 1994
36. Tu, Y., et al.: Towards identifying possible fault-tolerant advantage of quantum linear system algorithms in terms of space, time and energy, arXiv preprint arXiv:2502.11239, 2025
37. van Apeldoorn, J., Cornelissen, A., Gilyén, A., Nannicini, G.: Quantum tomography using state-preparation unitaries. In: Proceedings of the 2023 Annual ACM-SIAM Symposium on Discrete Algorithms (SODA). SIAM, pp. 1265–1318, 2023
38. van Apeldoorn, J., Gilyén, A.: "Improvements in quantum SDP-solving with applications," arXiv preprint arXiv:1804.05058, 2018
39. van den Brand, J.: "A deterministic linear program solver in current matrix multiplication time. In: Proceedings of the 2020 ACM-SIAM Symposium on Discrete Algorithms (SODA), pp. 259–278, 2020. https://epubs.siam.org/doi/abs/10.1137/1.9781611975994.16
40. Wossnig, L., Zhao, Z., Prakash, A.: Quantum linear system algorithm for dense matrices. Phys. Rev. Lett. **120**(5) (2018)
41. Wright, S.J.: Primal-Dual Interior-Point Methods. SIAM, 1997
42. Wu, Z., Mohammadisiahroudi, M., Augustino, B., Yang, X., Terlaky, T.: An inexact feasible quantum interior point method for linearly constrained quadratic optimization. Entropy **25**(2), 330 (2023)
43. Wu, Z., Sampourmahani, P., Mohammadisiahroudi, M., Terlaky, T.: A quantum dual logarithmic barrier method for linear optimization. INFORMS J. Optim. (2025). https://doi.org/10.1287/ijoo.2024.0062

Benchmarking Refined Quantum Linear Systems Algorithms

Adrian Harkness[1(✉)], Kate Saltovets[1], Mohammadhossein Mohammadisiahroudi[2], and Tamás Terlaky[1]

[1] Industrial and Systems Engineering, Lehigh University, Bethlehem, PA 18015, USA
adh323@lehigh.edu

[2] Mathematics and Statistics, University of Maryland, Baltimore County, College Park, MD 21250, USA

Abstract. Systems of linear equations are ubiquitous across science, engineering, machine learning, and finance. While classical methods can be prohibitively slow for large-scale problems, quantum linear systems algorithms offer the potential for exponential speedup in certain parameter regimes. However, a significant gap persists between this theoretical promise and practical implementation, as the advantages are often obscured by the substantial quantum resources and high sensitivity to noise inherent in current quantum hardware. One way to bridge this gap is through the use of Iterative Refinement, a classical post-processing scheme that can exponentially improve the accuracy to which a linear system of equations can be solved using low-precision arithmetic. In the context of quantum linear systems algorithms, such as the HHL algorithm proposed by Harrow, Hassidim, and Loyd, Iterative Refinement can greatly reduce the quantum resources required to calculate an accurate solution in terms of tomography cost, circuit volume, and fault-tolerant overhead. Here, we compute and benchmark highly precise solutions to linear systems of equations of up to eight variables by running HHL with Iterative Refinement on NISQ quantum computers. We also present our open-source implementation, emphasizing that our circuit is not tailored to specific problem instances, as most available implementations are.

Keywords: Quantum Linear System Algorithm · HHL Algorithm · Iterative Refinement · Benchmarking

1 Introduction

It is widely believed that quantum computers can efficiently solve some problems that do not admit efficient classical algorithms [11,14,15]. Recently, a considerable amount of attention has been devoted to quantum linear algebra [1,2,13]. A common class of methods use Hamiltonian simulation to prepare a quantum state $|x\rangle$ that is proportional to the solution of the linear system

$$Ax = b, \tag{1}$$

V. Aggarwal et al. (Eds.): PQAI 2025, CCIS 2724, pp. 140–147, 2026.
https://doi.org/10.1007/978-981-95-7829-0_12

for a given s-row sparse Hermitian matrix $A \in \mathbb{R}^{n \times n}$ (i.e., $A = A^{\dagger}$ has at most s nonzero entries per row) and a vector $b \in \mathbb{R}^n$. In the *quantum linear systems problem* (QLSP), one has input oracle access to the entries of A and the ability to prepare a quantum state $|b\rangle$ that is proportional to the right-hand side vector b. It is also standard to assume that $\|A\| \leq 1$, and that its non-zero eigenvalues lie in $[-1, -1/\kappa_A] \cup [1/\kappa_A, 1]$, where κ_A the (known) condition number of A.

Research into this subfield began with the work of Harrow, Hassidim, and Lloyd [6], who proposed what has come to be known as the HHL algorithm for solving the QLSP. In this seminal work, it was shown that a quantum computer could be used to solve a QLSP with a worst-case quantum complexity of

$$\mathcal{O}\left(\frac{s^2\kappa_A^2}{\epsilon} \cdot \text{polylog}(n)\right),$$

where $\epsilon > 0$ is the accuracy to which the solution is obtained. Although the HHL algorithm originally exhibited quadratic dependence on κ_A, its poly-logarithmic dependence on the dimension of the problem opened the doors for a potential exponential quantum speedup for solving linear systems of equations.

Many linear systems problems (for example, solving the Newton system within an interior point method algorithm for linear or semidefinite optimization [9]) require high-precision solutions, which can be challenging to compute on imperfect hardware. Iterative Refinement (IR) is a classical technique originally developed in response to this challenge on low-precision classical computers. IR can use an inexact or low-precision solution as a starting point and systematically improve its accuracy [5,12]. This idea has since been extended to compute highly accurate (and at times, *exact*) solutions to linear [3,4] and semidefinite optimization problems [7], and recent works have investigated their adaptability to quantum algorithms [8,10]. Given the speedups IR has facilitated for optimization problems, it is worthwhile to investigate how IR can be leveraged in combination with QLSAs. In this work, we show in practice that, by using a NISQ device to solve for the residual error term at each step, we can achieve high-precision final solutions while requiring significantly fewer resources from the quantum hardware.

The iterative refinement method for solving linear systems of equations [5] is described in Algorithm 1, and we summarize the core idea as follows. The algorithm commences from an initial solution $x^{(0)} \in \mathbb{R}^n$, and subsequently computes a refined solution $x^{(k+1)} \leftarrow x^{(k)} + u^{(k)}$ in iterations $k = 0, 1, 2, \ldots$, where $u^{(k)}$ acts as a correction of the error $r^{(k)} = z - Ax^{(k)}$ and is the solution to the *refining system* $Au^{(k)} = r^{(k)}$. These operations can all be carried out using the same level of accuracy or *fixed precision*.

1.1 Contributions

Our primary contribution is a benchmarking suite, available at https://github.com/QCOL-LU/QLSAs, for running the HHL algorithm with iterative refinement on quantum computer simulators/emulators as well as on real quantum computing hardware. The suite is integrated with the Quantinuum qnexus library for seamless execution on supported backends, including Quantinuum's own H-series hardware and high-performance emulators. Our implementation is generalized (not hard-coded to specific instances or sizes), unlike the majority of available implementations found online.

We also demonstrate that classical IR post-processing significantly lowers the quantum resource requirements for HHL. By iteratively refining a low-precision quantum solution, we reduce the need for high-precision quantum phase estimation (QPE). This directly translates to shallower circuits with fewer gates, fewer qubits required for the QPE register, and viable execution on current NISQ hardware, effectively enabling a fault-tolerant-era algorithm to provide meaningful results today. Iterative refinement also reduces statistical noise in the estimate of the solution vector during tomography. This reduces the number of shots required to estimate a quantum state to a given level of precision.

Lastly, we benchmark results for systems up to 4 variables on Quantinuum H2-2, and 8 variables on various emulators/simulators, quantifying the resource reductions and demonstrating exponential improvements in solution accuracy even on real noisy hardware.

2 Algorithm Implementation

Here, we discuss the details of our HHL-IR implementation.

2.1 Iterative Refinement

Our implementation follows the fixed-precision IR scheme detailed in and summarized in Algorithm 1, which alternates between a quantum solver and classical post-processing. In Algorithm 1, we use a low precision HHL+QTA oracle, denoted by $O_{LS}(\xi)$, iteratively to solve the refining problem. This approach uses the quantum computer only for the most challenging step (solving the linear system), while leveraging classical resources for all scaling and update operations.

2.2 HHL Circuit and Execution

The core quantum component is the HHL algorithm, which we implement in Qiskit. The circuit consists of three main stages: Quantum Phase Estimation (QPE) to find the eigenvalues of the matrix A, a controlled rotation based on the inverted eigenvalues, and an inverse QPE to uncompute the register. For hardware execution on Quantinuum systems, circuits are transpiled using pytket and submitted via the qnexus library, which handles the full compilation and job management workflow. While most existing HHL implementations currently available are hardcoded to a given linear system, this is not sufficient for our IR scheme because the right-hand side of the system changes at each iteration. Thus, our implementation allows us to define the requested precision and flexibly accepts input matrices and vectors just as classical linear solvers do.

Algorithm 1 Iterative Refinement for the Linear Systems Problem

Input: Coefficient matrix $A \in \mathbb{R}^{n\times n}$ with $\|A^{-1}\| \le \kappa_A$, right-hand side vector $b \in \mathbb{R}^n$, error tolerances $0 < \zeta \ll \xi < 1$.

Output: A classical solution vector $x \in \mathbb{R}^n$ satisfying $\left\|Ax - \frac{b}{\|b\|}\right\| \le \zeta$.

Let $\tilde{\alpha} = \max\{\|A\|_F, \|b\|\}$. Then, normalize the system $\tilde{b} \leftarrow \frac{b}{\tilde{\alpha}}$ and $\tilde{A} \leftarrow \frac{A}{\tilde{\alpha}}$.

Choose starting point $(x^{(0)}, r^{(0)}, \eta^{(0)}) \leftarrow (0, \tilde{b}, 1)$, $k \leftarrow 0$.

while $\|r^{(k)}\| > \frac{\zeta}{\tilde{\alpha}}$

1. $\tilde{u}^{(k)} \leftarrow$ **solve** $(A, \eta^{(k)} r^{(k)})$ using $O_{LS}(\xi)$
2. $x^{(k+1)} \leftarrow x^{(k)} + \frac{1}{\eta^{(k)}}\tilde{u}^{(k)}$
3. $r^{(k+1)} \leftarrow \tilde{b} - Ax^{(k+1)}$
4. $\eta^{(k+1)} \leftarrow \frac{1}{\|r^{(k+1)}\|}$
5. $k \leftarrow k + 1$

end

3 Experimental Results

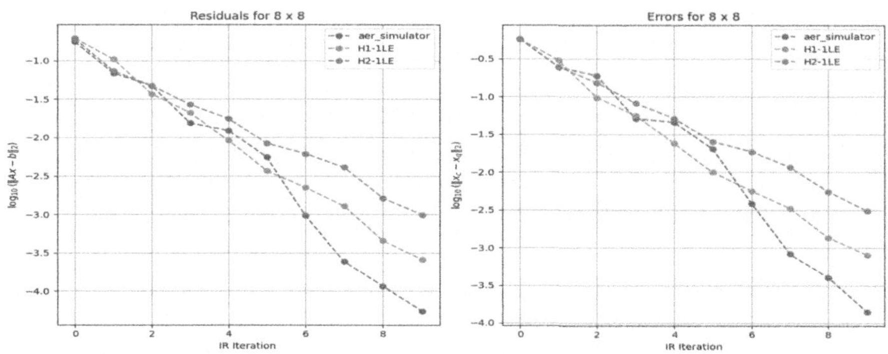

Fig. 1. Residuals and errors from an 8 variable linear system as a function of iterative refinement iteration on Quantinuum emulators and an IBM Qiskit simulator.

We run HHL-IR on various backends, including both emulators/simulators and quantum hardware. The results demonstrate that iterative refinement consistently and significantly improves the quality of the solutions obtained from the HHL algorithm.

Figure 1 shows the log-scale residuals and errors for an 8-variable linear system solved on Quantinuum emulators and an IBM Qiskit simulator. As the number of IR iterations increases, both the error and residual decrease exponentially, confirming the effectiveness of the refinement process in noiseless simulation.

This trend is successfully reproduced on NISQ hardware. Figure 2 plots the results for a 4-variable system executed on the Quantinuum H2-2 quantum computer. Despite the presence of hardware noise, IR still exponentially reduces the

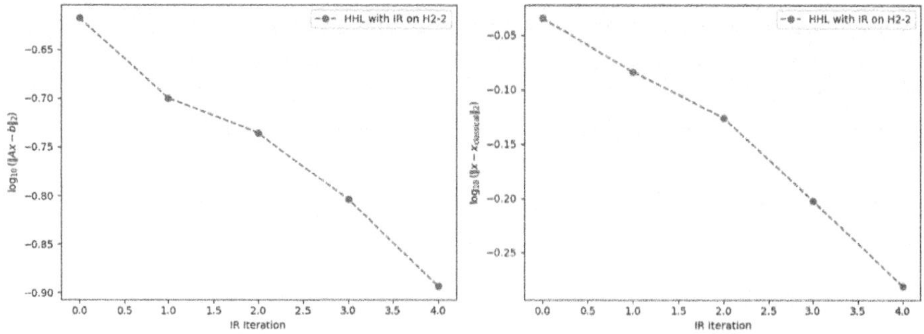

Fig. 2. Residuals and errors from a 4 variable linear system as a function of iterative refinement iteration on Quantinuum H2-2.

solution error and residual, improving the accuracy of the final result with each iteration.

A detailed summary of our benchmarking is presented in Table 1. This data quantifies the performance of HHL with and without IR across various problem sizes and backends. For every case, the application of IR yields a dramatic reduction in both the final solution error $||x_c - x_q||$ and the residual $||Ax - b||$. For example, for an 8-variable problem on the H1-1LE emulator, IR reduced the solution error by over three orders of magnitude, from approximately 0.85 to 0.0008. Note that we set a maximum of 10 iterations in these experiments, but in some cases, IR converges in fewer. These results empirically validate that IR is a crucial tool for obtaining high-precision solutions from quantum linear solvers on current and near-term hardware.

Table 1. Example statistics gathered in our QLSA benchmarking suite on a variety of problem sizes and backends.

Backend	Problem Size	Condition Number	Sparsity	Number of Qubits	Circuit Depth	Total Gates	$\|\|\mathbf{x}_c - \mathbf{x}_q\|\|$ without IR	$\|\|\mathbf{x}_c - \mathbf{x}_q\|\|$ with IR	$\|\|A\mathbf{x} - \mathbf{b}\|\|$ without IR	$\|\|A\mathbf{x} - \mathbf{b}\|\|$ with IR	Total Iterations of IR
aer_simulator	2 x 2	1.872340	0.500000	4	21	25	0.379778	0.000000	0.327757	0.000000	5
H1-1LE	2 x 2	1.872340	0.500000	4	12	22	0.255828	0.000050	0.121943	0.000042	10
H2-1LE	2 x 2	1.872340	0.500000	4	12	22	0.281076	0.000092	0.132106	0.000051	10
aer_simulator	4 x 4	2.571542	0.625000	7	107	132	0.547150	0.000030	0.141385	0.000009	9
H1-1LE	4 x 4	2.571542	0.625000	7	323	449	0.711828	0.096980	0.189305	0.024518	10
H2-1LE	4 x 4	2.571542	0.625000	7	323	449	0.765401	0.058367	0.204540	0.014552	10
aer_simulator	8 x 8	2.232571	0.750000	12	1007	1093	0.771855	0.000142	0.258965	0.000055	10
H1-1LE	8 x 8	2.232571	0.750000	12	2997	4310	0.848844	0.000807	0.305588	0.000260	10
H2-1LE	8 x 8	2.232571	0.750000	12	2997	4310	0.857951	0.003080	0.309835	0.000992	10

We also show that IR is the most resource-efficient way to obtain high precision solutions on current hardware with the HHL algorithm. Figure 3 shows the residuals as a function of both IR iteration and the number of qubits used for the Quantum Phase Estimation (QPE) subroutine. Every additional qubit used in the QPE clock register, in theory, exponentially improves the precision

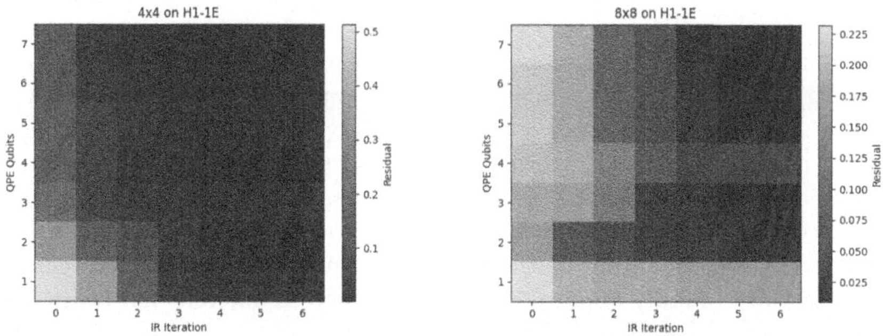

Fig. 3. Residual comparison for 4-variable (left) and 8-variable (right) systems across iterative refinement (IR) iterations and varying numbers of QPE qubits, obtained using the H1-1E noisy emulator.

to which the eigenvalues of the matrix A are estimated. However, circuit volume also increases exponentially due to the additional controlled rotation operations required by QPE. While a 4 by 4 linear system on a noisy emulator reproduces this theory, solution quality for an 8 by 8 linear system actually degrades with increasing QPE qubit count due to the accumulation of noise in the circuit. On both system sizes, IR improves precision as expected.

To emphasize the resource reductions enabled by using IR rather than larger QPE register sizes to compute high precision solutions, we note that on the 4-variable residual comparison in Fig. 3, we can achieve lower residuals with 3 iterations of IR and 1 QPE clock qubit than with 0 iterations of IR and 7 QPE clock qubits. However, the circuit for the 1 QPE qubit HHL circuit requires only 4 qubits, 48 2-qubit gates, and 170 gates total, while the larger 7 QPE qubit HHL circuit requires 10 qubits, 324 2-qubit gates, and 878 gates total. Similarly, for the 8-variable residual comparison, we can achieve much lower residuals with 3 iterations of IR and 3 QPE clock qubits than with 0 iterations of IR and 8 QPE clock qubits. However, the circuit for the 3 QPE qubit HHL circuit requires only 7 qubits, 714 2-qubit gates, and 1758 gates total, while the larger 7 QPE qubit HHL circuit requires 12 qubits, 1925 2-qubit gates, and 4604 gates total.

4 Conclusions

Quantum linear systems algorithms, such as HHL can provide speedups with respect to dimension over their classical counterparts. However, many instances of linear system problems require solutions calculated with high precision. Although the requirements for achieving these levels of precision directly with QLSAs like the HHL algorithm are out of reach on today's hardware, integrating QLSAs with iterative refinement can exponentially improve the quality of the solution with linear cost in time and no cost in space. Here, we take a step toward closing the gap between theory and practice, providing an open-source implementation with the state-of-the-art results in solution quality on current NISQ hardware.

Future work will focus on expanding this benchmarking suite to further enhance its practicality and scope. We plan to incorporate and evaluate other promising quantum linear systems algorithms, such as variational QLSAs, to provide a more comprehensive comparison of available methods. To continue reducing resource overhead, we will also implement more efficient variations of the HHL algorithm for specific problem instances, like using the Hadamard test to avoid costly tomography procedures. Finally, we will extend backend support to a wider array of quantum platforms beyond Quantinuum and IBM, ensuring our findings are broadly applicable and robust across different hardware architectures.

Acknowledgments. This research used resources of the Oak Ridge Leadership Computing Facility, which is a DOE Office of Science User Facility supported under Contract DE- AC05-00OR22725. This work is supported by Defense Advanced Research Projects Agency as part of the project W911NF2010022: *The Quantum Computing Revolution and Optimization: Challenges and Opportunities.*

Disclosure of Interests. The authors have no competing interests to declare that are relevant to the content of this article.

References

1. Chakraborty, S., Gilyén, A., Jeffery, S.: The power of block-encoded matrix powers: improved regression techniques via faster Hamiltonian simulation. arXiv preprint arXiv:1804.01973 (2018)
2. Childs, A.M., Kothari, R., Somma, R.D.: Quantum algorithm for systems of linear equations with exponentially improved dependence on precision. SIAM J. Comput. **46**(6), 1920–1950 (2017)
3. Gleixner, A., Steffy, D.E.: Linear programming using limited-precision oracles. Math. Program. **183**(1), 525–554 (2020)
4. Gleixner, A.M., Steffy, D.E., Wolter, K.: Improving the accuracy of linear programming solvers with iterative refinement. In: Proceedings of the 37th International Symposium on Symbolic and Algebraic Computation, pp. 187–194 (2012)
5. Golub, G.H., Van Loan, C.F.: Matrix Computations. JHU Press, Baltimore, MD, USA (2023)
6. Harrow, A.W., Hassidim, A., Lloyd, S.: Quantum algorithm for linear systems of equations. Phys. Rev. Lett. **103**(15) (2009). https://doi.org/10.1103/PhysRevLett.103.150502
7. Mohammadisiahroudi, M., Augustino, B., Sampourmahani, P., Terlaky, T.: Quantum computing inspired iterative refinement for semidefinite optimization. Math. Program. 1–40 (2025)
8. Mohammadisiahroudi, M., Fakhimi, R., Terlaky, T.: Efficient use of quantum linear system algorithms in inexact infeasible IPMs for linear optimization. J. Optim. Theory Appl. 1–38 (2024)
9. Mohammadisiahroudi, M., Wu, Z., Augustino, B., Carr, A., Terlaky, T.: Improvements to quantum interior point method for linear optimization. ACM Trans. Quantum Comput. **6**(1), 1–24 (2025)

10. Mohammadisiaroudi, M., Augustino, B., Fakhimi, R., Nannicini, G., Terlaky, T.: Exponentially more precise tomography for quantum linear system solutions via iterative refinement. Tech. rep., Lehigh University, 23T-007 (2023)
11. Tu, Y., et al.: Towards identifying possible fault-tolerant advantage of quantum linear system algorithms in terms of space, time and energy. arXiv preprint arXiv:2502.11239 (2025)
12. Wilkinson, J.H.: Rounding Errors in Algebraic Processes. SIAM Classics in Applied Mathematics (1994)
13. Wossnig, L., Zhao, Z., Prakash, A.: Quantum linear system algorithm for dense matrices. Phys. Rev. Lett. **120**(5) (2018)
14. Zheng, M., Chen, Y., Yang, X., Li, A.: Early exploration of a flexible framework for efficient quantum linear solvers in power systems. arXiv preprint arXiv:2402.08136 (2024)
15. Zheng, M., et al.: An early investigation of the HHL quantum linear solver for scientific applications. Algorithms **18**(8), 491 (2025)

Building Hybrid Quantum–Classical Computing Ecosystems: A Survey

Dhanvi Bharadwaj, Yuewen Hou, Alexander Knapen, Gino Kwun, Nikolai Zhitkov, and Gokul Subramanian Ravi(✉)

Computer Science and Engineering, University of Michigan, Ann Arbor, USA
gsravi@umich.edu

Abstract. Hybrid quantum–classical computing represents a pragmatic path toward harnessing the capabilities of quantum devices. Instead of treating quantum processors as standalone systems, they are integrated with powerful classical resources to form a complementary workflow. In this paradigm, each component of the computing stack plays a specific role: classical systems handle tasks that are scalable, but often complicated and data-intensive, while quantum hardware contributes non-classical primitives such as superposition and entanglement that are difficult to emulate efficiently. This review surveys the emerging quantum–classical ecosystem, emphasizing the workflows, system components, and application areas that define this paradigm. We structure the discussion around three key stages: pre-processing, which prepares quantum circuits before execution to reduce noise and minimize resource usage; co-processing, where dynamically adaptive methods leverage both application and device characteristics to enhance hybrid algorithms and enable scalable quantum error correction; and post-processing, which reconstructs high-fidelity outputs from noisy measurements to ensure reliable results.

1 Introduction

Quantum computing represents a disruptive technological paradigm with the potential to transform computing and society, with applications in critical domains such as chemistry [58], optimization [83], and machine learning [13]. Over the past three decades, its promise has grown stronger through advances in quantum algorithms and steady improvements in device technology. Nevertheless, current quantum systems remain far from achieving the scale and fidelity required to solve classically intractable problems with accuracy and efficiency. Meeting these challenges requires more than theoretical and technological progress in isolation: it necessitates the construction of an end-to-end ecosystem that integrates quantum capabilities with application requirements.

At its core, this ecosystem is inherently hybrid, comprising quantum devices that operate in concert with classical hardware and software, and that are often

Ordered alphabetically.

V. Aggarwal et al. (Eds.): PQAI 2025, CCIS 2724, pp. 148–169, 2026.
https://doi.org/10.1007/978-981-95-7829-0_13

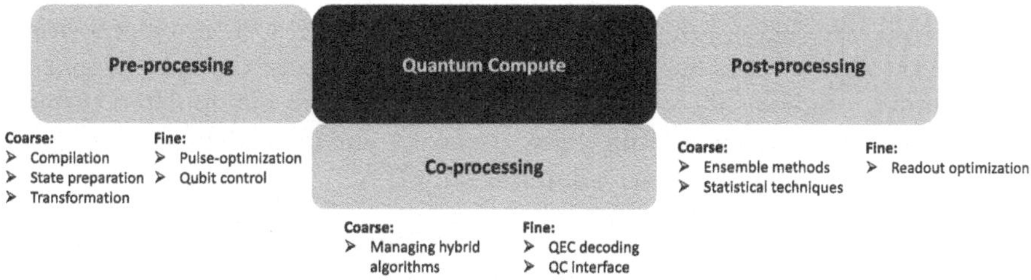

Fig. 1. Three core classes of the hybrid quantum–classical computing ecosystem, with representative topics in each, further categorized by 'coarse' and 'fine'-grained interactions between the quantum and classical components.

optimized through classically guided techniques. Progress therefore, depends on advancing quantum and classical components in a synergistic fashion. The needs and opportunities in this hybrid quantum–classical space can be broadly summarized through three key objectives:

1. Developing a scalable, modular, and adaptive **software stack** that harmonizes the interaction between diverse quantum and classical components, thereby enabling the execution of real-world applications on near-term noisy and rapidly evolving devices;
2. Designing and managing **classical hardware peripherals** that interact closely with quantum devices in resource-constrained environments, mitigating latency and bandwidth bottlenecks, and ultimately paving the way toward large-scale fault-tolerant systems;
3. Leveraging scalable **optimization and learning-based techniques**, adapted from classical computing where appropriate, to maximize the overall effectiveness of the hybrid hardware–software stack.

To advance these objectives, a variety of quantum–classical methods have been proposed. In this paper, we survey such methods, spanning software, hardware, and optimization, and categorize them into three classes based on how classical computation interacts with the quantum process (Fig. 1):

1. *Pre-processing:* Techniques that prepare an application's quantum circuits prior to execution, with the goal of reducing noise impact and minimizing quantum resource usage;
2. *Co-processing:* Dynamically adaptive methods that exploit application and device characteristics to enhance hybrid quantum–classical algorithms and enable scalable quantum error correction;
3. *Post-processing:* Techniques for reconstructing high-fidelity outputs from noisy quantum measurements.

2 Background

Quantum Computing: Quantum computers have the potential to boost computing capabilities in domains like chemistry [58], optimization [83], and machine learning [13]. Leveraging principles of quantum mechanics such as superposition, interference, and entanglement, they can tackle problems that are (exponentially) harder for classical computers. Quantum states are manipulated through unitary matrices called quantum gates, typically single- and two-qubit gates, which are sufficient for universal computation [87].

Noisy Intermediate-Scale Quantum Computing (NISQ): Near-term NISQ devices are error-prone and limited in qubit numbers [96], facing challenges like limited qubit coherence, state preparation and measurement (SPAM) errors, gate errors, and crosstalk [100]. These devices cannot execute large-scale quantum algorithms that require full fault-tolerance and millions of qubits [90]. However, they hold potential for quantum advantage in tasks such as variational quantum algorithms (VQAs) [37,95]. NISQ algorithms and devices benefit from error mitigation [9,15,28,33,44,70,103,114,119,120,126], pushing quantum systems towards practical utility [62].

Fault Tolerant Quantum Computing (FTQC): The long-term goal is to achieve quantum advantage in large-scale algorithms like Shor's Factoring [112] and Grover's Search [47]. Despite improvements, future quantum devices will still have qubit error rates too high for these applications, necessitating fault-tolerant systems [90] using quantum error correction (QEC) [38]. QEC improves qubit error rates by encoding logical qubits from multiple physical qubits and using redundancy and classical processing. Each logical qubit is encoded into a block of physical data qubits, with ancilla qubits entangled to detect errors without destroying the quantum state. Repeated measurements of ancilla qubits produce error signatures (syndromes), which a decoder analyzes to locate and correct errors. Accurate decoding allows continued error-free quantum execution, enabling large-scale quantum problem-solving.

3 Pre-processing

Pre-processing techniques aim to transform a quantum program or input prior to execution to reduce resource requirements and mitigate error effects. Typical examples include hardware-aware circuit compilation and hybrid classical–quantum workload partitioning.

3.1 Circuit Compilation and Optimization

Compilers form the backbone of quantum software stacks, bridging high-level algorithm design with the realities of hardware execution. For NISQ devices, compilers must prioritize aggressive optimizations to cope with limited qubit counts, noisy operations, and restricted connectivity. In contrast, fault-tolerant quantum systems require compilers that map algorithms into fault-tolerant gate sets while managing the overhead of quantum error-correcting codes. These contrasting requirements highlight the dual role of compilation: enabling near-term practicality while supporting long-term scalability.

Circuit compilation in the NISQ era extends beyond translating abstract algorithms into executable gate sequences. It now involves an ecosystem of optimization strategies that simultaneously address competing objectives: minimizing circuit depth to reduce decoherence, reducing gate count to limit error accumulation, accommodating hardware connectivity constraints, and adapting to device-specific noise profiles [2,41]. This multi-objective optimization becomes even more complex in hybrid quantum–classical workflows, where compilation decisions must also account for efficient classical–quantum data exchange and iterative feedback loops [117].

Compilation for hybrid systems is distinct from standalone quantum compilation due to the tight coupling between quantum subroutines and classical optimization routines. Recent approaches include noise-aware compilation using real-time calibration data [66], adaptive circuit optimization via machine learning [7], and dynamic strategies that adjust compilation based on evolving hardware states [110]. These methods recognize that static, hardware-agnostic approaches are insufficient; optimal NISQ compilation requires continuous adaptation to device-specific noise and performance characteristics.

A particularly promising direction is *approximate compilation*, which tolerates small approximation errors in exchange for substantial reductions in circuit depth. IBM's Approximate Quantum Compilation (AQC) exemplifies this approach by using tensor-network simulations to compress the initial layers of a quantum circuit into a shallower, equivalent subcircuit [55,76]. Introduced by Madden and Simonetto and later integrated into Qiskit, AQC finds a "target state" classically and replaces the first k layers of the circuit with a shorter subcircuit that prepares an approximation of that state [55,76]. This method yields shallower circuits that execute faster and more reliably, particularly for algorithms such as Trotterized time evolution, where early layers can be optimally merged [55].

While such approximate techniques are valuable for today's noisy devices, they do not directly extend to the demands of fault-tolerant quantum computing (FTQC). In the FTQC regime, compiler optimization must instead address the unique architectural constraints and steep resource requirements of error-corrected systems. Recent advances in FTQC compiler design demonstrate significant progress in mapping logical circuits to physical fault-tolerant implementations, underscoring the critical role of compilation in the fault-tolerant era.

3.2 Architecture-Aware Compilation and Hardware-Software Codesign

Modern FTQC compilers increasingly leverage hardware–software codesign principles to optimize resource utilization. The TACO (Transpiler–Architecture Codesign Optimization) framework exemplifies this approach, achieving a 91.7% reduction in Clifford gate overhead via bidirectional optimization between compiler and architecture [125]. Traditional metrics from the NISQ era, such as two-qubit gate count, can be suboptimal for FTQC: a single-qubit R_Z rotation may decompose into hundreds of Clifford+T gates, while two-qubit CX gates remain native Clifford operations [125]. Codesign allows architectures to reach single-gate-per-cycle throughput with only $1.5n+4$ logical qubit tiles, improving on earlier designs requiring $2n + \sqrt{8n+1}$ tiles [125].

Surface code compilation has similarly benefited from dependency-aware optimization. The DASCOT (Dependency-Aware Surface Code Optimization) algorithm formulates the mapping and routing problem as a discrete optimization task solvable via simulated annealing [82]. Optimal compilation requires simultaneous consideration of qubit mapping and routing on the constrained 2D lattice of surface codes. Edge-disjoint path algorithms enable constant-depth preparation of long-range Bell pairs, supporting parallel execution of long-range Clifford operations while maintaining asymptotic performance guarantees [11].

Lattice surgery compilation has been extended beyond surface codes to include color codes and folded surface codes, introducing *code substrates* as blueprints for topological code implementation [50]. Substrates cast compilation as a mapping and routing problem on macroscopic routing graphs while retaining substrate-specific constraints. Color codes, aided by advanced decoders such as VibeLSD, achieve performance parity with surface codes and enable more efficient compilation paths for certain circuits [64].

These advances mark a fundamental evolution in compilation methodology. Gate-count minimization alone has given way to holistic frameworks that account for error-correction overhead, real-time decoding, and hardware-specific features. The emerging metric of quantum logical operations per second (QLOPS) captures this broader perspective, integrating code rates, decoder accuracy, throughput, and latency. The continuum from NISQ to FTQC-oriented compilation illustrates how insights from near-term devices inform scalable fault-tolerant solutions.

3.3 Problem Transformation

Beyond architecture-aware co-design, another line of work focuses on problem transformation: restructuring quantum programs themselves so that more of the computational burden is shifted to classical processing. IBM's Operator Backpropagation (OBP) reduces circuit depth by *partitioning* a program as $U = U_C U_Q$, classically simulating the prefix U_C to back-propagate the observable $O' = U_C^{\dagger} O \, U_C$, and executing only the shorter suffix U_Q on hardware [39,56]. This trades quantum depth for classical computation by: (i) shifting a large gate prefix to classical evaluation, (ii) truncating negligible Pauli terms in O' to reduce measurements, and (iii) grouping remaining terms to minimize shots, all while preserving the target expectation within a user-defined error budget [39,56,104]. OBP exemplifies *problem transformation*, where the task is restructured so a simulable prefix is absorbed into the observable. Similarly, *Clapton* applies *Clifford-assisted problem transformation* to reframe Hamiltonians such that a classically identified high-quality state remains near-optimal under device noise [109].

3.4 Runtime Scheduling and Resource Management

Pre-processing also extends to *when* and *where* quantum kernels execute. Quantum cloud traces reveal long, variable queue times but predictable per-job runtimes in the NISQ era, motivating schedulers that jointly optimize latency, fidelity, and throughput [100]. User-centric orchestrators such as *QRIO* augment Kubernetes with quantum-aware descriptors (qubit count, topology, error/readout rates, fidelity targets) and score devices at submit time, enabling policy-driven placement and real-time selection across heterogeneous QPUs and attached CPUs/GPUs [23]. HPC–QC convergence stacks propose unified resource managers and gateways for simultaneous or interleaved quantum–classical workloads, event-driven execution (e.g., calibration/noise telemetry), and low-latency data movement, elevating scheduling, admission control, and measurement batching into first-class compile–runtime decisions [111].

Co-processing introduces additional challenges for circuit partitioning and distributed execution. Hierarchical compilation strategies decompose large circuits into smaller, hardware-manageable blocks, executed across multiple QPUs or via hybrid quantum–classical protocols [26, 36]. This improves scalability and enables targeted error mitigation, focusing on critical subcircuits while preserving overall algorithm fidelity.

Finally, integration of error mitigation directly into the compilation pipeline represents a major advance. Modern compilers embed techniques such as randomized compiling, zero-noise extrapolation, and Pauli twirling as intrinsic optimization passes [65, 127], ensuring that error suppression is co-designed with hardware constraints and algorithmic structure, maximizing quantum execution quality and downstream classical post-processing effectiveness.

4 Co-processing

In hybrid quantum–classical systems, *co-processing* refers to classical computations executed alongside quantum operations. In this section, we discuss two forms of co-processing: hybrid quantum–classical algorithms and quantum error correction (QEC) decoding. For hybrid quantum–classical algorithms, the classical processing occurs between quantum circuit executions, determining the next circuit or set of circuits to run. In this scenario, each quantum execution completes before the classical computation begins, so only one type of operation occurs at a time. This sequential interaction is inherent in near-term applications, such as variational quantum algorithms.

By contrast, in QEC decoding, classical processing occurs *concurrently* with quantum execution. Here, classical computations continuously monitor for errors affecting the ongoing quantum state and apply corrective operations to maintain the logical qubit state. This simultaneous interaction is essential for fault-tolerant quantum computing, enabling reliable execution of long-running quantum algorithms in the long-term future.

4.1 Variational Quantum Algorithms

In the current era of quantum computing, we primarily rely on noisy intermediate-scale quantum (NISQ) devices [96]. These platforms are constrained by limited qubit counts and imperfect reliability, with performance affected by short coherence times, SPAM errors, gate imperfections, and crosstalk [100]. As a result, current NISQ systems cannot yet execute large-scale quantum algorithms that require full fault tolerance and comprehensive error correction, which would necessitate millions of qubits [90]. Nevertheless, even with these limitations, NISQ devices can offer quantum advantage for specific, targeted applications, including optimization [37], physics simulations [62], and chemistry [95].

Certain algorithms developed for the NISQ era exhibit intrinsic robustness to noise, enabling them to produce meaningful results without full quantum error correction. Among these, variational quantum algorithms (VQAs) are particularly notable due to their broad applicability. VQAs leverage a parameterized quantum circuit that is iteratively optimized by a classical routine, exploring the quantum solution landscape to minimize an objective function and converge toward the problem's "ground state", i.e., the optimal solution [37,95]. Applications of VQAs include solving combinatorial optimization problems such as MaxCut [37] and estimating molecular electronic energies [95].

The hybrid structure of VQAs, alternating quantum circuit evaluations with noise-aware classical optimization, provides inherent resilience to errors [95]. Prominent VQA examples include the variational quantum eigensolver (VQE) [95], the quantum approximate optimization algorithm (QAOA) [37], and quantum machine learning algorithms [13]. VQE is widely used in molecular chemistry, condensed matter physics, and quantum many-body systems [122], while QAOA is specifically designed to produce approximate solutions to combinatorial optimization problems, particularly those formulated as Quadratic Unconstrained Binary Optimization (QUBO).

Each VQA iteration employs a parameterized quantum circuit, or *ansatz*, which defines the accessible state space and guides the optimization process. The ansatz typically comprises single-qubit rotations interleaved with fixed two-qubit entangling gates. A classical optimizer [68,116] iteratively updates the circuit parameters to minimize the objective function until convergence. The target problem is encoded as a Hamiltonian, expressed as a linear combination of Pauli operators, with the goal of estimating its ground state energy, i.e., the lowest eigenvalue [80]. In each iteration, the objective is evaluated by measuring the expectation value of the Hamiltonian using the current ansatz across the appropriate measurement bases.

Beyond VQE, Through Quantum Diagonalization. Beyond VQE, a class of *quantum diagonalization* algorithms replaces gradient-based classical optimization with classical eigenproblems defined on subspaces constructed from QPU-prepared states. *Krylov (Lanczos) quantum diagonalization* builds a Krylov basis on hardware (e.g., via short-time evolutions of a reference state) and solves a small generalized eigenproblem classically, yielding low-lying energies and properties with shallow circuits [89,130]. *Sample-based quantum diagonalization (SQD)* constructs a determinant or configuration subspace directly from computational-basis samples of a chemically informed trial state, followed by classical diagonalization; this approach has demonstrated scalability beyond exact diagonalization in recent benchmarks [34,102]. *Sample-based Krylov quantum diagonalization (SKQD)* combines these ideas by sampling Krylov states, avoiding Hadamard tests and large Pauli groupings while preserving Krylov convergence guarantees [131]. Finally, *quantum-selected configuration interaction (QSCI)* leverages a quantum state to *select* important configurations, then performs classical CI within the selected subspace; variants such as ADAPT-QSCI offer optimization-free or adaptive state-preparation pathways [59]. Although discussed in this co-processing section due to their relations to VQE, these methods shift circuit depth and optimizer sensitivity into co-, post-, and sometimes pre-processing steps (such as sampling, subspace construction, and diagonalization), providing shallow-circuit approaches to spectral estimation that complement the VQE workflow [95].

Application: Quantum Machine Learning for Generative Tasks. Quantum machine learning (QML) [13] leverages quantum computational properties to efficiently recognize and learn patterns in data. It has emerged as a versatile area of research within hybrid quantum–classical systems, both in near-term settings and beyond, often facilitated by variational quantum algorithms (VQAs) [22]. Among these, *generative QML* [135] has gained particular attention, extending classical generative modeling approaches, such as generative adversarial networks [45] and denoising diffusion probabilistic models [52], to the quantum domain. These frameworks can reproduce complex probability distributions and have applications in real-world tasks [53,129] as well as scientific domains including quantum physics [40] and drug discovery [69]. This section reviews current techniques and challenges in training QML models, with a focus on generative approaches.

To feed data into QML models, inputs must be encoded into the quantum domain, i.e., Hilbert space [135]. Quantum data naturally reside in this space, requiring only experimental adjustments for loading. Classical data, however, must be transformed into a suitable quantum representation or feature map. Efficient data encoding remains a critical bottleneck for demonstrating quantum advantage in QML [107]. Common approaches include basis encoding, amplitude encoding, angle encoding, and superposition encoding, with advanced methods developed for specific tasks and data structures [91,97,108].

In generative QML, initial states are often sampled from noisy distributions, such as quantum analogues of classical probability distributions [136], Haar-random states [132], or maximally mixed states [67,93], to efficiently produce high-quality classical or quantum data.

Generative QML models are inspired by classical generative approaches, including quantum Boltzmann machines (QBMs) [5], quantum generative adversarial networks (QGANs) [75], tensor-network models [27,123], and quantum denoising diffusion probabilistic models (QuDDPMs) [19,132]. Among these, QGANs and QuDDPMs are especially suited for hybrid quantum–classical implementations, thereby motivating intensive research and numerous applications [75,132].

QGANs [29,75] mimic classical generative adversarial networks, training a generator to produce samples indistinguishable from real data and a discriminator to precisely detect whether the given inputs are real or generated. VQAs with parameterized quantum circuits (PQCs) serve as generators, discriminators, or both. QGANs produce individual quantum states sampled from a target distribution, and their variants have been widely applied across classical and quantum domains [53,69,86,136].

QuDDPMs [132] adopt a stepwise approach to learn the inverse of noise injection processes, enabling quantum state ensemble generation. The forward process introduces noise to quantum data until a fully noisy distribution is obtained, while the backward process, implemented with PQCs and auxiliary qubits, is trained to remove the noise. Forward noise is applied via random scrambling unitaries in pure-state models [132] or depolarizing channels in mixed-state models [67,93]. Backward updates utilize projective measurements [132] or partial traces [24] to maintain output dimensions and handle non-unitary behaviors [124]. Recent improvements include temporal parameter sharing [133], joint probability learning for discrete diffusion models [25], and architectural advancement with randomized weak measurements and control Hamiltonian learning [73].

Training QML models relies on cost functions that guide optimization, often employing hardware-efficient ansätze [21,58]. For classical data, expectation values or statistics from repeated PQC executions are converted into classical metrics, e.g., cross-entropy loss or mean squared error for classification [54]. Other examples include maximum likelihood expectation for probabilistic unsupervised learning [94] and Kullback–Leibler divergence for comparing output and target distributions [134].

For quantum state generation, similarity measures such as fidelity or trace distance [87] assess output-target closeness. However, computing these experimentally may require many copies of the quantum state, particularly in noisy settings [18]. To address this issue, superfidelity-based metrics [67] provide efficient alternatives. Generative models employ loss functions constructed upon these measures: QGANs use adversarial or minimax losses [136], whereas QDMs leverage maximum mean discrepancy (MMD) [46] or the Wasserstein distance [88]. The model parameters are typically optimized via classical methods, such as gradient descent [80] or Newton's method [101].

Despite their promise, QML faces several challenges. Hardware limitations constrain the number of qubits and achievable fidelity. Model scalability is also an issue: as PQCs grow deeper and larger, barren plateaus [79] hinder gradient-based optimization, limiting effective depth and qubit utilization. Overcoming these issues may require architectural innovations, analogous to residual networks in classical deep learning [48], and theoretical developments to support robust training.

Nonetheless, QML remains a central area in hybrid quantum-classical computing, with the potential to handle and generate quantum data efficiently and to open new paradigms in information processing as hardware and experimental implementations continue to advance.

4.2 Quantum Error Correction Decoding

Solving practically useful problems with a quantum computer will require $10^5 - 10^6$ qubits with error rates near 10^{-15} [12]. However, due to qubit sensitivity to environmental noise, present-day systems are limited to just 10^2-10^3 qubits with error rates of $10^{-2} - 10^{-3}$. To both reduce error rates and scale the number of qubits to levels required for practical applications, future fault-tolerant quantum computing (FTQC) systems will rely on quantum error correction (QEC).

Analogous to classical error correction, QEC uses many noisy *physical* qubits to redundantly encode a robust *logical* qubit state with much lower error rates. Depending on the chosen error-correcting code, logical operations are performed on and between logical qubits to execute quantum algorithms. Each logical operation is interleaved with several rounds of error correction, during which errors on individual physical qubits are detected and corrected. Unlike classical error correction, however, QEC faces two unique challenges: (i) measuring a qubit collapses its state, so errors cannot be directly observed on the data qubits, and (ii) qubits suffer both bit-flip (X) and phase-flip (Z) errors, unlike classical bits which only experience bit flips.

To address these challenges, many QEC codes have been developed; here, we focus on the Calderbank-Shor-Steane (CSS) family of codes [20,118]. CSS codes partition the physical qubits into three groups: data qubits, X ancilla qubits, and Z ancilla qubits. Data qubits redundantly encode the logical qubit state, while X and Z ancilla qubits are used to indirectly extract information about Z-type and X-type errors, respectively. Each ancilla qubit is entangled with a subset of data qubits so that measuring it reveals the parity of the corresponding error type without collapsing the logical qubit state.

Measuring all ancilla qubits produces a binary string called an *error syndrome*, which a classical *decoder* uses to infer necessary corrections. These corrections are typically applied via Pauli frame updates [63] rather than directly to the data qubits. Decoders must operate in real-time for two primary reasons. First, non-Clifford logical gates such as T gates require corrections that depend on the history of decoded syndromes [38]. Second, if syndromes are generated faster than they can be decoded, the backlog grows exponentially, increasing the overall runtime [121]. The syndrome generation rate is determined by qubit readout latency, which can be as low as 1 μs for superconducting qubits [113].

Decoding algorithms typically operate on a weighted *decoding graph*, where vertices represent syndromes and edges (or hyperedges) represent error mechanisms that flip syndrome values. Edge weights reflect the probability of each error mechanism. Decoders infer corrections by matching active syndrome nodes (value 1) through a minimal set of error mechanism edges. Since ancilla qubits report only parity, multiple error configurations can yield identical syndromes, complicating decoding.

The decoding graph structure depends on noise modeling accuracy. Ancilla readout errors require multiple rounds of syndrome measurement, producing a 3D decoding graph with two spatial dimensions for the qubit layout and a third temporal dimension for multiple rounds. Imperfect single- and two-qubit gates entangling data and ancilla qubits introduce correlated errors, adding further complexity. While challenging computationally, this complexity is necessary to achieve the extremely low logical error rates required for future applications.

Matching algorithms vary with the chosen QEC code. For the surface code [38], several decoders trade off accuracy and resource complexity. For example, the Blossom algorithm [35] implements minimum-weight perfect matching with worst-case runtime $O(d^5)$, where d is the code distance. Modified algorithms [32,51,128] reduce latency at the cost of accuracy. Union-Find-based approaches [32] show promise for hardware acceleration [8,31,74], but typically require significant hardware resources. Due to these overheads, real-time QEC demonstrations remain scarce [1].

Achieving scalable, real-time QEC requires further algorithmic and architectural innovations. *Predecoding* [3,99,115] has been proposed as an initial preprocessing step. Leveraging the sparsity of syndromes, predecoders can locally infer the majority of corrections. The QEC stack can be arranged hierarchically: syndromes are first processed by a predecoder, which optionally passes complex cases to a second-level decoder.

Predecoders improve scalability in two ways. First, they can match local pairs of active syndromes, reducing the size of the problem passed to the second-level decoder [3,115]. The main design challenge is balancing *predecoder coverage* with logical error rate. Aggressive predecoding may increase coverage but also introduce logical errors, while conservative strategies reduce coverage and retain low logical error rates.

Second, predecoders benefit systems with classical electronics at room temperature (RT) and qubits at cryogenic temperatures, such as superconducting or solid-state spin qubits. High syndrome volume creates a bandwidth bottleneck across the cryogenic-to-RT interconnect. Cryogenic predecoders [99] use ultra-low-power logic to generate corrections for sparse syndromes and identify complex cases for second-level decoding, significantly reducing bandwidth requirements.

Looking forward, quantum low-density parity check (qLDPC) codes are gaining traction for their lower physical qubit overheads. For instance, bivariate bicycle (BB) codes [16] achieve surface code-level performance with $\sim$10$\times$ fewer qubits, and color codes [14] offer advantages in magic state preparation [43]. However, these codes increase decoding complexity: higher-weight or long-range qubit interactions convert edges into hyperedges, precluding traditional algorithms like minimum weight perfect matching (MWPM). Advanced decoders based on Belief Propagation or code-specific strategies [42,84,92] have been proposed, but their latency is typically higher than for surface codes, and few are viable for real-time implementation [84].

5 Post-processing

In hybrid quantum–classical workflows, post-processing converts analog measurement records into reliable classical information for downstream analysis, including calibration, error mitigation, and error correction. This section focuses on post-processing, particularly in the context of superconducting qubit platforms. We briefly review: (i) the signal-processing path from analog I/Q traces to discrete bit outcomes, (ii) single-shot state discrimination methods using statistically informed or learned filters, (iii) measurement-error mitigation on distributions via post-selection, and (iv) "soft-information" interfaces, where calibrated confidence values are passed to decoders instead of hard bits.

5.1 Converting Analog I/Q to Reliable Digital Bits

In dispersive readout, the qubit-state-dependent response of a readout resonator is probed, amplified, and digitized. Heterodyne demodulation then produces time series $I(t), Q(t)$ at a chosen intermediate frequency. A standard step is to compress the time record into a single complex statistic

$$Z = \int_0^T W(t)\,\big(I(t) + iQ(t)\big)\,dt,$$

where the (complex) weight function $W(t)$ is chosen to maximize the signal-to-noise ratio (SNR) for the qubit states being discriminated [57,105,106]. Commonly, W is chosen to be proportional to the difference between the class-mean trajectories, i.e., $W(t) \propto \bar{s}_0(t) - \bar{s}_1(t)$. This corresponds to a linear matched filter (MF): under the assumptions of stationary Gaussian noise and no state transitions during the readout, the matched filter maximizes both SNR and achievable assignment fidelity [105].

In practice, however, non-idealities such as state transitions during the readout window can bias this simple time-collapsed statistic [61]. To address these limitations, new techniques have emerged that operate directly on the raw I and Q traces; some of these methods will be discussed in the following subsection. In multi-qubit devices, readout is typically frequency-multiplexed: multiple tones share the same line and are digitally separated before per-qubit weighting and classification [49].

5.2 Single-Shot State Discrimination and Readout Optimization

A ubiquitous baseline for qubit readout is to compress the time-dependent I/Q record into a single complex value Z (as described earlier), which can be represented as a point in a two-dimensional plane. State assignment is then performed by projecting this 2D information onto a one-dimensional axis and applying a threshold. The optimal projection axis is the one that maximally separates the distributions corresponding to the $|g\rangle$ and $|e\rangle$ states, which can be systematically determined by maximizing the Fisher separation [71,72]. This reduces the two-dimensional problem to an effective single variable, significantly simplifying classification.

However, projecting two-dimensional data onto a line inevitably discards information. More expressive models such as two-dimensional Gaussian Mixture Models (GMMs) avoid this by directly modeling the full I/Q distribution for each qubit state. By fitting separate Gaussian distributions to each state's measurement data, GMMs enable nonlinear decision boundaries that better partition the measurement space and thus improve classification accuracy [49].

With frequency-multiplexed readout, per-qubit matched filtering (MF) or related compression strategies remain common baselines due to their simplicity, robustness, and straightforward FPGA implementation. Yet these methods ignore inter-qubit correlations and residual crosstalk. In contrast, some approaches omit matched filtering entirely, operating directly on the raw I/Q traces to further improve discrimination fidelity. Lightweight multivariate classifiers such as linear discriminants, support vector machines, or shallow neural networks trained on stacked channel features can capture such correlations and yield higher accuracy [71,78]. At larger qubit counts, however, neural networks face practical challenges: scaling requires more calibration data, greater computational resources, and more complex architectures to capture intricate correlations, which increases both training and deployment overhead. Hybrid pipelines that combine MF-based dimensionality reduction with compact neural network heads can balance fidelity with low on-chip cost and latency [78].

When relaxations ($|e\rangle \rightarrow |g\rangle$) or leakage ($|e\rangle \rightarrow |f\rangle$) occur during measurement, matched filters are suboptimal. Hidden-Markov or continuous-time Markov models that explicitly include these transitions provide a principled framework for separating overlapping distributions and scaling to multi-state discrimination [77]. Including the $|f\rangle$ manifold is especially important because excited-state histograms are frequently trimodal; ignoring this third cluster biases thresholds and corrupts calibration [71].

Recent "trainable post-processors" learn time-dependent linear filters directly from data, outperforming static matched filters under complex readout conditions such as high-power drive, while remaining FPGA-friendly [60]. Complementary efforts explore reservoir- and physics-aware post-processing to move portions of inference closer to the sensor, even into cryogenic environments, thereby reducing latency and I/O bottlenecks [6]. Collectively, these directions argue for *adaptive yet hardware-efficient* readout classifiers.

5.3 Measurement-Error Mitigation

Independent of the single-shot classifier, the observed bitstring distribution is affected by assignment errors and correlated crosstalk. Ensemble-level post-processing aims to invert a calibrated measurement channel. Two widely adopted techniques are: (i) *correlated/tensored mitigation*, which estimates a structured assignment matrix from calibration data and applies inversion to observed counts [17]; and (ii) *matrix-free measurement mitigation (M3)*, which iteratively solves the underlying linear problem without forming dense matrices, improving scalability [85]. These approaches are integrated into production toolchains and widely used in NISQ workflows, though they do not replace decoder-level fault tolerance. Some single-shot post-processing methods (e.g., multivariate classifiers) explicitly address correlated crosstalk at the feature level before any distributional mitigation [71].

5.4 Soft Information to QEC Decoders

Hardening measurements to $\{0, 1\}$ discards valuable confidence information. Passing *soft information*, such as log-likelihood ratios $\log P(z|1) - \log P(z|0)$, calibrated class probabilities, or leakage indicators, to decoders can significantly enhance logical performance. Early theory highlighted the benefits of soft decoding in qubit readout [30]. Recent experiments on superconducting platforms have demonstrated substantial reductions in logical error rates when soft readouts are fed into MWPM or neural-network-based decoders [4,81]. Hardware studies using soft-syndrome belief-propagation decoders on FPGAs showed that just 8-bit quantization of soft inputs preserves nearly full decoding performance, indicating that low-bit implementations are sufficient for resource-constrained, real-time stacks [98]. Learned decoders (e.g., transformer- or RNN-based) naturally exploit analog side information to outperform classical baselines, motivating readout pipelines that output calibrated confidence metrics rather than hard bits [10].

6 Conclusion

In conclusion, hybrid quantum–classical computing offers a practical route to harness the power of quantum devices. By combining quantum processors with classical resources, this paradigm forms a complementary workflow: classical systems add scalable, complex tasks, while quantum hardware provides non-classical capabilities such as superposition and entanglement. Achieving practical advantage requires an integrated ecosystem of scalable software stacks, low-latency classical peripherals, and optimization or learning-based methods to maximize overall performance. In this survey, we reviewed classical techniques that support quantum processes across key stages: pre-processing to efficiently prepare circuits, quantum execution on hardware or simulators, co-processing to adaptively enhance algorithms and enable fault tolerance, and post-processing to reconstruct high-fidelity outputs from noisy measurements. Together, these strategies define a hybrid quantum–classical ecosystem advancing toward practical, large-scale quantum computing.

Acknowledgment. This material is based upon work supported by the U.S. Department of Energy, Office of Science, Office of Advanced Scientific Computing Research, Accelerated Research in Quantum Computing under Award Number DE-SC0025633.

References

1. Quantum error correction below the surface code threshold. Nature **638**(8052), 920–926 (2025)
2. Academic Works CUNY: Survey on quantum circuit compilation for noisy intermediate-scale quantum devices (2021). https://academicworks.cuny.edu/ny_pubs/815/
3. Alavisamani, N., Vittal, S., Ayanzadeh, R., Das, P., Qureshi, M.: Promatch: extending the reach of real-time quantum error correction with adaptive predecoding. In: Proceedings of the 29th ACM International Conference on Architectural Support for Programming Languages and Operating Systems, vol. 3, pp. 818–833 (2024)
4. Ali, H., et al.: Reducing the error rate of a superconducting logical qubit using analog readout information. Phys. Rev. Appl. **22**, 044031 (2024). https://doi.org/10.1103/PhysRevApplied.22.044031
5. Amin, M.H., Andriyash, E., Rolfe, J., Kulchytskyy, B., Melko, R.: Quantum Boltzmann machine. Phys. Rev. X **8**, 021050 (2018). https://doi.org/10.1103/PhysRevX.8.021050
6. Angelatos, G., Khan, S.A., Türeci, H.E.: Reservoir computing approach to quantum state measurement. Phys. Rev. X **11**, 041062 (2021). https://doi.org/10.1103/PhysRevX.11.041062
7. Baird, L., Moin, A.: Orqstrator: an AI-powered framework for advanced quantum circuit optimization. arXiv preprint arXiv:2507.09682 (2025)
8. Barber, B., et al.: A real-time, scalable, fast and resource-efficient decoder for a quantum computer. Nat. Electron. **8**, 84–91 (2025). https://doi.org/10.1038/s41928-024-01319-5
9. Barron, G.S., Wood, C.J.: Measurement error mitigation for variational quantum algorithms (2020)
10. Bausch, J., et al.: Learning high-accuracy error decoding for quantum processors. Nature **635**(8040), 834–840 (2024). https://doi.org/10.1038/s41586-024-08148-8
11. Beverland, M., Kliuchnikov, V., Schoute, E.: Surface code compilation via edge-disjoint paths. PRX Quantum **3**, 020342 (2022). https://doi.org/10.1103/PRXQuantum.3.020342
12. Beverland, M.E., et al.: Assessing requirements to scale to practical quantum advantage (2022)
13. Biamonte, J., Wittek, P., Pancotti, N., Rebentrost, P., Wiebe, N., Lloyd, S.: Quantum machine learning. Nature **549**(7671), 195–202 (2017)
14. Bombin, H., Martin-Delgado, M.A.: Topological quantum distillation. Phys. Rev. Lett. **97**(18), 180501 (2006)
15. Botelho, L., Glos, A., Kundu, A., Miszczak, J.A., Salehi, Ö., Zimborás, Z.: Error mitigation for variational quantum algorithms through mid-circuit measurements (2021)
16. Bravyi, S., Cross, A.W., Gambetta, J.M., Maslov, D., Rall, P., Yoder, T.J.: High-threshold and low-overhead fault-tolerant quantum memory. Nature **627**, 778–782 (2024). https://doi.org/10.1038/s41586-024-07107-7

17. Bravyi, S., Sheldon, S., Kandala, A., Mckay, D.C., Gambetta, J.M.: Mitigating measurement errors in multiqubit experiments. Phys. Rev. A **103**, 042605 (2021). https://doi.org/10.1103/PhysRevA.103.042605
18. Bădescu, C., O'Donnell, R., Wright, J.: Quantum state certification. In: Proceedings of the 51st Annual ACM SIGACT Symposium on Theory of Computing, STOC 2019, pp. 503–514. Association for Computing Machinery, New York, NY, USA (2019). https://doi.org/10.1145/3313276.3316344
19. Cacioppo, A., Colantonio, L., Bordoni, S., Giagu, S.: Quantum diffusion models. arXiv:2311.15444 (2023)
20. Calderbank, A.R., Shor, P.W.: Good quantum error-correcting codes exist. Phys. Rev. A **54**(2), 1098 (1996)
21. Cao, S., Wossnig, L., Vlastakis, B., Leek, P., Grant, E.: Cost-function embedding and dataset encoding for machine learning with parametrized quantum circuits. Phys. Rev. A **101**, 052309 (2020). https://doi.org/10.1103/PhysRevA.101.052309
22. Cerezo, M., et al.: Variational quantum algorithms. Nat. Rev. Phys. **3**(9), 625–644 (2021). https://doi.org/10.1038/s42254-021-00348-9
23. Chakraborty, S., Hou, Y., Chen, A., Ravi, G.S.: Empowering the quantum cloud user with qrio (2024). https://arxiv.org/abs/2407.17676
24. Chen, C., Zhao, Q., Zhou, M., He, Z., Sun, Z., Situ, H.: Quantum generative diffusion model: a fully quantum-mechanical model for generating quantum state ensemble. arXiv preprint arXiv:2401.07039 (2024)
25. Chen, C., Zhao, Q., Zhou, M., Niyato, D., He, Z., Situ, H.: Overcoming dimensional factorization limits in discrete diffusion models through quantum joint distribution learning. arXiv preprint arXiv:2505.05151 (2025)
26. Chen, K.C., Xu, X., Burt, F., Liu, C.Y., Yu, S., Leung, K.K.: Noise-aware distributed quantum approximate optimization algorithm on near-term quantum hardware. In: 2024 IEEE International Conference on Quantum Computing and Engineering (QCE), vol. 2, pp. 144–149. IEEE (2024)
27. Cheng, S., Wang, L., Xiang, T., Zhang, P.: Tree tensor networks for generative modeling. Phys. Rev. B **99**, 155131 (2019). https://doi.org/10.1103/PhysRevB.99.155131
28. Czarnik, P., Arrasmith, A., Coles, P.J., Cincio, L.: Error mitigation with Clifford quantum-circuit data (2020)
29. Dallaire-Demers, P.L., Killoran, N.: Quantum generative adversarial networks. Phys. Rev. A **98**, 012324 (2018). https://doi.org/10.1103/PhysRevA.98.012324
30. D'Anjou, B., Coish, W.A.: Soft decoding of a qubit readout apparatus. Phys. Rev. Lett. **113**, 230402 (2014). https://doi.org/10.1103/PhysRevLett.113.230402
31. Das, P., et al.: Afs: accurate, fast, and scalable error-decoding for fault-tolerant quantum computers. In: 2022 IEEE International Symposium on High-Performance Computer Architecture (HPCA), pp. 259–273. IEEE (2022)
32. Delfosse, N., Nickerson, N.H.: Almost-linear time decoding algorithm for topological codes. Quantum **5**, 595 (2021)
33. Ding, Y., Gokhale, P., Lin, S.F., Rines, R., Propson, T., Chong, F.T.: Systematic crosstalk mitigation for superconducting qubits via frequency-aware compilation. arXiv preprint arXiv:2008.09503 (2020)
34. Duriez, A., et al.: Computing band gaps of periodic materials via sample-based quantum diagonalization. arXiv preprint arXiv:2503.10901 (2025)
35. Edmonds, J.: Paths, trees, and flowers. Can. J. Math. **17**, 449–467 (1965)
36. EPiQC Research Group: Scalable quantum circuit optimization (2018). https://www.epiqc.cs.uchicago.edu/scalable-quantum-circuit-optimization

37. Farhi, E., Goldstone, J., Gutmann, S.: A quantum approximate optimization algorithm (2014)
38. Fowler, A.G., Mariantoni, M., Martinis, J.M., Cleland, A.N.: Surface codes: towards practical large-scale quantum computation. Phys. Rev. A **86**(3), 032324 (2012)
39. Fuller, B., et al.: Improved quantum computation using operator backpropagation. arXiv preprint (2025). https://doi.org/10.48550/arXiv.2502.01897
40. Gao, X., Anschuetz, E.R., Wang, S.T., Cirac, J.I., Lukin, M.D.: Enhancing generative models via quantum correlations. Phys. Rev. X **12**(2), 021037 (2022)
41. Ge, Y., et al.: Quantum circuit synthesis and compilation optimization: overview and prospects. arXiv preprint arXiv:2407.00736 (2024)
42. Gidney, C., Jones, C.: New circuits and an open source decoder for the color code. arXiv preprint arXiv:2312.08813 (2023)
43. Gidney, C., Shutty, N., Jones, C.: Magic state cultivation: growing t states as cheap as cnot gates. arXiv preprint arXiv:2409.17595 (2024)
44. Giurgica-Tiron, T., Hindy, Y., LaRose, R., Mari, A., Zeng, W.J.: Digital zero noise extrapolation for quantum error mitigation. In: 2020 IEEE International Conference on Quantum Computing and Engineering (QCE), pp. 306–316. IEEE (2020)
45. Goodfellow, I., et al.: Generative adversarial networks. Commun. ACM **63**(11), 139–144 (2020)
46. Gretton, A., Borgwardt, K.M., Rasch, M.J., Schölkopf, B., Smola, A.: A kernel two-sample test. J. Mach. Learn. Res. **13**(25), 723–773 (2012). http://jmlr.org/papers/v13/gretton12a.html
47. Grover, L.K.: A fast quantum mechanical algorithm for database search. In: Annual ACM Symposium on Theory of Computing, pp. 212–219. ACM (1996)
48. He, K., Zhang, X., Ren, S., Sun, J.: Deep residual learning for image recognition. In: Proceedings of the IEEE Conference on Computer Vision and Pattern Recognition, pp. 770–778 (2016)
49. Heinsoo, J., et al.: Rapid high-fidelity multiplexed readout of superconducting qubits. Phys. Rev. Appl. **10**, 034040 (2018). https://doi.org/10.1103/PhysRevApplied.10.034040
50. Herzog, L.S., Berent, L., Kubica, A., Wille, R.: Lattice surgery compilation beyond the surface code. arXiv preprint arXiv:2504.10591 (2025)
51. Higgott, O., Gidney, C.: Sparse blossom: correcting a million errors per core second with minimum-weight matching. Quantum **9**, 1600 (2025)
52. Ho, J., Jain, A., Abbeel, P.: Denoising diffusion probabilistic models. In: Proceedings of the 34th International Conference on Neural Information Processing Systems, NIPS 2020. Curran Associates, Red Hook, NY, USA (2020)
53. Huang, H.L., et al.: Experimental quantum generative adversarial networks for image generation. Phys. Rev. Appl. **16**, 024051 (2021). https://doi.org/10.1103/PhysRevApplied.16.024051
54. Hur, T., Kim, L., Park, D.K.: Quantum convolutional neural network for classical data classification. Quantum Mach. Intell. **4**(1), 3 (2022)
55. IBM Quantum: Approximate quantum compiler (AQC) – API reference (qiskit 0.41) (2025). https://quantum.cloud.ibm.com/docs/api/qiskit/0.41/synthesis_aqc. Accessed 03 Sept 2025
56. IBM Quantum: Operator backpropagation (qiskit-addon-obp) – guide (2025). https://quantum.cloud.ibm.com/docs/guides/qiskit-addons-obp. Accessed 03 Sept 2025

57. Jeffrey, E., et al.: Fast accurate state measurement with superconducting qubits. Ph.D. thesis (2014). https://doi.org/10.1103/PhysRevLett.112.190504
58. Kandala, A., Mezzacapo, A., Temme, K., Takita, M., Brink, M., Chow, J.M., Gambetta, J.M.: Hardware-efficient variational quantum eigensolver for small molecules and quantum magnets. Nature **549**(7671), 242–246 (2017)
59. Kanno, K., et al.: Quantum-selected configuration interaction: classical diagonalization of hamiltonians in subspaces selected by quantum computers. arXiv preprint arXiv:2302.11320 (2023)
60. Khan, S.A., Kaufman, R., Mesits, B., Hatridge, M., Türeci, H.E.: Practical trainable temporal postprocessor for multistate quantum measurement. PRX Quantum **5**, 020364 (2024). https://doi.org/10.1103/PRXQuantum.5.020364
61. Khezri, M., et al.: Measurement-induced state transitions in a superconducting qubit: within the rotating-wave approximation. Phys. Rev. Appl. **20**, 054008 (2023). https://doi.org/10.1103/PhysRevApplied.20.054008
62. Kim, Y., et al.: Evidence for the utility of quantum computing before fault tolerance. Nature **618**(7965), 500–505 (2023). https://doi.org/10.1038/s41586-023-06096-3
63. Knill, E.: Quantum computing with realistically noisy devices. Nature **434**(7029), 39–44 (2005)
64. Koutsioumpas, S., Noszko, T., Sayginel, H., Webster, M., Roffe, J.: Colour codes reach surface code performance using vibe decoding. arXiv preprint arXiv:2508.15743 (2025)
65. Kurita, T., Qassim, H., Ishii, M., Oshima, H., Sato, S., Emerson, J.: Synergetic quantum error mitigation by randomized compiling and zero-noise extrapolation for the variational quantum eigensolver. Quantum **7**, 1184 (2023). https://doi.org/10.22331/q-2023-11-20-1184
66. Kurniawan, H., Rodríguez-Soriano, L., Cuomo, D., Almudever, C.G., Herrero, F.G.: On the use of calibration data in error-aware compilation techniques for nisq devices. In: 2024 IEEE International Conference on Quantum Computing and Engineering (QCE), vol. 1, pp. 338–348. IEEE (2024)
67. Kwun, G., Zhang, B., Zhuang, Q.: Mixed-state quantum denoising diffusion probabilistic model. Phys. Rev. A **111**, 032610 (2025). https://doi.org/10.1103/PhysRevA.111.032610
68. Lavrijsen, W., Tudor, A., Müller, J., Iancu, C., de Jong, W.: Classical optimizers for noisy intermediate-scale quantum devices. In: 2020 IEEE International Conference on Quantum Computing and Engineering (QCE), pp. 267–277 (2020). https://doi.org/10.1109/QCE49297.2020.00041
69. Li, J., Topaloglu, R.O., Ghosh, S.: Quantum generative models for small molecule drug discovery. IEEE Trans. Quantum Eng. **2**, 1–8 (2021)
70. Li, Y., Benjamin, S.C.: Efficient variational quantum simulator incorporating active error minimization. Phys. Rev. X **7**, 021050 (2017). https://doi.org/10.1103/PhysRevX.7.021050
71. Lienhard, B.: Machine learning assisted superconducting qubit readout. Ph.D. dissertation, Massachusetts Institute of Technology, Cambridge, MA (2021). https://hdl.handle.net/1721.1/140024
72. Lienhard, B., et al.: Deep-neural-network discrimination of multiplexed superconducting-qubit states. Phys. Rev. Appl. **17**, 014024 (2022). https://doi.org/10.1103/PhysRevApplied.17.014024
73. Liu, X., Zhuang, J., Hou, W., You, Y.Z.: Measurement-based quantum diffusion models. arXiv preprint arXiv:2508.08799 (2025)

74. Liyanage, N., Wu, Y., Deters, A., Zhong, L.: Scalable quantum error correction for surface codes using FPGA. In: 2023 IEEE International Conference on Quantum Computing and Engineering (QCE), vol. 1, pp. 916–927. IEEE (2023)
75. Lloyd, S., Weedbrook, C.: Quantum generative adversarial learning. Phys. Rev. Lett. **121**, 040502 (2018). https://doi.org/10.1103/PhysRevLett.121.040502
76. Madden, L., Simonetto, A.: Best approximate quantum compiling problems. ACM Trans. Quantum Comput. **3**(2), 1–29 (2022). https://doi.org/10.1145/3505181
77. Martinez, L.A., Rosen, Y.J., DuBois, J.L.: Improving qubit readout with hidden Markov models. Phys. Rev. A **102**, 062426 (2020). https://doi.org/10.1103/PhysRevA.102.062426
78. Maurya, S., Mude, C.N., Oliver, W.D., Lienhard, B., Tannu, S.: Scaling qubit readout with hardware efficient machine learning architectures. In: Proceedings of the 50th Annual International Symposium on Computer Architecture, ISCA 2023. Association for Computing Machinery, New York, NY, USA (2023). https://doi.org/10.1145/3579371.3589042
79. McClean, J.R., Boixo, S., Smelyanskiy, V.N., Babbush, R., Neven, H.: Barren plateaus in quantum neural network training landscapes. Nat. Commun. **9**(1), 4812 (2018)
80. McClean, J.R., Romero, J., Babbush, R., Aspuru-Guzik, A.: The theory of variational hybrid quantum-classical algorithms. New J. Phys. **18**(2), 023023 (2016)
81. Miao, K.C., et al.: Overcoming leakage in quantum error correction. Nat. Phys. **19**(12), 1780–1786 (2023). https://doi.org/10.1038/s41567-023-02226-w
82. Molavi, A., Xu, A., Tannu, S., Albarghouthi, A.: Dependency-aware compilation for surface code quantum architectures. arXiv preprint arXiv:2311.18042 (2023), full version of OOPSLA 2025 paper
83. Moll, N., et al.: Quantum optimization using variational algorithms on near-term quantum devices. Quantum Sci. Technol. **3**(3), 030503 (2018)
84. Müller, T., et al.: Improved belief propagation is sufficient for real-time decoding of quantum memory. arXiv preprint arXiv:2506.01779 (2025)
85. Nation, P.D., Kang, H., Sundaresan, N., Gambetta, J.M.: Scalable mitigation of measurement errors on quantum computers. PRX Quantum **2**, 040326 (2021). https://doi.org/10.1103/PRXQuantum.2.040326
86. Ngo, T.A., Nguyen, T., Thang, T.C.: A survey of recent advances in quantum generative adversarial networks. Electronics **12**(4), 856 (2023)
87. Nielsen, M.A., Chuang, I.L.: Quantum computation and quantum information. Phys. Today **54**(2), 60 (2001)
88. Oreshkov, O., Calsamiglia, J.: Distinguishability measures between ensembles of quantum states. Phys. Rev. A **79**, 032336 (2009). https://doi.org/10.1103/PhysRevA.79.032336
89. Oumarou, O., Ollitrault, P.J., Cortes, C.L., Scheurer, M., Parrish, R.M., Gogolin, C.: Molecular properties from quantum Krylov subspace diagonalization. arXiv preprint arXiv:2501.05286 (2025)
90. O'Gorman, J., Campbell, E.T.: Quantum computation with realistic magic-state factories. Phys. Rev. A **95**(3) (2017). https://doi.org/10.1103/physreva.95.032338
91. Pagni, V., Huber, S., Epping, M., Felderer, M.: Fast quantum amplitude encoding of typical classical data. arXiv preprint arXiv:2503.17113 (2025)
92. Panteleev, P., Kalachev, G.: Degenerate quantum LDPC codes with good finite length performance. Quantum **5**, 585 (2021)
93. Parigi, M., Martina, S., Caruso, F.: Quantum-noise-driven generative diffusion models. Advanced Quantum Technol. 2300401 (2024)

94. Perdomo-Ortiz, A., Benedetti, M., Realpe-Gómez, J., Biswas, R.: Opportunities and challenges for quantum-assisted machine learning in near-term quantum computers. Quantum Sci. Technol. **3**(3), 030502 (2018)
95. Peruzzo, A., et al.: A variational eigenvalue solver on a photonic quantum processor. Nat. Commun. **5**, 4213 (2014)
96. Preskill, J.: Quantum computing in the nisq era and beyond. Quantum **2**, 79 (2018)
97. Rath, M., Date, H.: Quantum data encoding: a comparative analysis of classical-to-quantum mapping techniques and their impact on machine learning accuracy. EPJ Quantum Technol. **11**(1), 72 (2024)
98. Raveendran, N., Valls, J., Pradhan, A.K., Rengaswamy, N., Garcia-Herrero, F., Vasić, B.: Soft syndrome iterative decoding of quantum LDPC codes and hardware architectures. EPJ Quantum Technol. **10**(1) (2023). https://doi.org/10.1140/epjqt/s40507-023-00201-1
99. Ravi, G.S., et al.: Better than worst-case decoding for quantum error correction. In: Proceedings of the 28th ACM International Conference on Architectural Support for Programming Languages and Operating Systems, vol. 2, pp. 88–102 (2023)
100. Ravi, G.S., Smith, K.N., Gokhale, P., Chong, F.T.: Quantum computing in the cloud: analyzing job and machine characteristics. In: 2021 IEEE International Symposium on Workload Characterization (IISWC), pp. 39–50. IEEE (2021). https://doi.org/10.1109/IISWC53511.2021.00015
101. Rebentrost, P., Schuld, M., Wossnig, L., Petruccione, F., Lloyd, S.: Quantum gradient descent and newton's method for constrained polynomial optimization. New J. Phys. **21**(7), 073023 (2019)
102. Robledo-Moreno, J., et al.: Chemistry beyond the scale of exact diagonalization on a quantum-centric supercomputing architecture. Sci. Adv. **11**(25), eadu9991 (2025). https://doi.org/10.1126/sciadv.adu9991
103. Rosenberg, E., Ginsparg, P., McMahon, P.L.: Experimental error mitigation using linear rescaling for variational quantum eigensolving with up to 20 qubits. Quantum Sci. Technol. (2021). https://doi.org/10.1088/2058-9565/ac3b37
104. Rudolph, et al.: Pauli propagation for efficient classical simulation of noisy quantum circuits. Phys. Rev. Lett. (2025), pauli propagation methods
105. Ryan, C.A., et al.: Tomography via correlation of noisy measurement records. Phys. Rev. A **91**, 022118 (2015). https://doi.org/10.1103/PhysRevA.91.022118
106. Sank, D., et al.: System characterization of dispersive readout in superconducting qubits. Phys. Rev. Appl. **23**, 024055 (2025). https://doi.org/10.1103/PhysRevApplied.23.024055
107. Schuld, M.: Supervised quantum machine learning models are kernel methods. arXiv preprint arXiv:2101.11020 (2021)
108. Schuld, M., Sweke, R., Meyer, J.J.: Effect of data encoding on the expressive power of variational quantum-machine-learning models. Phys. Rev. A **103**, 032430 (2021). https://doi.org/10.1103/PhysRevA.103.032430
109. Seifert, L.M., Dangwal, S., Chong, F.T., Ravi, G.S.: Clapton: Clifford-assisted problem transformation for error mitigation in variational quantum algorithms. In: Proceedings of ASPLOS, pp. 47–62 (2025). https://doi.org/10.1145/3622781.3674178
110. Sharma, A., Kumar, K.: Adaptive quantum circuit compilation for nisq devices: a dynamic programming approach. Int. J. Res. Appl. Sci. Eng. Technol. **13**, 351–358 (2025). https://doi.org/10.22214/ijraset.2025.66859

111. Shehata, A., et al.: Bridging paradigms: designing for HPC-quantum convergence (2025). https://arxiv.org/abs/2503.01787
112. Shor, P.W.: Polynomial-time algorithms for prime factorization and discrete logarithms on a quantum computer. SIAM J. Comput. **26**(5), 1484–1509 (1997). https://doi.org/10.1137/s0097539795293172
113. Skoric, L., Browne, D.E., Barnes, K.M., Gillespie, N.I., Campbell, E.T.: Parallel window decoding enables scalable fault tolerant quantum computation. Nat. Commun. **14**(1), 7040 (2023)
114. Smith, K.N., et al.: Error mitigation in quantum computers through instruction scheduling. arXiv preprint arXiv:2105.01760 (2021)
115. Smith, S.C., Brown, B.J., Bartlett, S.D.: Local predecoder to reduce the bandwidth and latency of quantum error correction. Phys. Rev. Appl. **19**(3), 034050 (2023)
116. Spall, J.: An overview of the simultaneous perturbation method for efficient optimization (2001)
117. SpinQ Technology: Hybrid quantum-classical algorithms: The future of computing (2025). https://www.spinquanta.com/news-detail/hybrid-quantum-classical-algorithms-the-future-of-computing20250123075527. Accessed 15 Sept 2025
118. Steane, A.M.: Error correcting codes in quantum theory. Phys. Rev. Lett. **77**(5), 793 (1996)
119. Takagi, R., Endo, S., Minagawa, S., Gu, M.: Fundamental limits of quantum error mitigation (2021)
120. Temme, K., Bravyi, S., Gambetta, J.M.: Error mitigation for short-depth quantum circuits. Phys. Rev. Lett. **119**(18), 180509 (2017)
121. Terhal, B.M.: Quantum error correction for quantum memories. Rev. Mod. Phys. **87**(2), 307–346 (2015). https://doi.org/10.1103/revmodphys.87.307
122. Tilly, J., et al.: The variational quantum eigensolver: a review of methods and best practices (2021)
123. Wall, M.L., Abernathy, M.R., Quiroz, G.: Generative machine learning with tensor networks: benchmarks on near-term quantum computers. Phys. Rev. Res. **3**(2), 023010 (2021)
124. Wang, H., Wu, L.A., Liu, Y., Nori, F.: Measurement-based quantum phase estimation algorithm for finding eigenvalues of non-unitary matrices. Phys. Rev. A **82**, 062303 (2010). https://doi.org/10.1103/PhysRevA.82.062303
125. Wang, M., et al.: Optimizing FTQC programs through qec transpiler and architecture codesign. arXiv preprint arXiv:2412.15434 (2024)
126. Wang, S., Czarnik, P., Arrasmith, A., Cerezo, M., Cincio, L., Coles, P.J.: Can error mitigation improve trainability of noisy variational quantum algorithms? (2021)
127. Wang, S.P., Mao, J., Sakk, E.: Compilation, optimization, error mitigation, and machine learning in quantum algorithms. arXiv preprint arXiv:2506.15760 (2025)
128. Wu, Y., Zhong, L.: Fusion blossom: fast MWPM decoders for qec. In: 2023 IEEE International Conference on Quantum Computing and Engineering (QCE), vol. 1, pp. 928–938. IEEE (2023)
129. Yeter-Aydeniz, K., Bauer, N.M., Jain, P., Masnick, M.: Hybrid quantum-classical latent diffusion models for medical image generation. arXiv preprint arXiv:2508.09903 (2025)
130. Yoshioka, N., et al.: Krylov diagonalization of large many-body Hamiltonians on a quantum processor. Nat. Commun. **16**, 5014 (2025). https://doi.org/10.1038/s41467-025-59716-z

131. Yu, J., et al.: Sample-based krylov quantum diagonalization. arXiv preprint arXiv:2501.09702 (2025)
132. Zhang, B., Xu, P., Chen, X., Zhuang, Q.: Generative quantum machine learning via denoising diffusion probabilistic models. Phys. Rev. Lett. **132**(10), 100602 (2024)
133. Zhang, X., Chen, C.: Parameter-efficient quantum denoising diffusion probabilistic models with temporal encoding. Futur. Gener. Comput. Syst. **174**, 107981 (2026)
134. Zhu, D., et al.: Training of quantum circuits on a hybrid quantum computer. Sci. Adv. **5**(10), eaaw9918 (2019)
135. Zoufal, C.: Generative quantum machine learning (2021). https://arxiv.org/abs/2111.12738
136. Zoufal, C., Lucchi, A., Woerner, S.: Quantum generative adversarial networks for learning and loading random distributions. NPJ Quantum Inf. **5**(1), 103 (2019)

Quantum Optimization via Gradient-Based Hamiltonian Descent

Jiaqi Leng[1,2(✉)] and Bin Shi[3,4]

[1] Simons Institute for the Theory of Computing, University of California, Berkeley, USA
jiaqil@berkeley.edu

[2] Department of Mathematics, University of California, Berkeley, USA

[3] Center for Mathematics and Interdisciplinary Sciences, Fudan University, Shanghai, China
binshi@fudan.edu.cn

[4] Shanghai Institute for Mathematics and Interdisciplinary Sciences, Shanghai, China

Abstract. With rapid advancements in machine learning, first-order algorithms have emerged as the backbone of modern optimization techniques, owing to their computational efficiency and low memory requirements. Recently, the connection between accelerated gradient methods and damped heavy-ball motion, particularly within the framework of Hamiltonian dynamics, has inspired the development of novel quantum algorithms for continuous optimization. One such algorithm, Quantum Hamiltonian Descent (QHD), leverages quantum tunneling to escape saddle points and local minima, facilitating the discovery of global solutions in complex optimization landscapes. However, QHD faces several challenges, including slower convergence rates compared to classical gradient methods and limited robustness in highly non-convex problems due to the non-local nature of quantum states. Furthermore, the original QHD formulation primarily relies on function value information, which limits its effectiveness. Inspired by insights from high-resolution differential equations that have elucidated the acceleration mechanisms in classical methods, we propose an enhancement to QHD by incorporating gradient information, leading to what we call gradient-based QHD. Gradient-based QHD achieves faster convergence and significantly increases the likelihood of identifying global solutions. Numerical simulations on challenging problem instances demonstrate that gradient-based QHD outperforms existing quantum and classical methods by at least an order of magnitude.

Keywords: Quantum Optimization · Quantum Dynamics · Gradient-Based Optimization

In modern machine learning, a central challenge lies in unconstrained optimization, particularly the task of minimizing a continuous objective function without any constraints. Mathematically, this problem is formulated as:

$$\min_{x\in\mathbb{R}^d} f(x).$$

V. Aggarwal et al. (Eds.): PQAI 2025, CCIS 2724, pp. 170–177, 2026.
https://doi.org/10.1007/978-981-95-7829-0_14

Efficiently solving such optimization problems is fundamental to a wide range of machine learning applications. First-order optimization algorithms have emerged as the cornerstone of this endeavor due to their computational efficiency and low memory requirements. In the early 1980s, a groundbreaking advancement was introduced by [19]: the accelerated gradient method, now widely known as Nesterov's accelerated gradient descent method (NAG). This method revolutionized first-order optimization by achieving a faster convergence rate compared to vanilla gradient descent. The iterative update rules for NAG are as follows:

$$x_k = y_{k-1} - s\nabla f(y_{k-1}), \quad y_k = x_k + \frac{k-1}{k+2}(x_k - x_{k-1}), \tag{1}$$

where $s > 0$ is the step size. The key innovation of NAG is the introduction of momentum to the vanilla gradient descent, which effectively reduces oscillations and speeds up progress in the optimization trajectory.

Recent advancements have shed light on the mechanisms underlying the acceleration of NAG. In 2016, Wibisono et al. [25] introduced a variational perspective on the acceleration phenomenon, known as the Bregman Lagrangian framework. Specifically, the continuous limit of NAG can be recast as the Euler-Lagrange equation of the following Lagrangian function:

$$\mathcal{L}(X, V, t) = \frac{1}{2}t^3\|V\|^2 - t^3 f(X), \tag{2}$$

where X and V represent the position and velocity variables, respectively. Furthermore, employing the Legendre transformation, we can convert this Lagrangian into its Hamiltonian form:

$$H(X, P, t) = \frac{1}{2t^3}\|P\|^2 + t^3 f(X), \tag{3}$$

which paves the way for extending the analysis from classical dynamics to quantum dynamics.

By transforming the classical momentum variable P to the quantum momentum operator $-i\nabla$ within the Hamiltonian (3), Leng et al. [12] have pioneered a groundbreaking algorithm, known as Quantum Hamiltonian Descent (QHD), which defines the quantum dynamics through a Schrödinger equation:

$$i\partial_t \Psi(t, x) = \hat{H}(t)\Psi(t, x), \tag{4}$$

where the time-dependent Hamiltonian is articulated as[1]:

$$\hat{H}(t) = \frac{1}{2}\left\|t^{-3/2}(-i\nabla)\right\|^2 + t^3 f(x). \tag{5}$$

[1] The original formulation of QHD allows a more general Hamiltonian: $\hat{H}(t) = e^{\alpha_t - \gamma_t}(-\Delta/2) + e^{\alpha_t + \beta_t + \chi_t} f(x)$, where the Laplacian $\Delta = \nabla \cdot \nabla$, as given in Eq. (A.24) of [12]. For simplicity, we specialize to the parameter choices corresponding to the classical NAG, namely $\alpha_t = -\log(t)$ and $\beta_t = \gamma_t = 2\log(t)$.

Let $\Psi(t,x)\colon [0,\infty) \times \mathbb{R}^d \to \mathbb{C}$ denote a quantum wave function, whose squared modulus $|\Psi(t,x)|^2$ represents the probability distribution of a hypothetical quantum particle in $\mathbb{R}^d$ at any time $t \geq 0$. For sufficiently large evolution time t, the probability distribution is expected to concentrate near the low-energy configurations of the potential f, particularly around its global minimum. Measuring the quantum state in the computational basis at such times yields a random vector $X \sim |\Psi(t,x)|^2$, which is likely to lie close to the global minimizer of f, thereby approximately solving the associated optimization problem.

As a quantum algorithm, QHD is implemented by simulating the time-dependent Hamiltonian (5), which relies only on oracle access to the function values of f. Thus, QHD can be viewed as a quantum *zeroth-order* method. A natural extension of QHD is to develop its higher-order variants that leverage additional information such as the gradient of f, and to analyze whether such extensions can enhance QHD's efficiency on various continuous optimization problems.

Inspired by the high-resolution ODE framework introduced by [22], where the Lyapunov function is conceptualized as a form of energy or Hamiltonian involving the interplay of kinetic energy and gradient, we propose a novel time-dependent Hamiltonian as

$$\hat{H}(t) = \frac{1}{2}\left\| t^{-3/2}(-i\nabla) + \alpha t^{3/2}\nabla f \right\|^2 + \frac{\beta}{2}\|t^{3/2}\nabla f\|^2 + (t^3 + \gamma t^2) f(x). \tag{6}$$

In this paper, we mainly investigate the Schrödinger equation (4) with the gradient-based Hamtiltonian (6), termed as **gradient-based QHD**.

Convergence Analysis

We theoretically establish the convergence of the newly derived quantum dynamics (6). Without loss of generality, we assume $f(x^*) = 0$ and $x^* = 0$. The first result proves that the quantum dynamics converge to the global minimum of $f(x)$.

Theorem 1 (Informal version of Theorem 1, [14]). *Let $\beta = 0$ and $\gamma \geq \max(3\alpha, 0)$ for any $\alpha \in \mathbb{R}$. For any $1/\alpha \geq T_0 > 0$, we denote $\Psi(t,x)$ as the solution to the PDE (6) for $t \geq T_0$. Let X_t be a random variable distributed according to the probability density $|\Psi(t,x)|^2$. Then, for a convex and continuously differentiable f, we have $\mathbb{E}[f(X_t)] \leq O(t^{-2})$.*

The second result proves that the quantum dynamics converge to the first-order stationary points of $f(x)$ under certain conditions. In what follows, we denote the function $G(x) = \|\nabla f(x)\|^2$ as the square of the gradient norm of f.

Theorem 2 (Informal version of Theorem 2, [14]). *Let $\gamma \geq \max(3\alpha, 0)$ and $\beta > 0$. For any $\min(1/\alpha, \sqrt{2/\beta}) \geq T_0 > 0$, we denote $\Psi(t,x)$ as the solution to the PDE (6). Let X_t be a random variable distributed according to the probability density $|\Psi(t,x)|^2$. Then, for a convex and continuously differentiable f such that its gradient norm satisfies the following identity $G(x) - \nabla G(x)^\top x \leq 0$, we have $\mathbb{E}[\|\nabla f(X_t)\|^2] \leq O(t^{-2})$.*

Algorithm 1. Gradient-based QHD with fixed step size

Classical Input: Hamiltonian parameters α, β, γ, step size h, number of iterations K.
Quantum Input: an initial guess state $|\Psi_0\rangle$
Output: a classical point $\xi \in \mathbb{R}^d$.

Initialize the quantum register to $|\Psi_0\rangle$.
for $k = 1$ **to** K **do**
  Determine $t_k = kh$.
  Implement a quantum circuit U_k as described in (7).
  Compute $|\Psi_k\rangle = U_k|\Psi_{k-1}\rangle$.
end for
Measure the final quantum state $|\Psi_K\rangle$ with the position observable $\hat{x}$ to obtain a sample point $\xi \in \mathbb{R}^d$.

Both results are proven via a Lyapunov function approach, where we establish the time-monotonicity of a carefully designed Lyapunov function using the commutation relation of quantum operators.

Time Discretization and Complexity Analysis

Note that the gradient-based QHD Hamiltonian can be decomposed in the form $\hat{H}(t_k) = H_{k,1} + H_{k,2} + H_{k,3}$, where

$$H_{k,1} = -\frac{1}{2t_k^3}\Delta, \quad H_{k,2} = \frac{\alpha}{2}\{-i\nabla, \nabla f\},$$
$$H_{k,3} = \frac{(\alpha^2 + \beta)}{2}t^3\|\nabla f\|^2 + (t^3 + \gamma t^2)f.$$

Therefore, we can further decompose a short-time evolution step using the product formula (i.e., operator splitting):

$$e^{-ih\hat{H}(t_k)} \approx e^{-ihH_{k,1}}e^{-ihH_{k,2}}e^{-ihH_{k,3}}. \tag{7}$$

Since all the Hamiltonians $H_{k,1}$, $H_{k,2}$, and $H_{k,3}$ can be efficiently simulated using a quantum computer, we obtain a quantum algorithm that implements gradient-based QHD to solve large-scale optimization problems, as summarized in Algorithm 1.

A crucial step in Algorithm 1 is to implement the quantum unitary operator U_k based on the operator splitting formula (7). We note that the sub-Hamiltonians $H_{k,1}$ and $H_{k,3}$ are fast-forwardable, and the operator $H_{k,2}$ can be simulated by invoking Quantum Singular Value Transformation (QSVT). Combining these technical results together, we end up with the overall complexity of the quantum algorithm, as summarized in Theorem 3.

Theorem 3. *Let f be L-Lipschitz and $|\Psi_0\rangle$ be a sufficiently smooth function. Then, we can implement Algorithm 1 for K iterations using $\mathcal{O}(K)$ queries to the quantum zeroth-order oracle O_f and $\widetilde{\mathcal{O}}(\alpha dhKL)$ queries to the quantum first-order oracle $O_{\nabla f}$ and its inverses.*[2]

Numerical Experiments

We conduct extensive numerical experiments to evaluate the performance of gradient-based QHD and compare it with other prominent optimization algorithms. The source code of the experiments is available at https://github.com/jiaqileng/Gradient-Based-QHD.

Convex Optimization. We conduct a numerical comparison of gradient-based QHD against three baseline algorithms, including SGDM, NAG, and QHD, for convex optimization. The test function used is

$$f(x, y) = \frac{(x+y)^4}{256} + \frac{(x-y)^4}{128}, \tag{8}$$

which is a convex yet non-strongly convex function, with a singular Hessian at its unique minimum $(0, 0)$. Notably, the gradient of this function does not satisfy the Lipschitz continuity condition. This flat geometry presents significant challenges for classical methods that rely heavily on curvature information, making it a suitable benchmark for comparative evaluation. The performance of these optimization algorithms is visualized in Fig. 1, where two key metrics are employed to access convergence: the average function values $\mathbb{E}[f(X_k)]$ (depicted in the left subplot) and the average gradient norm $\mathbb{E}[\|\nabla f(X_k)\|^2]$ (depicted in the right subplot). Both quantities are tracked over iterations $1 \le k \le 25$. The

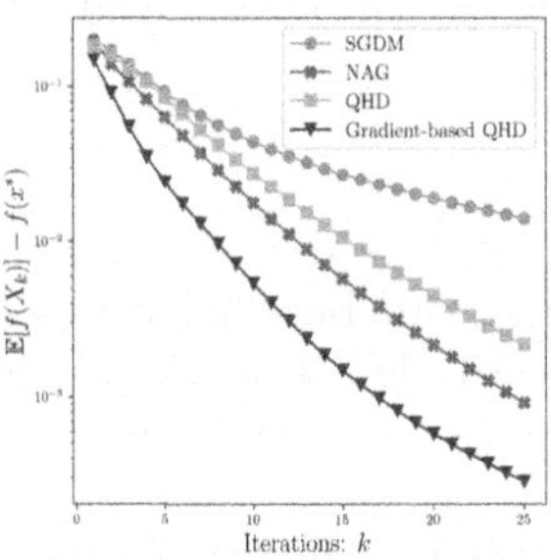

(a) Function value

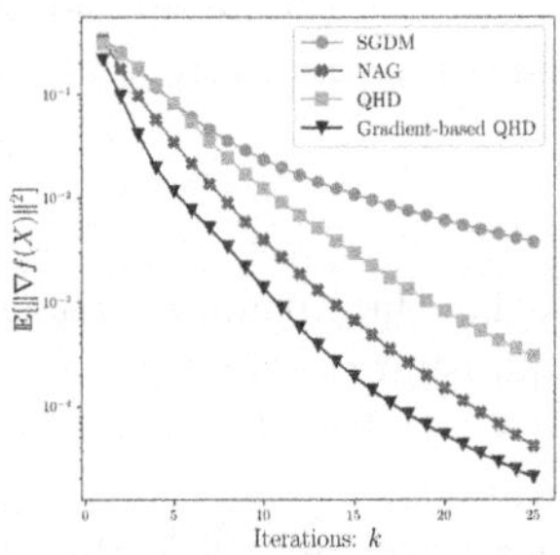

(b) Gradient norm

Fig. 1. Numerical comparison of various optimization algorithms on the convex objective function (8), including function values and success probability.

[2] Here, the $\widetilde{\mathcal{O}}$ notation suppresses poly-logarithmic factors in the error parameter ϵ. The parameter $\epsilon > 0$ represents the error budget in the Hamiltonian simulation.

results reveal distinct convergence behaviors. While the (classical) QHD exhibits a slower convergence rate compared to NAG, the gradient-based QHD stands out by achieving a remarkably faster convergence rate, outperforming all other algorithms. This superior performance highlights the effectiveness of incorporating gradient-based techniques into QHD, particularly for challenging optimization landscapes.

Non-convex Optimization. Next, we compare gradient-based QHD against three baseline algorithms, including SGDM, NAG, and QHD, in non-convex optimization settings. Non-convexity introduces significant challenges for classical first-order methods, as local gradient information alone is often insufficient to distinguish the global minimum from other spurious local optima. We evaluate a variety of non-convex optimization problem instances characterized by diverse landscape features. In Fig. 2, we illustrate the behaviors of four optimization algorithms for two problem instances, namely, the Michalewicz function and the Rastrigin function.

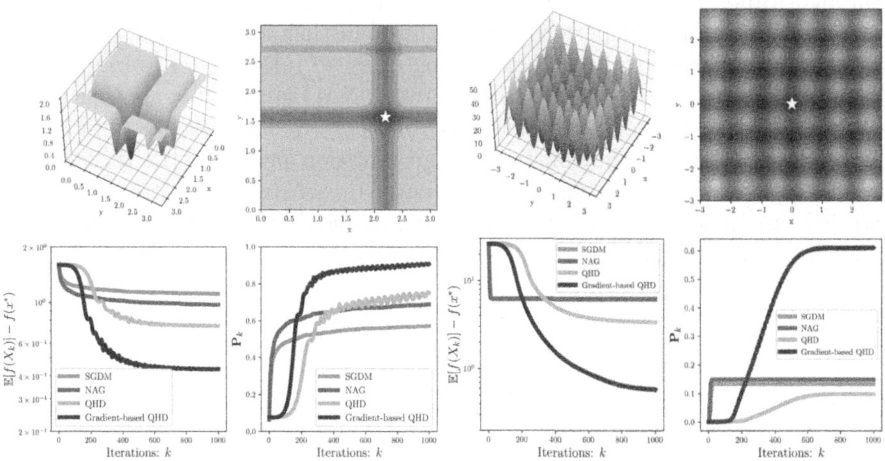

(a) Michalewicz function (b) Rastrigin function

Fig. 2. Numerical comparison of various optimization algorithms, including functions values and success probability.

Despite the diversity of non-convex test problems, gradient-based QHD consistently delivers robust and favorable performance. Compared to both QHD and the classical algorithms, it achieves a significantly faster convergence rate and yields notably lower terminal objective function values. In the Rastrigin function, for instance, the final objective value obtained by gradient-based QHD is nearly an order of magnitude lower than that of QHD and two orders of magnitude lower than those achieved by SGDM and NAG.

Further numerical analysis highlights that gradient-based QHD attains a higher success probability across all problem instances, indicating that its final states are tightly concentrated around the global minimizer. In summary, by leveraging gradient information within the quantum Hamiltonian framework, gradient-based QHD demonstrates enhanced *global* convergence properties, outperforming QHD and classical optimization methods.

Related Work

There has been a long history of analyzing NAG-related optimization algorithms [6,20]. [24] sheds new light on the understanding and design of NAG using an ODE perspective. In [3,25,26], a Lagrangian (or Hamiltonian) framework is used to describe a larger class of ODEs that provides a unified perspective for the acceleration phenomenon in first-order optimization. Notably, accelerated gradient descent has been investigated in non-Euclidean settings, including mirror descent [10,16] and more generally, Riemannian manifolds [1,7,9,23].

Using quantum computers to accelerate bottleneck steps in classical optimization algorithms has shown promise in achieving quantum advantage [8,17,21]. However, their practical performance requires further investigation due to the non-trivial overhead involved in extracting classical information from quantum states [27]. Motivated by the interplay between NAG and ODEs, another line of research proposes leveraging quantum Hamiltonian dynamics as an algorithmic surrogate for addressing unconstrained optimization problems [12,18,28], with recent extensions to open quantum systems [4], constrained optimization [2], and discrete optimization [5]. This approach is particularly effective for highly non-convex problems [15] and well-suited for hardware implementation [11,13].

References

1. Ahn, K., Sra, S.: From Nesterov's estimate sequence to Riemannian acceleration. In: Conference on Learning Theory, pp. 84–118. PMLR (2020)
2. Augustino, B., Leng, J., Nannicini, G., Terlaky, T., Wu, X.: A quantum central path algorithm for linear optimization. arXiv preprint arXiv:2311.03977 (2023)
3. Betancourt, M., Jordan, M.I., Wilson, A.C.: On symplectic optimization. arXiv preprint arXiv:1802.03653 (2018)
4. Chen, Z., Lu, Y., Wang, H., Liu, Y., Li, T.: Quantum Langevin dynamics for optimization. Commun. Math. Phys. **406**(3), 52 (2025)
5. Cheng, J., Zhou, R., Gan, Y., Qian, C., Liu, J.: Quantum Hamiltonian descent for graph partition. arXiv preprint arXiv:2411.14696 (2024)
6. Giselsson, P., Boyd, S.: Monotonicity and restart in fast gradient methods. In: 53rd IEEE Conference on Decision and Control, pp. 5058–5063. IEEE (2014)
7. Han, A., Mishra, B., Jawanpuria, P., Gao, J.: Riemannian accelerated gradient methods via extrapolation. In: International Conference on Artificial Intelligence and Statistics, pp. 1554–1585. PMLR (2023)
8. Kerenidis, I., Prakash, A.: Quantum gradient descent for linear systems and least squares. Phys. Rev. A **101**(2), 022316 (2020)

9. Kong, L., Tao, M.: Quantitative convergences of Lie group momentum optimizers. arXiv preprint arXiv:2405.20390 (2024)
10. Krichene, W., Bayen, A., Bartlett, P.L.: Accelerated mirror descent in continuous and discrete time. In: Advances in Neural Information Processing Systems, vol. 28 (2015)
11. Kushnir, S., Leng, J., Peng, Y., Fan, L., Wu, X.: QHDOPT: a software for nonlinear optimization with quantum Hamiltonian descent. INFORMS J. Comput. (2024)
12. Leng, J., Hickman, E., Li, J., Wu, X.: Quantum Hamiltonian descent. arXiv preprint arXiv:2303.01471 (2023)
13. Leng, J., Li, J., Peng, Y., Wu, X.: Expanding hardware-efficiently manipulable hilbert space via Hamiltonian embedding. arXiv preprint arXiv:2401.08550 (2024)
14. Leng, J., Shi, B.: Quantum optimization via gradient-based Hamiltonian descent. In: Forty-Second International Conference on Machine Learning (ICML) (2025)
15. Leng, J., Zheng, Y., Wu, X.: A quantum-classical performance separation in nonconvex optimization. arXiv preprint arXiv:2311.00811 (2023)
16. Lin, T., Ho, N., Jordan, M.: On efficient optimal transport: an analysis of greedy and accelerated mirror descent algorithms. In: International Conference on Machine Learning, pp. 3982–3991. PMLR (2019)
17. Liu, J., et al.: Towards provably efficient quantum algorithms for large-scale machine-learning models. Nat. Commun. **15**(1), 434 (2024)
18. Liu, Y., Su, W.J., Li, T.: On quantum speedups for nonconvex optimization via quantum tunneling walks. Quantum **7**, 1030 (2023)
19. Nesterov, Y.: A method for solving the convex programming problem with convergence rate $O(1/k^2)$. In: Dokl Akad Nauk Sssr, vol. 269, p. 543 (1983)
20. O'donoghue, B., Candes, E.: Adaptive restart for accelerated gradient schemes. Found. Comput. Math. **15**, 715–732 (2015)
21. Rebentrost, P., Schuld, M., Wossnig, L., Petruccione, F., Lloyd, S.: Quantum gradient descent and Newton's method for constrained polynomial optimization. New J. Phys. **21**(7), 073023 (2019)
22. Shi, B., Du, S.S., Jordan, M.I., Su, W.J.: Understanding the acceleration phenomenon via high-resolution differential equations. Math. Program. 1–70 (2022)
23. Siegel, J.W.: Accelerated optimization with orthogonality constraints. arXiv preprint arXiv:1903.05204 (2019)
24. Su, W., Boyd, S., Candes, E.J.: A differential equation for modeling Nesterov's accelerated gradient method: theory and insights. J. Mach. Learn. Res. **17**(153), 1–43 (2016)
25. Wibisono, A., Wilson, A.C., Jordan, M.I.: A variational perspective on accelerated methods in optimization. Proc. Natl. Acad. Sci. **113**(47), E7351–E7358 (2016)
26. Wilson, A.C., Recht, B., Jordan, M.I.: A Lyapunov analysis of accelerated methods in optimization. J. Mach. Learn. Res. **22**(113), 1–34 (2021)
27. Yuen, H.: An improved sample complexity lower bound for (fidelity) quantum state tomography. Quantum **7**, 890 (2023)
28. Zhang, C., Leng, J., Li, T.: Quantum algorithms for escaping from saddle points. Quantum **5**, 529 (2021)

Quantum Cognition and Human-Centered Computing

Human Factors of Quantum Modeling and Computing

Peiran Liu and Denny Yu(✉)

Purdue University, West Lafayette, IN 47907, USA
{liu3820,dennyyu}@purdue.edu

Abstract. Quantum mechanics and computing has potential for advancing human factors and ergonomics (HFE) research by providing novel theories and methods for modeling cognition and human-centered systems. This paper presents a comprehensive review of quantum modeling and computing within HFE on both theory and application. This review first discuss quantum cognition, where the uncertainty principle and quantum probability theory refines models of human decision-making under uncertainty. Building on these foundations, the review discusses human–computer interaction (HCI) on quantum systems, demonstrating current challenges in usability. The concept of quantum digital twins (QDTs) is introduced as an extension of digital twin technology, integrating human centric data for real-time healthcare. Finally, the paper surveys the role of quantum artificial intelligence (QAI) in augmenting human work and analyzing human-centric data with potential for computational acceleration and improved predictive performance. Together, these advances illustrate a bidirectional pathway, HCI drive innovations in quantum technologies while integrating quantum frameworks into reshapes cognitive models along with expanding human-centered AI, and human digital twin. This review establishes a conceptual baseline for future interdisciplinary research of quantum science and HFE.

150–250 words.

Keywords: Quantum Computing · Quantum AI · Quantum Digital Twin · Quantum Cognition · Human factors · Ergonomics

1 Introduction

Quantum mechanism and computing has substantial potential for accelerating human-centric research. Specifically, the quantum uncertainty principle provides a robust framework for modeling complex human decision-making and cognition under uncertainty. Furthermore, the fundamental mechanics of quantum computation, which utilize qubits in superposition will improve the efficiency of information transmission and computing. This review examines the quantum modeling or quantum-inspired modeling of human factors and ergonomics,

V. Aggarwal et al. (Eds.): PQAI 2025, CCIS 2724, pp. 181–196, 2026.
https://doi.org/10.1007/978-981-95-7829-0_15

the other theme is the Quantum computing and quantum AI applications in the domain. This manuscript is structured as follows: Sect. 3 discusses the role of quantum cognition and brain modeling under uncertainty, Sect. 4 reviews advancements in quantum human–computer interaction, Sect. 5 introduces the current developments in quantum digital twins of human healthcare, finally in Sect. 6, we summarizes potential application of quantum AI augmenting human work and human-centric data. As shown in Fig. 1, this paper reviews the conceptual pathway linking fundamental quantum principles and technology to applications in human factors and ergonomics. Quantum mechanism and computing has substantial potential for accelerating human well-being. Specifically, the quantum uncertainty principle models complex human decision-making and cognition under uncertainty. One benefit is it offers robust framework for capacity to natively represent uncertainty as intrinsic, context-sensitive, and trade-offs between incompatible mental states [3]. Furthermore, the fundamental mechanics of quantum computation, which utilize qubits in superposition will improve the efficiency of human-centric data transmission and computing. This review examines the quantum modeling or quantum-inspired modeling of human factors and ergonomics, the other theme is the human centric applications of Quantum computing and quantum AI. This manuscript is structured as follows: Sect. 3 discusses the role of quantum cognition and brain modeling under uncertainty, Sect. 4 reviews advancements in quantum human–computer interaction, Sect. 5 introduces the current developments in quantum digital twins of human healthcare, finally in Sect. 6, we summarizes potential application of quantum AI augmenting human work and human-centric data. As shown in Fig. 1, this paper reviews the conceptual pathway linking fundamental quantum principles and technology to applications in human factors and ergonomics.

2 Human Factors and Ergonomics

Human factors is a discipline optimizes human well-being and system performance by understanding interactions between people and system entity [86] [81]. Academic research in human factors is broadly classified into three areas of research track: cognitive factors, physical factors, and organizational factors [36].

As for cognitive factors, human-computer interaction (HCI) remains one of the most extensively investigated areas. This is because even the most sophisticated technological systems have zero value unless they can be operated properly by human users [69]. Human-Computer Interaction (HCI) is guided by two core objectives. The first is usability, which create or optimize the systems that are easy to learn, efficient to use, error-tolerant, and pleasing [27]. The second objective is User Experience (UX), which extends beyond functional usability to include the full spectrum of a user's interaction, including emotions, perceptions, beliefs, and preferences [50].

Physical factors focus on the physical and physiological responses of the human body to physical work. In application, ergonomists design workplaces, tools, and tasks to fit the human body's capabilities to maximizing performance

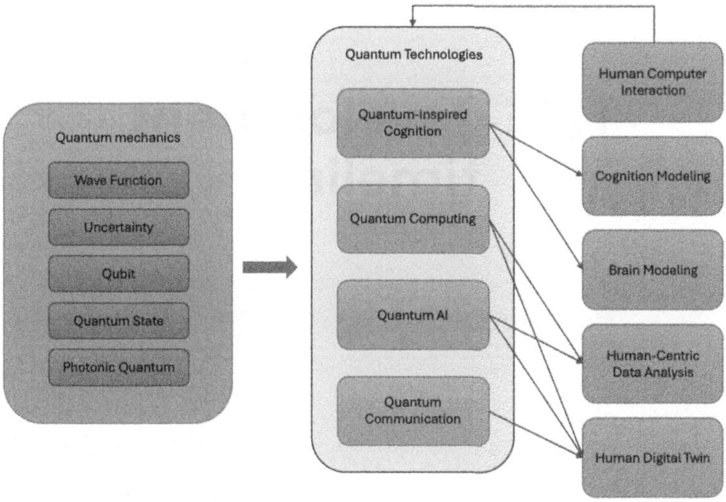

Fig. 1. A Figure Review of Quantum Modeling and Computing for Human Factors and Ergonomics, From Physics to Quantum Technologies to Human-Centric Applications

while minimizing discomfort and the risk of injury [55]. The primary physical factors examined in the field includes: posture, repetitive movements, forceful exertions, and manual material handling [54]. Over the past decade, the community observed a growing adoption of artificial intelligence methods in the prevention of musculoskeletal disorders [46,47,87].

2.1 History of Human Factors and Ergonomics

Starting 1900s, Frederick Taylor, Frank Gilbreth and Lillian Gilbreth started to investigate the work design that improve the industrial efficiency, whose theory was also referred as 'Taylorism' [7]. After the 'process engineering' era, the formal subject of modern human factors emerged during World War II, when researchers realized complexity of new military technology (aircrafts, radar systems) reveals that highly trained operators were making catastrophic errors [79]. Since 1980s, The digital revolution soon propelled the focus to Human-Computer Interaction (HCI), which focusing on usability studies, user interfaces (UI), and user experience (UX) [65]. In the past decade, the application of machine learning in human factors and ergonomics has significantly expanded dramatically. In 2025, empirical evidence of HFE efficacy can be applied in diverse engineering applications such as healthcare (medical devices to prevent errors) [6,11,74], automotive driving (cybersecurity of autonomous vehicles) [17,45,82], and manufacturing. Recently, quantum AI have received increasing attention in analyzing human-centric data [57], meanwhile, theoretical and cognitive science communities have proposed a increasing number of quantum-inspired cognitive models. The development timeline of human factors and ergonomics is illustrated in Fig. 2.

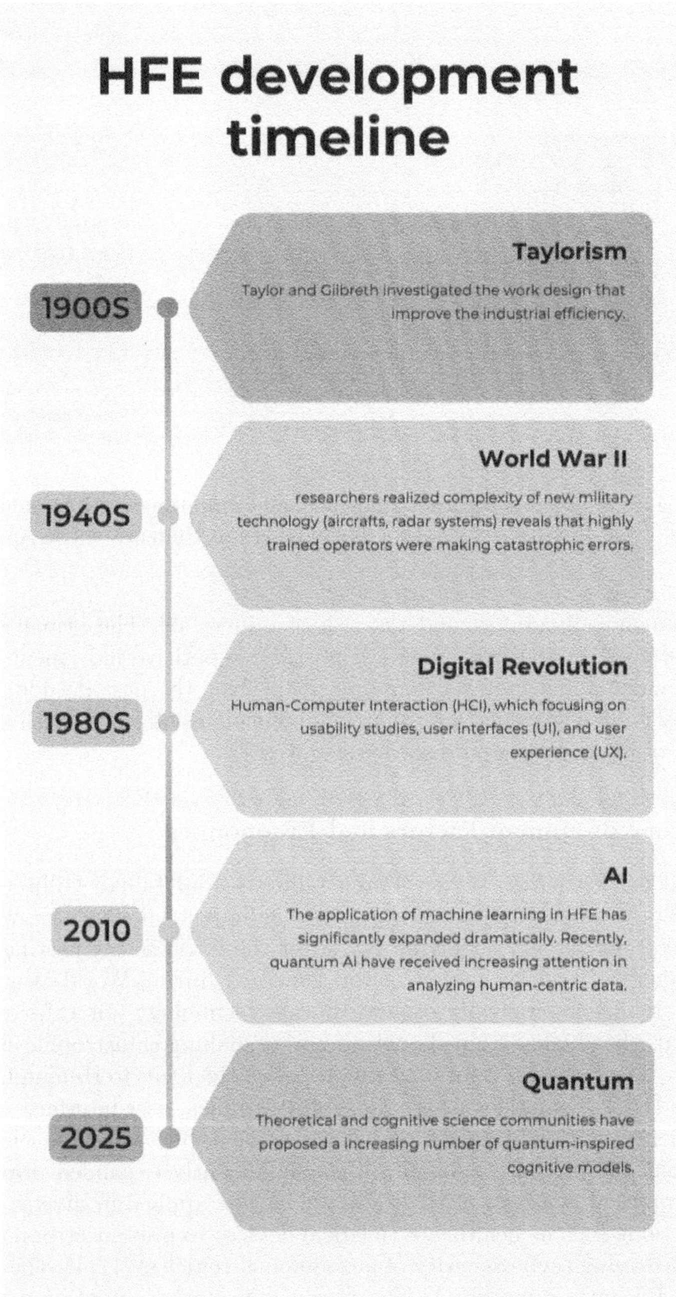

Fig. 2. A Development Timeline of Human Factors and Ergonomics

3 Quantum Cognition and Brain Modeling with Uncertainty

Cognition represents one of the most active research areas in human factors. Inheriting frameworks from systems engineering, the field generally decomposes cognition into constituent components [5,64]. Standard model segments cognitive processes into four core elements: perception, attention and mental workload, memory, and decision-making [34]. The first three components are often studied through biological and physiological mechanisms. Decision-making, however, is more complicated, research in psychology and cognitive science has studied human decision-making under conditions of uncertainty [10]. Indeed, a significant proportion of daily decisions involve uncertainty. Within this field, psychologists have identified and delineated several key scopes or dimensions of uncertainty, including reward expectancy, reward probability, ambiguity, and risk [3]. Thus, Amy Bland concluded three types of uncertainty in human decision making [8]. The first type is the expected uncertainty, week predictions of future result based on reinforcement learning from previous event and practice with the recognized and established unreliability [26]. The second type is the unexpected uncertainty, which invalidates the current reinforcement learning, since the established rules of action and outcome become obsolete, forcing the model incapable of forecasting the results of future actions. The last type is volatility that periodic environmental changes involve a continuous revising of the reinforcement learning model [62].

3.1 Uncertainty Principle

Addressing uncertainty in human cognition can use quantum uncertainty principle. The uncertainty principle is one of the most fundamental concept of quantum mechanics, reflecting the limits of simultaneously knowing conjugate observables (position and momentum), from the non-commuting algebra of operators representing physical quantities [29,58].

Let A, B be Hermitian operators corresponding to two observables, and let $|\psi\rangle$ be a normalized quantum state [80]. We define shifted operators $\tilde{A} = A - \langle A\rangle$ and $\tilde{B} = B - \langle B\rangle$, where $\langle A\rangle = \langle\psi|A|\psi\rangle$.

Their variances are

$$(\Delta A)^2 = \langle\psi|\tilde{A}^2|\psi\rangle, \qquad (\Delta B)^2 = \langle\psi|\tilde{B}^2|\psi\rangle.$$

Applying the Cauchy–Schwarz inequality [40] to $\tilde{A}|\psi\rangle$ and $\tilde{B}|\psi\rangle$ yields

$$\langle\tilde{A}^2\rangle\langle\tilde{B}^2\rangle \geq |\langle\tilde{A}\tilde{B}\rangle|^2.$$

The expectation $\langle\tilde{A}\tilde{B}\rangle$ can be split into commutator and anticommutator terms:

$$\langle\tilde{A}\tilde{B}\rangle = \tfrac{1}{2}\langle\{\tilde{A}, \tilde{B}\}\rangle + \tfrac{1}{2}\langle[A, B]\rangle.$$

Taking the imaginary part leads to the inequality:

$$\Delta A\,\Delta B \geq \tfrac{1}{2}\big|\langle[A, B]\rangle\big|.$$

For the canonical pair $A = x$ and $B = p$, we have $[x, p] = i\hbar$. Therefore, the Heisenberg uncertainty [67] relation becomes:

$$\Delta x\, \Delta p \geq \frac{\hbar}{2}.$$

This math formulation indicates that quantum states cannot be simultaneously localized in both position and momentum beyond this limit [28].

3.2 Uncertainty in Quantum Cognition

Quantum cognition uses the mathematical framework of quantum probability as a better model for how the cognition works with uncertain, ambiguous information [63]. As a limitation of classic framework of cognitive science, its persistent difficulty in explaining common fallacies, but this is a natural consequence of how human judgment operates. In fact, psychology researchers have discovered more and more paradoxical cognitive phenomena that contradict the predictions of classical probability theory, and these fallacies in human decision-making are rooted in processes such as memory, perception, causal reasoning, and probability judgments [9,16,70].

Conjunction Fallacy with Uncertainty. An example explaining this is the conjunction fallacy. The often-cited example Linda Problem, was introduced by Amos Tversky and Daniel Kahneman is typically presented as follows [19,37,77]:

> *Linda is 31 years old, single, outspoken, and very bright. She majored in philosophy. As a student, she was deeply concerned with issues of discrimination and social justice, and also participated in anti-nuclear demonstrations. Which of the following is more probable?*
> *1. Linda is a bank teller.*
> *2. Linda is a bank teller and is active in the feminist movement.*

Human participants consistently believe that the likelihood of the conjunction (a feminist and a bank teller) as higher than that of its constituent event (a bank teller), according to extensive experiments. However, in probability theory formulation, we know this inequality must be true for this type of problem $P(A \wedge B) \leq P(A)$ for arbitrary two events A and B.

In the quantum model, a person's belief about "what Linda is" is not in a definite state while they are considering the question [32]. Instead, it's in a superposition of multiple possibilities: bank teller, feminist, environmentalist, etc. The description creates a specific mental state vector. The key is that the questions "Is Linda a bank teller?" and "Is Linda a feminist?" are not compatible. Therefore, the order matters, asking one question disturbs the system (the mental representation of Linda) and changes the context for asking the next one.

The first question provides context: When you are first asked, "Is Linda a feminist?", the description is perfectly aligned with this hypothesis. The probability of a "yes" is very high. This question collapses the mental superposition toward a state compatible with being a feminist. However, the next question in that Context: Now, from within that context, you are asked about bank teller? While being a feminist bank teller is still unlikely, it's now more thinkable than it was before. The context has been changed by the first question. If the question was asked in the reverse order: "Is Linda a bank teller?" This is highly unlikely given the description. The mental state collapses to a firm "no." Now, asking about feminist from within that context feels even more unlikely. The probability of the conjunction in cognition is very low, which is correct.

Or in classic probability based cognitive modeling, the fallacy occurs because the two options are not evaluated independently [24]. The conjunction is evaluated as a single, whole concept. The vivid, story-consistent concept "feminist bank teller" feels more plausible and has a higher projection from the initial mental state than the bland, story-inconsistent concept "bank teller" alone. In fact, either classic cognition models with uncertainty or quantum cognition modeling, instead of providing completely contradictory results, provide identical outcomes in most cases, as researchers have proven [9], although their approaches and derivations are no way near each other.

3.3 Quantum Brain Modeling

Neurobiological evidence supports the uncertainty of brain structure, because the release of synaptic vesicles from trillions of neruons in the brain is independently random, and the quantum mechanical state of the entire brain evolves into a "fuzzy cloud" containing macroscopic possibilities [68]. These possibilities represent a variety of plans with uncertainty. Similar to the conclusion from neurobiology, previous research in neuromodulation has identified two key chemical neurotransmitters responsible for modulating uncertainty processing in the human brain: acetylcholine (ACh) and noradrenaline (NA) [4]. Acetylcholine signals expected uncertainty, which arises from the known unreliability of a familiar environment or task [18]. In contrast, noradrenaline is linked to signaling unexpected uncertainty, which occurs from sudden, fundamental changes in environmental contingencies.

Beyond uncertainty, some researchers has proposed that human brain systems may function as intrinsic quantum-information processors [68]. By the second law of thermodynamics, the entropy of an isolated system tends to increase, but brain systems can maintain and create highly ordered states [30]. Resisting the entropy tendency towards disorder, non-Markovian quantum dynamics suggest a camel-like shape stage in quantum entropy, where it first increases and then decreases, as human adapt to their environment [41]. Furthermore, brain systems can maintain global order and stability through quantum entanglement, even when their subsystems experience an increase in entropy [1]. The electrochemical states of neurons turns into quantum information states and describes cognitive processes using open quantum system dynamics [76].

4 Quantum Human–Computer Interaction

Current quantum computing research and engineering presents a high barrier to entry for researchers and practitioners [56]. This challenge is multifaceted, from the complexity of quantum mechanics, lack of user-friendly application programming interfaces (APIs), and a lack of comprehensive documentation. This accessibility gap becomes evident when contrasted with the current progress of artificial intelligence (AI) development. Frameworks like PyTorch and TensorFlow are supported by extensive documentation and APIs. Moreover, a vast ecosystem of educational resources, millions of online tutorials, which have accelerated research in AI.

This challenge has also attracted the attention of IBM's quantum computer team, a leading research group in the field [71]. Their work has identified four research areas for Human-Computer Interaction (HCI) of quantum computing in the future [2]. The first area is the analysis of user workflows, since current quantum computing systems are not yet fully aligned with user demand. This primarily beginning with a classification of the distinct requirements of researchers, students, and developers who interact with the APIs. The second research area focuses on coding tools. Present frameworks like IBM's Qiskit and Cirq from Google heavily rely on expert knowledge, which created a significant barrier to entry. Furthermore, an example of unique constraints is the no-cloning theorem prevents conventional collaborative debugging workflow, due to quantum states cannot be copied and shared [72]. Although features like autocomplete, documentation, and usage analysis are a well-established discipline in classical computing API usability study, it remains a critical outlook for development in quantum software. The third area addresses the visualization of quantum states. Given the difficulty of intuitively representing high-dimensional quantum states, the IBM team suggested that research should progress from improving single-state representations to further visualizing multiqubit entanglement and eventually to tackling the visual compression of exponential state spaces [83]. From a human-computer interaction (HCI) perspective, an objective is to avoid reinventing wheels and mitigating challenges, for instance manually transfer classical computing logic into quantum gate operations and analysis of raw bit-string outputs. A recent study proposes three major principles to address these specific issues in quantum computing: (1) linking conceptual ideas directly to quantum information theory, (2) providing multi-level computational support, and (3) applying established usability standards directly to quantum computing tools [42].

5 Quantum Digital Twins of Human

The digital twin is a concept widely used in manufacturing and transportation to create cyber representations of physical entities [49,78]. The development of a human digital twin presents greater complexity than industrial digital twins. While an industrial digital twin typically represents a single physical asset, a human digital twin exist as either a high-fidelity digital representation of an individual (mesh and skeleton data) or as a collective group model, which requires the intricate modeling of social and interactions between humans [44]. Moreover, the human digital twin is multifaceted; beyond the physical features modeled in machines, it integrates a complex array of physiological signals, cognitive, behavioral, physical, and personality features [11,52,61].

Quantum entanglement is an objective phenomenon when a pair of particles are interacted or share proximity that the quantum state of each particle cannot be described independently of the state of another, the condition holds true with the particles are separated by arbitrary distance. As one particle is observed, the other particle instantly percepts, this can be treat as a telecommunication mechanism and system, which is analogy to the bidirectional data exchange of the cyber and physical communication synchronizing in digital twin. Moreover, recent studies proved that the information transfer with quantum entanglement is continuous, instead of instantaneous [53]. Therefore, some scientists consider quantum entanglement as a manifestation of Digital Twins itself [48]. Researchers suggest future work to integrating data across micro to macro scales, from quantum particles to cosmic systems, for Digital Twins achieve flexibility and a unified dimensions.

However, as current telecommunication technology approaches the foundational limits of Moore's Law, it raises concerns about the technology improvement of research and development for future generation communication [20,22]. Hence, researchers suggest Quantum Optics would increasing the ability of telecommunication [38]. Quantum Optics (QO) utilizes the quantum mechanical rules of photons and its interactions with atoms based on the quantum properties of photons. Quantum optics has demonstrated its application in the development of a next-generation digital twin, which Manzalini proposed as the Quantum Optical Twin [51]. The Quantum Optical Twin is conceptualized as a quantum cyber twin of physical entities, created through the principles of quantum optics. In information theory, data is transmitted and encoded as digital binary bits. However, in quantum communication, information is encoded in qubits. Similarly, the Quantum Optical Twin technology utilizes decoherence-free photons to encode and decode cyber-physical information, thereby constructing a cyber twin of the physical world [75].

More efficient encoding of information will also improve the efficiency for application in healthcare digital twin.

Efficient quantum encoding of information improves the efficiency of digital twin applications in healthcare. In healthcare, a Quantum Digital Twin (QDT) represents a technology where the increasing complexity of patient care requires multi-constraint optimization across diagnostics, treatment planning, and ultimately resource allocation in almost real time level, since generating accurate predictions of the patient is the biggest obstacle of healthcare digital twins [85]. In personalized healthcare, each patient is supported by a quantum digital twin that continuously integrates multimodal health data, especially genomics, medical imaging, and wearable sensors to offer real-time monitoring for disease detection or therapy adjustment [13,84]. While these local decisions elevate personalized care, they also have to fully align with system-wide objective, hospital resource management. The QDT framework could enable region-level optimization for clinicians and administrators to balance individual diagnosis with hospital-wise healthcare efficiency with Quantum Digital Twin (QDT) to optimize healthcare resources globally.

6 Quantum AI Augmenting Human Work and Human-Centric Data

In quantum mechanics, the wavefunction is the fundamental mathematical representation of a quantum system's state [25]. The wavefunction, denoted by $\psi(\mathbf{r}, t)$, encapsulates all possible information about a particle's position, momentum, and all observables at time t. Unlike classical mechanics where states are described deterministically, the wavefunction is a probabilistic model derived from the Born rule [23]. Specifically, the squared modulus of the wavefunction yields the probability density:

$$P(\mathbf{r}, t) = |\psi(\mathbf{r}, t)|^2,$$

which represents the likelihood of a particle position $\mathbf{r}$ and time t. The wavefunction shifts according to the time-dependent Schrödinger equation [21]:

$$i\hbar\frac{\partial}{\partial t}\psi(\mathbf{r}, t) = \hat{H}\psi(\mathbf{r}, t),$$

where $\hat{H}$ is the Hamiltonian operator. Fundamentally, quantum machine learning (QML) utilizes the principles of quantum mechanics for information processing, the information is encoded into the state of qubits, and this quantum state is described mathematically by a wave function [15]. In other words, wave functions are the medium in which quantum machine learning algorithms operate and store information. This framework uses quantum phenomena: superposition, entanglement, and interference to process data with higher efficiency over the classical method [31].

When applied to classical data, quantum computing can function as a powerful accelerator for classic machine learning and deep learning algorithms [39,60]. This is achieved by representing classical information as quantum states, or wave functions, and use quantum mechanics for computation. An example is the quantum algorithm for solving linear systems of equations. By encoding the matrix A and the vector b into quantum states, the algorithm can prepare a final quantum state corresponding to the solution with a potential exponential speedup over classical methods. Thus, measurement of this final state provides the solution. Another example application is in Quantum Kernel Methods. In this approach, classical data points are mapped to quantum states (wave functions) via a quantum feature map to embed the data into a high-dimensional Hilbert space. This process allows for the computation of a quantum kernel, which can capture complex patterns and non-linear relationships that are difficult for classical methods to enhance higher performance with more powerful classification or regression capabilities.

Healthcare represents a critical application for human-centric data, offering potential to improve patient diagnosis [14]. Currently, quantum computing becomes as a Monumental tool for analyzing complex human-centric data. Recent reports have identified three potential applications where quantum computing can be applied to healthcare with human-centric data to drive innovation [73]. An application is Quantum EEG, which integrates quantum computing with neural signal analysis. The conventional EEG analysis pipeline, which focuses on extracting and classifying features across channels to mathematically formed as a signal process problem. By reformulating this challenge for quantum computation, it becomes possible to use quantum machine learning (QML) to enhanced performance. Specifically, quantum Recurrent Neural Networks (RNNs) and quantum Support Vector Machines (SVMs) outperform the classical Kalman filtering method in multiple tasks. The second application is Radiology. Studies have shown that a qutrit model more effectively corresponds to grayscale imaging data for brain image segmentation tasks when utilizing a quantum deep neural network [43]. Furthermore, quantum machine learning are validated to be applied classifying over 120 different types of brain tumors using MRI scan data [12]. Another application is Genomics, where quantum annealing machines are used to predict protein fold and DNA sequence reconstruction [59,66]. Beyond healthcare, the quantum walk model also demonstrated its application in predicting the inherent and obscure human minds, particularly emotions and ideas, employing controlled unitary operations and quantum superpositions mechanism in the serial reproduction of narrative data [33].

7 Conclusion

In this study, we present a comprehensive review of quantum modeling and computing in human factors and ergonomics. Within our proposed quantum framework for human factors and ergonomics, we observe a bidirectional pathway of innovation. First, advancements in fundamental quantum mechanisms are pushing the advancements of quantum Ftechnologies, which in turn enable novel applications and enhance human factors and ergonomics research. Meanwhile, fundamental physics principles are actively advancing the development of cognitive and brain modeling. Likewise, innovations in Human-Computer Interaction (HCI) are assisting to visualize quantum states and develop interfaces for quantum programming. Nevertheless, the unique characteristics, along with advantages, of quantum systems create novel challenges that traditional HFE is only beginning to address. Quantum programming languages, like Qiskit, and their integrated development environments (IDEs) are in their early stage from a usability study [35]. Another research is control room design for quantum computer labs. For engineers onsite working with the physical quantum computers, the control systems are immensely complex. Designing control rooms and interfaces that allow for effective monitoring and manipulation of these delicate systems is a classic human factors challenge. A constraint within human factors is the acquisition of sufficient data to fully leverage the capabilities of quantum computing. This challenge of data scarcity is not unique to quantum computing but is also a well-documented in the application of machine learning.

References

1. Adams, B., Petruccione, F.: Quantum effects in the brain: a review. AVS Quantum Sci. **2**(2) (2020)
2. Ashktorab, Z., Weisz, J.D., Ashoori, M.: Thinking too classically: research topics in human-quantum computer interaction. In: Proceedings of the 2019 CHI Conference on Human Factors in Computing Systems, pp. 1–12 (2019)
3. Atmanspacher, H., et al.: What are the appropriate axioms of rationality for reasoning under uncertainty with resource-constrained systems? Behav. Brain Sci. **43** (2020)
4. Avery, M.C., Nitz, D.A., Chiba, A.A., Krichmar, J.L.: Simulation of cholinergic and noradrenergic modulation of behavior in uncertain environments. Front. Comput. Neurosci. **6**, 5 (2012)
5. Azuma, R., Daily, M., Furmanski, C.: A review of time critical decision making models and human cognitive processes. In: 2006 IEEE Aerospace Conference, p. 9. IEEE (2006)
6. Barragan, J.A., Yang, J., Yu, D., Wachs, J.P.: A neurotechnological aid for semi-autonomous suction in robotic-assisted surgery. Sci. Rep. **12**(1), 4504 (2022)
7. Björkman, T.: The rationalisation movement in perspective and some ergonomic implications. Appl. Ergon. **27**(2), 111–117 (1996)
8. Bland, A.R., Schaefer, A.: Different varieties of uncertainty in human decision-making. Front. Neurosci. **6**, 85 (2012)
9. Bruza, P.D., Wang, Z., Busemeyer, J.R.: Quantum cognition: a new theoretical approach to psychology. Trends Cogn. Sci. **19**(7), 383–393 (2015)
10. Busemeyer, J., Zhang, Q., Balakrishnan, S., Wang, Z.: Application of quantum–Markov open system models to human cognition and decision. Entropy **22**(9), 990 (2020)
11. Cha, J.S., Athanasiadis, D., Anton, N.E., Stefanidis, D., Yu, D.: Measurement of nontechnical skills during robotic-assisted surgery using sensor-based communication and proximity metrics. JAMA Netw. Open **4**(11), e2132209–e2132209 (2021)
12. Chakrabarty, S., Sotiras, A., Milchenko, M., LaMontagne, P., Hileman, M., Marcus, D.: Mri-based identification and classification of major intracranial tumor types by using a 3d convolutional neural network: a retrospective multi-institutional analysis. Radiology: Artif. Intell. **3**(5), e200301 (2021)
13. Chen, H., Anton, N.E., Zheng, B., Hornbeck, T., Scanlon, M., Yu, D.: Visual cue processing and scanning efficiency in decision-making of expert and novice nurses during rapid patient deterioration. In: Proceedings of the Human Factors and Ergonomics Society Annual Meeting. p. 10711813251372097. SAGE Publications Sage CA: Los Angeles, CA (2025)
14. Chen, H., Liu, P., Zhou, G., Lu, M.L., Yu, D.: Computer vision and tactile glove: a multimodal model in lifting task risk assessment. Appl. Ergon. **127**, 104513 (2025)
15. Cheng, G., Guo, J.: Modeling correlated-noise in silicon spin qubit device. APL Quantum **2**(1) (2025)

16. Cools, R., D'Esposito, M.: Inverted-u-shaped dopamine actions on human working memory and cognitive control. Biol. Psychiat. **69**(12), e113–e125 (2011)
17. Cui, C., et al.: Large language models for autonomous driving (llm4ad): Concept, benchmark, experiments, and challenges. arXiv preprint arXiv:2410.15281 (2024)
18. Dayan, P., Yu, A.J.: Expected and unexpected uncertainty: Ach and ne in the neocortex. Adv. Neural Inf. Process. Syst. **15** (2002)
19. Donovan, S., Epstein, S.: The difficulty of the linda conjunction problem can be attributed to its simultaneous concrete and unnatural representation, and not to conversational implicature. J. Exp. Soc. Psychol. **33**(1), 1–20 (1997)
20. Eldering, C.A., Sylla, M.L., Eisenach, J.A.: Is there a moore's law for bandwidth? IEEE Commun. Mag. **37**(10), 117–121 (1999)
21. Feit, M., Fleck, J., Jr., Steiger, A.: Solution of the schrödinger equation by a spectral method. J. Comput. Phys. **47**(3), 412–433 (1982)
22. Gao, K., Bhuyan, A.K., Lee, M.H., Biswas, S.: Network-disjointed asymmetric autoencoders for data compression in IoT based healthcare applications. IEEE Internet Things Mag. (2025)
23. Gao, S.: The Meaning of the Wave Function: In Search of the Ontology of Quantum Mechanics. Cambridge University Press (2017)
24. Gigerenzer, G.: The bounded rationality of probabilistic mental models. In: Rationality: Psychological and philosophical perspectives, pp. 284–313. Routledge (1993)
25. Goldstein, S., Zanghì, N.: Reality and the role of the wave function in quantum theory. The wave function: Essays on the metaphysics of quantum mechanics, pp. 91–109 (2013)
26. Gordon, K.: The impermanence of being: Toward a psychology of uncertainty. J. Humanist. Psychol. **43**(2), 96–117 (2003)
27. Gunesekera, A.I., Bao, Y., Kibelloh, M.: The role of usability on e-learning user interactions and satisfaction: a literature review. J. Syst. Inf. Technol. **21**(3), 368–394 (2019)
28. Heller, E.J.: Quantum localization and the rate of exploration of phase space. Phys. Rev. A **35**(3), 1360 (1987)
29. Hestenes, D.: Spin and uncertainty in the interpretation of quantum mechanics. Am. J. Phys. **47**(5), 399–415 (1979)
30. Hilbert, S., Hänggi, P., Dunkel, J.: Thermodynamic laws in isolated systems. Phys. Rev. E **90**(6), 062116 (2014)
31. Horodecki, R., Horodecki, P., Horodecki, M., Horodecki, K.: Quantum entanglement. Revi. Modern Phys. **81**(2), 865–942 (2009)
32. Huang, J., Busemeyer, J.R., Ebelt, Z., Pothos, E.M.: Bridging the gap between subjective probability and probability judgments: the quantum sequential sampler. Psychol. Rev. (2024)
33. Huang, J., Zhang, Q., Xie, X., Breithaupt, F., Busemeyer, J.: A quantum model for serial reproduction. Cognition **266**, 106266 (2026)
34. Jalote-Parmar, A., Badke-Schaub, P., Ali, W., Samset, E.: Cognitive processes as integrative component for developing expert decision-making systems: a workflow centered framework. J. Biomed. Inform. **43**(1), 60–74 (2010)
35. Jiménez-Navajas, L., Bühler, F., Leymann, F., Pérez-Castillo, R., Piattini, M., Vietz, D.: Quantum software development: a survey. Quantum Inf. Comput. **24**(7&8), 609–642 (2024)
36. Kadir, B.A., Broberg, O., da Conceição, C.S.: Current research and future perspectives on human factors and ergonomics in industry 4.0. Comput. Ind. Eng. **137**, 106004 (2019)

37. Kahneman, D., Tversky, A.: Choices, values, and frames. Am. Psychol. **39**(4), 341 (1984)
38. Kamdjou, H.M., Baudry, D., Havard, V., Ouchani, S.: Resource-constrained extended reality operated with digital twin in industrial internet of things. IEEE Open J. Commun. Soc. **5**, 928–950 (2024)
39. Khan, T.M., Robles-Kelly, A.: Machine learning: quantum vs classical. IEEE Access **8**, 219275–219294 (2020)
40. Kheruntsyan, K., et al.: Violation of the cauchy-schwarz inequality with matter waves. Phys. Rev. Lett. **108**(26), 260401 (2012)
41. Khrennikov, A.: Open systems, quantum probability, and logic for quantum-like modeling in biology, cognition, and decision-making. Entropy **25**(6), 886 (2023)
42. Kim, H., Jeng, M.J., Smith, K.N.: Toward human-quantum computer interaction: Interface techniques for usable quantum computing. In: Proceedings of the 2025 CHI Conference on Human Factors in Computing Systems, pp. 1–18 (2025)
43. Konar, D., Bhattacharyya, S., Panigrahi, B.K., Behrman, E.C.: Qutrit-inspired fully self-supervised shallow quantum learning network for brain tumor segmentation. IEEE Trans. Neural Netw. Learn. Syst. **33**(11), 6331–6345 (2021)
44. Lin, Y., et al.: Human digital twin: a survey. J. Cloud Comput. **13**(1), 131 (2024)
45. Linkov, V., Zámečník, P., Havlíčková, D., Pai, C.W.: Human factors in the cybersecurity of autonomous vehicles: trends in current research. Front. Psychol. **10**, 995 (2019)
46. Liu, P., Chen, H., Yu, D.: Computer vision embedded post-processing algorithm on lifting risks. In: Proceedings of the Human Factors and Ergonomics Society Annual Meeting. vol. 68, pp. 1372–1373. SAGE Publications Sage CA: Los Angeles, CA (2024)
47. Liu, P., Chen, H., Zhou, G., Yu, D.: Are llms the new ergonomists? Predicting lifting risks with llms and machine learning. Proc. Human Factors Ergonomics Soc. Ann. Meet. **0**(0), 10711813251370730 (2025). https://doi.org/10.1177/10711813251370730
48. Lv, Z., Xie, S., Li, Y., Hossain, M.S., El Saddik, A.: Building the metaverse using digital twins at all scales, states, and relations. Virtual Reality Intell. Hardware **4**(6), 459–470 (2022)
49. Ma, Y., et al.: Lampilot: An open benchmark dataset for autonomous driving with language model programs. In: Proceedings of the IEEE/CVF Conference on Computer Vision and Pattern Recognition, pp. 15141–15151 (2024)
50. MacKenzie, I.S.: Human-computer interaction: an empirical research perspective. Elsevier (2024)
51. Manzalini, A.: Quantum communications in future networks and services. Quantum Rep. **2**(1), 221–232 (2020)
52. Miller, M.E., Spatz, E.: A unified view of a human digital twin. Human-Intell. Syst. Integ. **4**(1), 23–33 (2022)
53. Minev, Z.K., et al.: To catch and reverse a quantum jump mid-flight. Nature **570**(7760), 200–204 (2019)
54. Muggleton, J., Allen, R., Chappell, P.: Hand and arm injuries associated with repetitive manual work in industry: a review of disorders, risk factors and preventive measures. Ergonomics **42**(5), 714–739 (1999)
55. Niu, S.: Ergonomics and occupational safety and health: an ilo perspective. Appl. Ergon. **41**(6), 744–753 (2010)
56. Nofer, M., Bauer, K., Hinz, O., van der Aalst, W., Weinhardt, C.: Quantum computing. Bus. Inf. Syst. Eng. **65**(4), 361–367 (2023)

57. Olorunsogo, T., Jacks, B.S., Ajala, O.A.: Leveraging quantum computing for inclusive and responsible ai development: a conceptual and review framework. Comput. Sci. IT Res. J. **5**(3), 671–680 (2024)
58. Oppenheim, J., Wehner, S.: The uncertainty principle determines the nonlocality of quantum mechanics. Science **330**(6007), 1072–1074 (2010)
59. Outeiral, C., Morris, G.M., Shi, J., Strahm, M., Benjamin, S.C., Deane, C.M.: Investigating the potential for a limited quantum speedup on protein lattice problems. New J. Phys. **23**(10), 103030 (2021)
60. Pan, X., et al.: Segan: semi-supervised learning approach for missing data imputation. arXiv preprint arXiv:2405.13089 (2024)
61. Peng, Y., Anton, N.E., Cha, J., Mizota, T., Hennings, J.M., Stambro, R., Rendina, M.A., Stanton-Maxey, K.J., Stefanidis, D., Yu, D.: Objective measures of communication behavior predict clinical performance. J. Surg. Educ. **76**(5), 1337–1347 (2019)
62. Pidgeon, N.F., Beattie, J.: The psychology of risk and uncertainty. Handbook of Environmental Risk Assessment and Management, pp. 289–318 (1998)
63. Pothos, E.M., Busemeyer, J.R.: Quantum cognition. Annu. Rev. Psychol. **73**(1), 749–778 (2022)
64. Pylyshyn, Z.W.: Computation and cognition: issues in the foundations of cognitive science. Behav. Brain Sci. **3**(1), 111–132 (1980)
65. Rogers, Y.: New theoretical approaches for Hci. Ann. Rev. Inf. Sci. Technol. **38**(1), 87–143 (2004)
66. Sarkar, A., Al-Ars, Z., Bertels, K.: Quaser: quantum accelerated de novo dna sequence reconstruction. PLoS ONE **16**(4), e0249850 (2021)
67. Schrödinger, E.: About heisenberg uncertainty relation. arXiv preprint quant-ph/9903100 **10** (1999)
68. Schwartz, J.M., Stapp, H.P., Beauregard, M.: Quantum physics in neuroscience and psychology: a neurophysical model of mind-brain interaction. Philosop. Trans. Royal Soc. B Biol. Sci. **360**(1458), 1309–1327 (2005)
69. Sinha, G., Shahi, R., Shankar, M.: Human computer interaction. In: 2010 3rd International Conference on Emerging Trends in Engineering and Technology, pp. 1–4. IEEE (2010)
70. Stacy, A.W., Wiers, R.W.: Implicit cognition and addiction: a tool for explaining paradoxical behavior. Annu. Rev. Clin. Psychol. **6**, 551–575 (2010)
71. Steffen, M., DiVincenzo, D.P., Chow, J.M., Theis, T.N., Ketchen, M.B.: Quantum computing: an ibm perspective. IBM J. Res. Dev. **55**(5), 13-1 (2011)
72. Streltsov, A., Zurek, W.H.: Quantum discord cannot be shared. Phys. Rev. Lett. **111**(4), 040401 (2013)
73. Swan, M., dos Santos, R.P., Witte, F.: Quantum neurobiology. Quantum Rep. **4**(1), 107–126 (2022)
74. Toombs, A., Gray, C., Zhou, G., Light, A.: Appropriated or inauthentic care in gig-economy platforms: a psycho-linguistic analysis of uber and lyft. In: Extended Abstracts of the 2018 CHI Conference on Human Factors in Computing Systems, pp. 1–6 (2018)
75. Tran, D.H., et al.: Network digital twin for 6g and beyond: an end-to-end view across multi-domain network ecosystems. IEEE Open J. Commun. Soc. (2025)
76. Tuszynski, J.A.: From quantum chemistry to quantum biology: a path toward consciousness. J. Integr. Neurosci. **19**(4), 687–700 (2020)
77. Tversky, A., Kahneman, D.: Judgment under uncertainty: Heuristics and biases: biases in judgments reveal some heuristics of thinking under uncertainty. Science **185**(4157), 1124–1131 (1974)

78. Wang, Z., et al.: Mobility digital twin: Concept, architecture, case study, and future challenges. IEEE Internet Things J. **9**(18), 17452–17467 (2022)
79. Waterson, P.: World war ii and other historical influences on the formation of the ergonomics research society. Ergonomics **54**(12), 1111–1129 (2011)
80. Widom, H.: On the eigenvalues of certain hermitian operators. Trans. Am. Math. Soc. **88**(2), 491–522 (1958)
81. Wilson, J.R.: Fundamentals of ergonomics in theory and practice. Appl. Ergon. **31**(6), 557–567 (2000)
82. Xiao, X., Ma, X., McDonald, A.D., Mehta, R.K.: What leads to reliance on automated vehicles? an inferential analysis of responses to variable av performance. Appl. Ergon. **128**, 104511 (2025)
83. Yang, Y., Sun, Z.Z., Ran, S.J., Su, G.: Visualizing quantum phases and identifying quantum phase transitions by nonlinear dimensional reduction. Phys. Rev. B **103**(7), 075106 (2021)
84. Yi, S., Lu, M., Busemeyer, J.: Application of quantum cognition to judgments for medical decisions. Quantum Rep. **4**(2) (2022)
85. Zhang, K., et al.: Concepts and applications of digital twins in healthcare and medicine. Patterns **5**(8) (2024)
86. Zheng, B., Chen, H., Zhou, X., Kim, S.Y., Yu, D.: Passive exoskeletons during live surgeries: supporting forward head postures among veterinary surgeons. IISE Trans. Occup. Ergonomics Human Factors **13**(1), 32–39 (2025)
87. Zhou, G., Aggarwal, V., Yin, M., Yu, D.: A computer vision approach for estimating lifting load contributors to injury risk. IEEE Trans. Human-Mach. Syst. **52**(2), 207–219 (2022). https://doi.org/10.1109/THMS.2022.3148339

Q-XAI: Towards Explainable Quantum Neural Networks for Human-Centered Intelligence

Richard Jiang[1(✉)], Yijie Zhu[1], Plamen Angelov[1], Yuri Pashkin[2], and Vaneet Aggarwal[3]

[1] LIRA Centre, Lancaster University, Lancaster LA1 4YW, UK
r.jiang2@lancaster.ac.uk
[2] Quantum Technology Centre, Lancaster University, Lancaster LA1 4YW, UK
[3] Purdue University, West Lafayette, IN 47907, USA

Abstract. Artificial intelligence (AI) is transforming society, from healthcare diagnostics to financial decision-making, yet its rapid growth has outpaced our ability to understand its reasoning. Complex models, particularly deep neural networks, operate largely as "black boxes," raising critical concerns about trust, accountability, and fairness. Explainable AI (XAI) has emerged to illuminate these opaque systems, enabling human-understandable explanations and fostering reliable deployment. At the same time, quantum computing introduces Quantum Neural Networks (QNNs), which exploit superposition, entanglement, and nonlocal correlations to process information in fundamentally new ways, promising breakthroughs in speed, scalability, and modeling complexity. However, these quantum advantages come with unique interpretability challenges: probabilistic outputs, entangled states, and complex causal dependencies make traditional XAI approaches insufficient. In this position paper, we propose Q-XAI, a comprehensive framework that integrates explainability into quantum machine learning. We discuss its core principles, architectural components, and applications across healthcare, finance, and neuroscience, highlighting its potential to bridge human understanding with quantum intelligence. We further outline challenges, ethical considerations, and future research directions. By providing a roadmap for interpretable quantum AI, this work addresses the urgent need for trustworthy, transparent, and human-centered quantum intelligence.

Keywords: Quantum Neural Computing · Quantum Informatics · Explainable Quantum AI · Deep Neural Networks

---

Y. Zhu, P. Angelov, Y. Pashkin and V. Aggarwal—These authors contributed equally to this work.

V. Aggarwal et al. (Eds.): PQAI 2025, CCIS 2724, pp. 197–209, 2026.
https://doi.org/10.1007/978-981-95-7829-0_16

1 Introduction

In 2016, the United States Defense Advanced Research Projects Agency (DARPA) initiated the Explainable Artificial Intelligence (XAI) program, aiming to develop AI systems whose actions can be understood by human users [1]. This move was propelled by growing concerns over the opacity of AI decision-making processes, particularly in high-stakes domains such as healthcare, finance, and criminal justice. The inability of traditional AI models to provide transparent reasoning has led to a crisis of trust, hindering their adoption and raising ethical questions about accountability and fairness [2,3].

1.1 The Rise of Explainable AI

The journey towards explainable AI began with expert systems in the 1970s, which were designed to mimic human expertise in specific domains. These systems provided rule-based explanations for their decisions, offering a glimpse into the "why" behind AI actions. However, with the advent of deep learning in the 2000s, models became increasingly complex, operating as "black boxes" with millions of parameters and non-linear decision boundaries. This complexity made it challenging to interpret their internal workings and understand the rationale behind their predictions [2,4].

In response, the field of XAI emerged, focusing on developing methods to make AI systems more transparent and interpretable. Techniques such as Local Interpretable Model-agnostic Explanations (LIME) and SHapley Additive exPlanations (SHAP) have been proposed to approximate model behavior locally and provide feature importance scores, respectively. Despite these advancements, challenges remain in ensuring that explanations are faithful, consistent, and accessible to end-users [5].

1.2 Quantum Machine Learning: A New Frontier

Parallel to the developments in classical AI, quantum computing has introduced new paradigms in information processing. Quantum Machine Learning (QML) leverages quantum mechanical phenomena, such as superposition and entanglement, to process information in ways that classical computers cannot. Quantum Neural Networks (QNNs) are a class of models that utilize quantum bits (qubits) and quantum gates to perform computations, potentially offering exponential speedups for certain tasks.

However, the integration of quantum principles into machine learning introduces unique challenges. The probabilistic nature of quantum measurements means that QNNs inherently produce outputs with associated uncertainties. Additionally, the non-local correlations in quantum systems can lead to complex interdependencies between variables, complicating the interpretation of model behavior. These factors necessitate the development of novel interpretability techniques tailored to the quantum domain [6].

1.3 The Need for a Roadmap in Explainable Quantum Neural Networks

Given the complexities inherent in QNNs, there is an urgent need for a structured approach to their explainability. Traditional XAI methods may not be directly applicable to quantum models due to differences in computational paradigms and the unique properties of quantum systems. Therefore, a dedicated roadmap is essential to guide the development of interpretability techniques for QNNs.

This roadmap should encompass the identification of key challenges in quantum explainability, the adaptation of existing XAI methods to the quantum context, and the creation of new methodologies that leverage quantum-specific characteristics. Furthermore, it should address the ethical implications of deploying quantum-enhanced AI systems, ensuring that they are aligned with human values and societal norms.

1.4 Contributions of This Paper

In this position paper, we propose the Q-XAI framework, which integrates quantum computing principles with explainable AI methodologies to enhance the interpretability of quantum machine learning models. We outline the core principles of this framework, including quantum superposition and interference, entanglement and non-locality, and causal inference, and discuss their implications for model transparency.

Additionally, we present the architectural components of Q-XAI, comprising Quantum Neural Networks, causal inference modules, and interpretability layers, and explore their roles in facilitating human-understandable explanations. We also examine the applications of Q-XAI in various domains, such as healthcare, finance, and neuroscience, highlighting its potential to provide insights into complex systems.

Finally, we address the challenges and future directions in the field, emphasizing the need for scalable quantum algorithms, integration with classical systems, and ethical considerations in the deployment of quantum-enhanced AI. Through this paper, we aim to stimulate discourse on the importance of explainability in quantum machine learning and advocate for the development of comprehensive frameworks that ensure the responsible and transparent use of these advanced technologies.

2 Background and Challenges

2.1 Classical XAI: A Brief Survey

Explainable Artificial Intelligence (XAI) has become a critical research area aiming to interpret, visualize, and explain the decision-making of AI models. Classical XAI approaches are typically categorized into post-hoc and intrinsic methods. Post-hoc methods, such as feature attribution (e.g., SHAP [7], LIME [8]), aim to interpret model outputs after training. Intrinsic approaches involve designing inherently interpretable models, such as attention-based architectures [9] or rule-based neural networks [10].

Another emerging direction involves generative explanations, where models produce textual or visual explanations alongside predictions [11]. Explainable AI has also advanced in large-scale models, including Transformers and LLMs, with attention visualization and counterfactual methods providing human-understandable insights [12].

While these methods have improved transparency, limitations remain. Feature attribution can be noisy, attention weights do not always align with causal importance, and post-hoc methods may introduce approximation errors. Despite these challenges, classical XAI provides foundational tools and principles that inspire the design of XAI in the quantum domain [13].

2.2 Challenges from Quantum Neural Networks

Quantum Neural Networks (QNNs) introduce unique challenges for interpretability. QNNs leverage qubits and quantum gates [14–16], which exist in superposition and can be entangled, creating highly non-classical computation spaces [17]. This high-dimensional Hilbert space allows QNNs to represent complex patterns and correlations that classical models cannot, but it simultaneously complicates human interpretation.

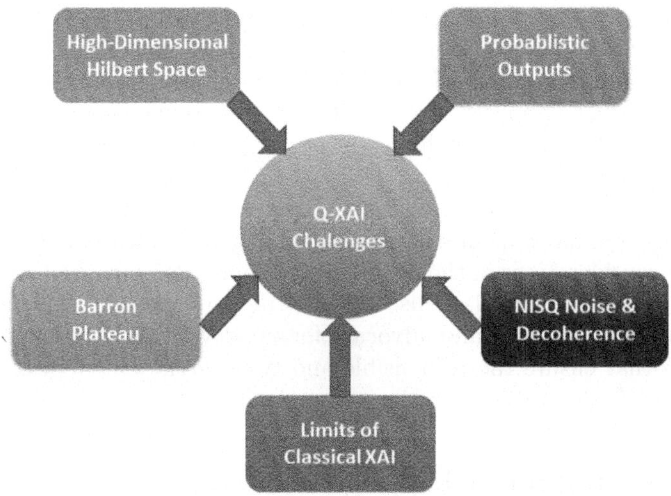

Fig. 1. Challenges for Q-XAI to address.

Measurement in quantum systems introduces probabilistic outputs; thus, QNNs do not produce deterministic predictions like classical ANNs, making standard post-hoc explanations less reliable [18]. Moreover, parameter optimization in QNNs often suffers from barren plateaus, where gradients vanish exponentially with system size, hindering both training and interpretability [19].

Additionally, QNNs are implemented on NISQ (Noisy Intermediate-Scale Quantum) devices, where noise and decoherence can significantly affect model behavior [20]. This adds uncertainty to explanations and complicates the causal attribution of outputs to specific parameters or quantum gates.

Finally, classical XAI methods do not trivially extend to QNNs due to quantum phenomena such as superposition, entanglement, and non-commutativity. Novel interpretability frameworks are needed that respect quantum mechanics while producing explanations intelligible to human users [17].

Overall, the interpretability of QNNs faces multiple interrelated challenges, including high-dimensional quantum state spaces, probabilistic outputs, barren plateaus in training, and the impact of noise and decoherence on NISQ devices. As illustrated in Fig. 1, these factors collectively create a complex landscape that classical XAI approaches cannot fully address. Developing Quantum-XAI frameworks that integrate quantum-aware interpretability strategies is therefore essential to bridge the gap between quantum model complexity and human-understandable explanations.

3 XAI for Human-Like Intelligence

3.1 What XAI Can Really Do

Recent advances in XAI demonstrate its capacity to move beyond simple explanations. Generative models, such as variational autoencoders, can create counterfactual examples to illustrate model decisions [11]. Large Language Models (LLMs) now offer textual explanations of reasoning chains, supporting factual XAI [21]. Furthermore, probabilistic and causal XAI approaches allow models to communicate uncertainty and causal relationships, aligning AI reasoning closer to human understanding [2,22–24].

3.2 XAI2Brain: Bridging Bio-Inspired Neural Networks and Human Cognition

XAI can serve as a bridge between artificial neural networks and human cognitive processes, translating complex computations into human-understandable causal explanations [25]. Biological neural systems differ significantly from classical ANNs: neurons in the brain operate with low-precision, stochastic synapses, and the network is wired in 3D with recurrent long-range connections, resembling Reservoir Computing [26].

By applying XAI to bio-inspired architectures, we can identify causal pathways and map representations in AI to neural activity in the brain [27]. This bidirectional approach helps us understand human cognition while improving AI interpretability. Causal reasoning modules, feature attribution, and attention-based explanations provide insights into high-level decision processes, fostering human-like intelligence in AI systems [28].

3.3 Quantum Minds: Drawing Parallels Between QNNs and Human Consciousness

QNNs leverage quantum superposition, entanglement, and interference to encode information in ways classical ANNs cannot, potentially mirroring the brain's parallel processing capabilities [29]. The human brain is a complex, three-dimensional network with recurrent loops and long-range connectivity [30]. Penrose and Hameroff hypothesize that consciousness arises from quantum processes in microtubules, involving non-computable effects that could underlie cognition [31].

QNNs, by simulating entangled and superposed states, offer a framework to explore such phenomena in artificial systems [32]. As illustrated in Fig. 2, integrating XAI with QNNs can make these quantum computations interpretable, allowing humans to understand and trust quantum AI reasoning while providing potential insights into neural computation [33].

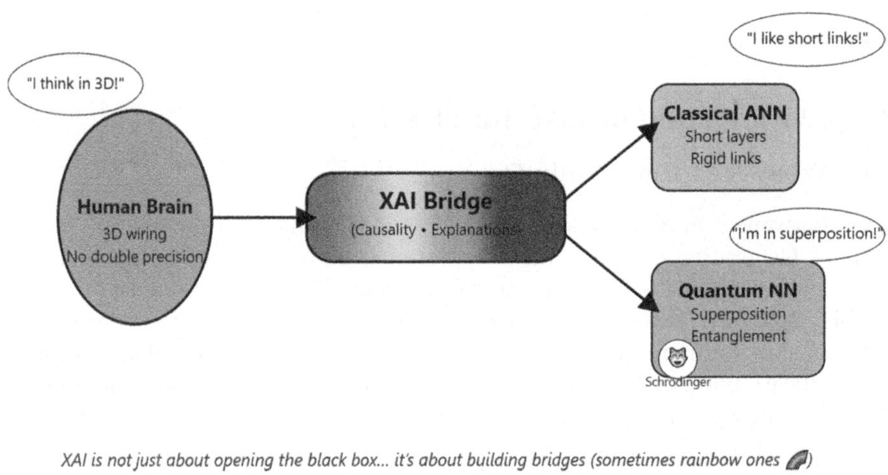

Fig. 2. XAI for Human-Like Intelligence: From Brains to Quantum Minds.

4 Q-XAI Framework: Bridging Quantum Computing and Explainable AI

The integration of quantum computing with Explainable Artificial Intelligence (XAI) has given rise to the Q-XAI framework, a novel paradigm that seeks to harness the unique capabilities of quantum mechanics to enhance the interpretability of AI models. This approach addresses the inherent opacity of complex AI systems by providing explanations that are accurate and comprehensible to human users [34].

4.1 Core Principles of Q-XAI

The Q-XAI framework is grounded in several core principles derived from quantum mechanics, each contributing to the interpretability and human-alignment of AI models:

- **Quantum Superposition and Interference:** Quantum systems can exist in multiple states simultaneously, a property known as superposition. This allows quantum models to process and represent a vast array of possibilities in parallel. Interference effects enable the combination of these possibilities to model complex decision-making processes, capturing nuanced patterns that classical models may overlook. By utilizing these properties, Q-XAI can provide insights into how different factors collectively influence model predictions [17].
- **Entanglement and Non-locality:** Quantum entanglement describes a phenomenon where particles become interconnected such that the state of one instantaneously influences the state of another, regardless of spatial separation. This non-locality allows Q-XAI to model intricate dependencies within data, capturing correlations that may be hidden or difficult to represent in classical architectures. In interpretability, this principle enables the identification of subtle feature interactions and emergent behaviors that contribute to model decisions [35].
- **Causal Inference:** Understanding cause-and-effect relationships is crucial for generating explanations that humans can comprehend. Quantum causal inference extends classical causal models by incorporating quantum phenomena, allowing analysis of complex causal structures with probabilistic and non-commutative effects. This approach facilitates the identification of direct and indirect influences within AI models, producing explanations that reflect the true underlying decision logic [33].

Together, these principles provide a foundation for Q-XAI to produce interpretable, human-aligned insights from quantum neural networks, as illustarted in Fig. 3. By leveraging the parallelism of superposition, the deep correlations of entanglement, and the rigor of causal inference, Q-XAI offers a framework capable of uncovering both global and local patterns in model behavior. This integration of quantum mechanics with explainable AI principles positions Q-XAI to bridge the gap between complex quantum computations and human-understandable reasoning, ultimately fostering trust and transparency in next-generation AI systems.

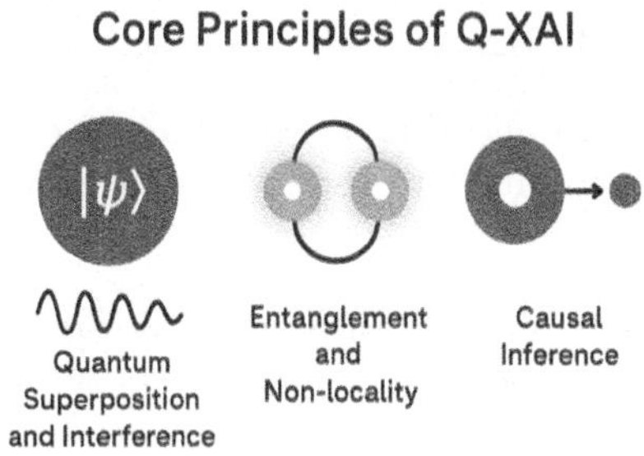

Fig. 3. The illustration of the core principles needed by the Quantum XAI framework.

4.2 Architectural Components of Q-XAI

The Q-XAI framework integrates several key architectural components that function cohesively to enhance the interpretability of quantum neural networks while maintaining computational power. At the core are the **Quantum Neural Networks (QNNs)**, which utilize qubits and quantum gates to process information in ways inaccessible to classical models. By leveraging superposition, QNNs can represent multiple potential states simultaneously, and entanglement enables capturing complex correlations across features. This intrinsic capability allows QNNs to encode rich, high-dimensional patterns and relationships, making them particularly effective for tasks requiring deep abstraction and nuanced pattern recognition [29]. However, this complexity also poses challenges for transparency, necessitating additional components for interpretability.

To address this, **Causal Inference Modules** are integrated into the architecture. These modules extend classical causal reasoning into the quantum domain, enabling the identification of cause-and-effect relationships within the model. By analyzing how perturbations in one qubit or parameter propagate through the network, these modules elucidate the internal logic of decision-making processes. This is particularly valuable in sensitive domains, such as healthcare and finance, where understanding the underlying rationale behind predictions is essential for trust and accountability [33].

Complementing these components are the **Interpretability Layers**, which translate quantum computations into human-understandable insights. These layers implement techniques such as feature attribution, decision boundary visualization, and attention mapping, bridging the gap between abstract quantum representations and intuitive explanations. By providing interpretable outputs, these layers empower users to comprehend how input variations influence model predictions, fostering trust and enabling informed decision-making [17].

Overeall, these components form an integrated Q-XAI architecture that balances the computational advantages of QNNs with the need for transparency and human-aligned reasoning. As illustrated in Fig. 4, the framework provides a roadmap for future research, highlighting how quantum computation, causal analysis, and interpretability layers can be synergistically combined. By establishing this guideline, Q-XAI serves not only as a technical framework but also as a community-oriented reference for developing interpretable and trustworthy quantum AI systems.

4.3 Applications and Implications

The Q-XAI framework holds transformative potential across a wide range of domains where interpretability and trust are paramount. In **healthcare**, AI models increasingly assist in medical diagnostics and treatment planning [36, 37]. However, the opacity of many AI systems has hindered their adoption in clinical settings. By providing causal, human-understandable explanations for model predictions, Q-XAI allows clinicians to comprehend the rationale behind AI-driven recommendations. This transparency not only fosters trust but also facilitates critical validation and collaborative decision-making, which is essential in high-stakes medical contexts [33].

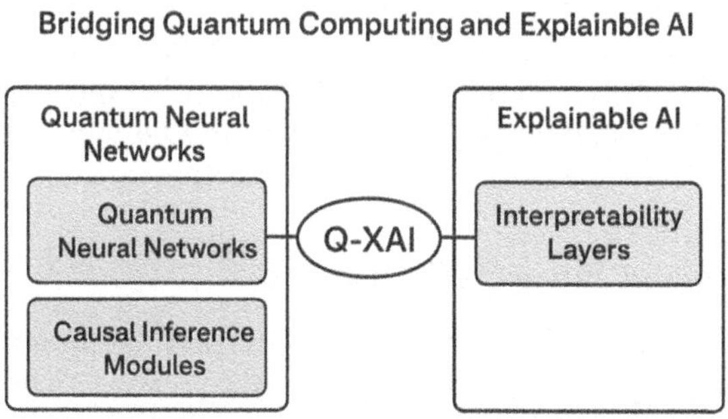

Fig. 4. The proposed Q-XAI framework.

In the **financial sector**, AI is extensively applied for credit scoring, risk assessment, and fraud detection. Regulatory requirements and the need for stakeholder confidence demand interpretable models. Q-XAI enhances transparency by elucidating how complex patterns in financial data influence predictions. This enables both regulators and decision-makers to understand model behavior, detect potential biases, and ensure compliance with fairness and accountability standards [35]. Beyond compliance, interpretable insights can inform strategic business decisions and risk mitigation strategies, bridging the gap between predictive power and human oversight.

Neuroscience represents another frontier for Q-XAI applications. AI models are employed to analyze neural activity and brain imaging data, yet understanding the underlying cognitive processes remains a formidable challenge [30,31]. Q-XAI can provide interpretable mappings between neural inputs and model outputs, helping researchers to uncover causal relationships and latent representations in neural signals. This approach not only aids in understanding complex brain functions but also opens pathways for developing AI models inspired by biological cognition, potentially advancing studies in neurological disorders and cognitive neuroscience.

Ethical and legal domains also benefit from Q-XAI. In environments where AI systems make high-stakes decisions, such as autonomous systems, criminal justice, or policy recommendations, the ability to provide clear, justified explanations is critical. Q-XAI ensures that the reasoning behind model outputs is transparent and accountable, facilitating ethical oversight and stakeholder confidence. By integrating interpretability with quantum-enhanced computation, Q-XAI provides a foundation for responsible AI that aligns performance with societal values [33].

Across these domains, Q-XAI exemplifies a convergence of advanced quantum computation and explainable AI, offering interpretable insights without compromising on predictive capability. Figure 4 presents an overview of the framework, highlighting how QNNs, causal inference modules, and interpretability layers collectively support applications ranging from healthcare to finance and neuroscience. Looking forward, the development of Q-XAI encourages a broader community effort, motivating researchers to explore not only practical deployments but also the ethical, regulatory, and cognitive implications of interpretable quantum AI. As quantum technologies mature, Q-XAI could serve as a guiding blueprint for building trustworthy, transparent, and human-aligned AI systems in increasingly complex and high-stakes environments.

4.4 Challenges and Future Directions

While Q-XAI offers substantial promise for interpretable quantum AI, it faces several critical challenges that must be addressed to realize its full potential. One major hurdle is the **scalability of quantum algorithms**. Current quantum neural networks are often limited by the number of qubits and gate fidelity, which constrains their ability to process large datasets or complex high-dimensional tasks [29]. Overcoming this limitation will require both advancements in quantum hardware and the development of efficient, noise-resilient quantum algorithms capable of scaling without compromising interpretability.

Another significant challenge is the **integration of quantum components with classical AI systems**. Most practical AI applications rely on classical computing infrastructure, and embedding quantum modules within these systems necessitates hybrid architectures that efficiently manage resources and maintain computational coherence [35]. Ensuring seamless interaction between classical and quantum layers is essential not only for performance but also for providing interpretable outputs that are meaningful to human users accustomed to classical AI paradigms.

Ethical considerations constitute a further layer of complexity. As Q-XAI models are deployed in domains with high societal impact, such as healthcare, finance, and autonomous systems, it is imperative that their explanations are both accurate and equitable [28]. Biases inherent in training data or model design can propagate through quantum computations, potentially leading to misleading explanations or unfair decision-making. Establishing standards for responsible Q-XAI deployment will be necessary to ensure transparency, fairness, and accountability.

Looking ahead, future research directions encompass the development of robust quantum algorithms capable of efficiently handling large and complex datasets [29], alongside enhanced interpretability techniques that can translate intricate quantum computations into explanations comprehensible to human stakeholders [17]. Additionally, the formulation of ethical frameworks and community standards will be crucial for responsible and trustworthy deployment of Q-XAI [28].

Overall, the challenges and future directions highlight the intertwined nature of technical, interpretability, and ethical considerations in Q-XAI. As illustrated in Fig. 4, addressing these challenges requires a synergistic approach, combining scalable quantum architectures, hybrid integration strategies, and principled ethical guidelines. The ongoing evolution of quantum technologies offers a unique opportunity for the research community to pioneer methods that are not only computationally powerful but also human-aligned, interpretable, and socially responsible. Future work should continue to explore these intersections, positioning Q-XAI as both a scientific framework and a guiding paradigm for trustworthy quantum AI research.

4.5 Overview

Q-XAI represents a significant step toward transparent and interpretable AI systems. By leveraging quantum mechanics, it provides new avenues for understanding AI decision-making. While challenges remain, the potential benefits in enhancing trust, accountability, and human-like reasoning are substantial. Continued research and development will be key to realizing the full potential of Q-XAI across sectors.

5 Conclusion

The convergence of quantum computing and artificial intelligence presents unprecedented opportunities to model complex systems and accelerate decision-making. Yet, the interpretability of Quantum Neural Networks (QNNs) remains a critical bottleneck, limiting their adoption in high-stakes applications where trust, accountability, and human understanding are essential. Traditional XAI techniques are insufficient for the probabilistic, entangled, and non-local nature of quantum models, highlighting the urgent need for specialized frameworks.

In this position paper, we introduced the Q-XAI framework, a roadmap for embedding explainability into quantum machine learning. By leveraging principles such as superposition, entanglement, and quantum causal inference, Q-XAI enables transparent and human-comprehensible insights into QNN behavior. We discussed architectural components, applications across healthcare, finance, and neuroscience, and the technical and ethical challenges that accompany this emerging field.

Looking forward, Q-XAI paves the way for human-centered quantum intelligence, bridging the gap between abstract quantum computations and meaningful explanations. It provides a foundation for developing scalable, ethical, and interpretable quantum AI systems, fostering both scientific discovery and societal trust. By establishing a structured roadmap, this work emphasizes the transformative potential of explainable quantum AI and the imperative to guide its responsible evolution.

References

1. Agency, D.A.R.P.: Explainable Artificial Intelligence (XAI) Program (2016). https://www.darpa.mil/program/explainable-artificial-intelligence
2. Angelov, P.P., Soares, E.A., Jiang, R., Arnold, N.I., Atkinson, P.M.: Explainable artificial intelligence: an analytical review. Wiley Interdisc. Rev. Data Min. Knowl. Discov. **11**(5), 1424 (2021)
3. Zhang, Z., Aggarwal, V., Angelov, P., Jiang, R.: Modeling brain aging with explainable triamese vit: towards deeper insights into autism disorder. IEEE J. Biomed. Health Inf. (2025)
4. Longo, L., et al.: Explainable artificial intelligence (xai) 2.0: A manifesto of open challenges and interdisciplinary research directions. Inf. Fusion **106**, 102301 (2024)
5. Saeed, W., Omlin, C.: Explainable ai (xai): a systematic meta-survey of current challenges and future opportunities. arXiv preprint arXiv:2111.06420 (2023)
6. Gil-Fuster, E., et al.: Opportunities and limitations of explaining quantum machine learning. arXiv preprint arXiv:2412.14753 (2024)
7. Lundberg, S.M., Lee, S.-I.: A unified approach to interpreting model predictions. Adv. Neural Inf. Process. Syst. (2017)
8. Ribeiro, M.T., Singh, S., Guestrin, C.: Why should I trust you?: explaining the predictions of any classifier. ACM SIGKDD Int. Conf. Knowl. Discov. Data Min. (2016)
9. Selvaraju, R.R., et al.: Grad-cam: visual explanations from deep networks via gradient-based localization. ICCV (2017)
10. Sethi, N., et al.: Interpretable neural networks via rule-based methods. J. AI Res. (2021)
11. Hase, P., Bansal, M.: Evaluating explainable AI: which algorithmic explanations help users predict model behavior? ACL (2020)
12. Hohman, F., al.: Visual analytics in deep learning: An interrogative survey. IEEE Trans. Visualization Comput. Graph. (2020)
13. Molnar, C.: Interpretable machine learning (2020)
14. Zhu, Y., et al.: Quantum face recognition with multigate quantum convolutional neural network. IEEE Trans. Artif. Intell. **5**(12), 6330–6341 (2024)
15. Zhu, Y., Jiang, R., Ni, Q., Bouridane, A.: Enable quantum graph neural networks on a single qubit with quantum walk. IEEE Trans. Artif. Intell. (2025)
16. Easom-McCaldin, P., Bouridane, A., Belatreche, A., Jiang, R., Al-Maadeed, S.: Efficient quantum image classification using single qubit encoding. IEEE Trans. Neural Netw. Learn. Syst. **35**(2), 1472–1486 (2022)
17. Beer, K.e.a.: Training deep quantum neural networks. Nat. Commun. **11**(1), 14454 (2020). https://doi.org/10.1038/s41467-020-14454-2
18. Schuld, M., Sinayskiy, I., Petruccione, F.: The quest for a quantum neural network. Quantum Inf. Process. (2014)

19. McClean, J.R.: Barren plateaus in quantum neural network training landscapes. Nat. Commun. **9**, 4812 (2018)
20. Bharti, K.e.a.: Noisy intermediate-scale quantum algorithms. Rev. Modern Phys. **94**(1), 015004 (2022)
21. Bubeck, S., et al.: Sparks of artificial general intelligence: early experiments with gpt-4. arXiv preprint arXiv:2303.12712 (2023)
22. Pearl, J., Mackenzie, D.: The Book of Why: the New Science of Cause and Effect. Penguin Books (2019)
23. Bartle, A.-S., Jiang, Z., Jiang, R., Bouridane, A., Almaadeed, S.: A critical appraisal on deep neural networks: bridge the gap between deep learning and neuroscience via xai. HANDBOOK ON COMPUTER LEARNING AND INTELLIGENCE: Volume 2: Deep Learning, Intelligent Control and Evolutionary Computation, pp. 619–634 (2022)
24. Zhou, Y., Jiang, R.: Advancing explainable ai toward human-like intelligence: Forging the path to artificial brain. arXiv preprint arXiv:2402.06673 (2024)
25. Doerig, A.e.a.: High-level visual representations in the human brain are predicted by large language model embeddings. Nat. Mach. Intell. **7**(5), 313–322 (2025) https://doi.org/10.1038/s42256-025-01072-0
26. Goodwin, N.L.e.a.: Simple behavioral analysis (simba) as a platform for explainable machine learning in behavioral neuroscience. Nat. Neurosci. **27**(7), 1411–1424 (2024). https://doi.org/10.1038/s41593-024-01649-9
27. Wang, E.Y.e.a.: Foundation model of neural activity predicts response to naturalistic stimuli. Nature **617**(7955), 1–6 (2025). https://doi.org/10.1038/s41586-025-08829-y
28. Hutson, M.: How does chatgpt 'think'? psychology and neuroscience crack open ai large language models. Nature **614**(7949), 22–25 (2024). https://doi.org/10.1038/d41586-024-01314-y
29. Zhang, X.e.a.: Deep quantum neural networks on a superconducting processor. Nat. Commun. **14**, 3412 (2023). https://doi.org/10.1038/s41467-023-39246-x
30. Koch, C., Hepp, K.: Quantum mechanics in the brain. Nat. **440**, 611 (2006). https://doi.org/10.1038/440611a
31. Penrose, R., Hameroff, S.: Consciousness, microtubules, and quantum computation. J. Conscious. Stud. **3**(1), 36–53 (1996)
32. Ho, J.K.W.e.a.: Quantum affective processes for multidimensional decision making. Sci. Rep. **12**(1), 22855 (2022) https://doi.org/10.1038/s41598-022-22855-0
33. Brody, D.C.: Quantum formalism for the dynamics of cognitive psychology. Sci. Rep. **13**(1), 43403 (2023). https://doi.org/10.1038/s41598-023-43403-4
34. Zhang, Z., Jiang, R.: User-centric democratization towards social value aligned medical ai services. In: IJCAI, pp. 6326–6334 (2023)
35. Long, C.e.a.: Hybrid quantum-classical-quantum convolutional neural networks. Sci. Rep. **15**(1), 13417 (2025) https://doi.org/10.1038/s41598-025-13417-1
36. Konar, D., Sreekumar, N., Jiang, R., Aggarwal, V.: Alz-qnet: A quantum regression network for studying alzheimer's gene interactions. Comput. Biol. Med. **196**, 110837 (2025)
37. Jiang, R., Chazot, P., Pavese, N., Crookes, D., Bouridane, A., Celebi, M.E.: Private facial prediagnosis as an edge service for Parkinson's dbs treatment valuation. IEEE J. Biomed. Health Inform. **26**(6), 2703–2713 (2022)

QToMNet: A Quantum-Enhanced Machine Theory of Mind

Sabine Brunswicker(✉) and Mudit Gaur

Purdue University, West-Lafayette, USA
{sbrunswi,mgaur}@purdue.edu

Abstract. Theory of Mind (ToM) broadly refers to a human's cognitive capacity to attribute *mental states*, such as beliefs, to oneself and others. The emerging stream of machine ToM seeks to equip artificial agents with this ability using machine learning, including neural networks. In this paper, we build upon evidence on the quantum dynamics of human decision making and hypothesize that quantum embeddings are a promising direction for enhancing machine ToM neural networks, such as ToMNet and its extensions. Such meta-learning models learn a small number of behavioral observations to make rich predictions about other agents' characteristics and mental states. Quantum embeddings map outputs of ToMNet's character and mental networks into parameterized superimposed and entangled quantum states, and provide a mathematical structure that allows the network to better approximate the uncertainty inherent to partially observable mental states. We outline a quantum-enhanced version of ToMNet, called QToMNet, which uses quantum embeddings to encode past episodes and current trajectories into quantum circuits, with a classical prediction head producing standard ToMNet outputs. Further, we present an experimental plan to evaluate QToMNet's predictive power using grid-world settings with random and deep reinforcement learning (RL) agents. With this position paper, we aim to catalyze rigorous experiments that examine whether quantum embeddings can offer practical, drop-in gains in performance of ToM algorithms, and hope to nurture a new stream of literature on quantum-enhanced ToM networks.

1 Introduction

Theory of Mind (ToM) broadly refers to a human's cognitive capacity to attribute *mental states*, such as beliefs, desires or intentions (sometimes abbreviated with BDIs), to oneself and others, and recognize that others' mental states may differ from ones own [1–4]. We as humans have the ability to approximate and make predictions about such latent informational states that are difficult to access and observe, even under challenging conditions. For example, even when we interact with strangers we are able to make good guesses about what they think and how they act based on how they behave, even though we cannot draw upon intense prior experiences and interactions with them [5]. Cognitive

V. Aggarwal et al. (Eds.): PQAI 2025, CCIS 2724, pp. 210–228, 2026.
https://doi.org/10.1007/978-981-95-7829-0_17

psychologists argue that we humans develop a theory of the mind during early childhood allowing us to engage in social reasoning using a mental model of others' mind. This model relies on abstract representations of BDI states rather than the underlying complex reasoning processes and the physical mechanisms in the brain [3]. We use this abstract model in generative way to simulate others' behaviors in our mind, and to predict current and future states of ourself, as well as those of others [6].

Scholars in the field of artificial intelligence (AI) and machine learning (ML), have recently increased their interest in theorizing and implementing computational models that imbue artificial agents with such human-like abilities of mindfulness so that they can infer BDIs related to humans as well as other artificial (potentially human-like) agents. Some of such efforts seek to account for the limited rationality of human agents to simulate mental states (BDIs) and predict human behaviors that result from such states [7,8]. Currently used approaches range from hand-crafted models of agents as planners, such as Bayesian Theory of Mind models that formalize behaviors as noisy-rational beliefs and plans that are guided by the agent's desire to maximize gains and minimize loss [8–11], model-based inverse reinforcement learning [5], or epistemic approaches that infer BDIs based on epistemic logic [12]. In contrast to such models that seek to invert a model-based approach and assert a generative model of the agents' behavior, some scholars turn to deep learning approaches to make weak assumptions about the generative capacity of the agent. Instead, the goal is to learn latent higher-level mental states directly from data using neural networks. The ToMNet, proposed by Rabinowitz et al. in 2018 [13], produced impressive results, and triggered a new stream of research on deep learning (DL) focused ToM research, framing ToM learning as a meta-learning problem. Despite the significance of this stream of research on a new field on machine Theory of Mind, this DL stream of work is still in its early stages of development, compared to the work on human ToM.

One of the key challenges in this stream of research is that existing neural network architectures like ToMNet struggle with approximating the partially observable mental states that condition human behaviors. Specifically, they fail to properly account for the uncertainty inherent to the dynamics of mental states and associated decisions. Researchers in the field of mathematical psychology, neuroscience and cognitive sciences have provided empirical evidence that our actions and mental states are not always in line with Bayes' theorem in the sense that we carefully reflect on prior observations and experiences (priors) to form future beliefs (posterior). Research suggests that quantum mechanics offers a formal way to model the unique uncertainties and dynamics of human beliefs[1] and decision making [14]. In other words, just like a quantum system can exist in multiple states simultaneously, with specific probability amplitudes, humans can be in two or multiple belief states concurrently, with varying probabilities

[1] We will use belief and mental states interchangeable throughout the paper since beliefs are informational states central for ToM.

of emphasis. Such forms of *superposition* of beliefs may exist until we make a decision or take an action (and our beliefs become observable and measurable).

In this position paper, we seek to make a first step to enhance the emerging stream of literature on deep learning research on ToM. Specifically, we extend the influential work from Rabinowitz et al. [13], and propose a quantum-enhanced ToMNet, in which we leverage quantum embeddings to characterize an agent's behavior and mental states through representations in a Hilbert space. Our proposed architecture is complemented with an experimentation plan and a set of hypotheses that guide QToMNet's future evaluation in established ToM task settings (e.g. 2-grid search). Our vision is to nurture a new stream of research on quantum-enhanced machine Theory of Mind that rigorously evaluates the performance benefits of quantum embeddings in ToMNet and its extension, as well as complementary deep learning models for machine ToM.

Our paper is structured as follows: In Sect. 2, we provide a brief review of the literature on deep learning focused ToM. In Sect. 3 we depart from the recent work in psychology, economics, and neuroscience, and introduce a quantum-inspired view of human decision making, and formally represent human beliefs and their dynamics as quantum states in a Hilbert Space. Then we introduce our QToMNet framework in Sect. 4. Afterwards (Sect. 5) we discuss a detailed experimental plan evaluating this network in rigorous grid-world ToM experiments to inspire future work, and eventually draw final conclusions (see Sect. 6).

2 Deep Learning (DL) for a Machine Theory of Mind: ToMNet and Its Extensions

Deep learning (DL) has recently emerged as a promising avenue to develop machine ToM models. DL-based approaches complement hand-crafted models of ToM agents acting as planners, such as Bayesian Theory of Mind models that formalize behaviors as noisy-rational beliefs and plans that are guided by the agent's desire to maximize gains and minimize loss [8,9] or model-based inverse reinforcement learning [5] that uses Bayesian inference to produce human-like judgment [12][2]. DL architectures are argued to offer new opportunities to infer abstract and only partially observable mental states using neural networks. The influential ToMNet, proposed by Rabinowitz et al. [13], set the foundation for a range of models and frameworks that leverage neural networks for learning higher-level representations from lower-level and complex feature spaces constituting the agent's environment, including 2-D, 3-D grid worlds as well as real-world scene information (images and videos). ToMNet is based on the idea of meta-learning [15]. Essentially, at test time, ToMNet should be able to have a strong and rich prior about an agent's future behavior, even though there were no or little prior encounters with that particular agent. ToMNet formulates a task for an observer, who, in each iteration/episode, gets access to a set of behavioral traces of such a novel agent and predicts future behaviors. Essentially, the observer learns priors based on a known family of agents to predict

[2] For a recent literature review see Mao et al. [7].

the behavior of agents whose rewards, parameterizations, and policies may vary considerably. The architecture seeks to implicitly capture commonalities across different agents, given only limited data about the agent's behavior. ToMNet has been extended in various ways. Table 1 provides an excerpt of such extensions, including those that explicitly represent social relationships in graphs and multi-agent settings, or account for fact that one's belief inference is conditioned by other agents' beliefs (see ToMCAT). The original ToMNet consists of 3 modules that were further modified by its extension over the last years: A *character net*, that processes trajectories from past episodes of an observer agent, i, and computes a character embedding, that is input the *mental net*, that focuses on the current episode's trajectory so far to provide the necessary context to infer an agent's current *mental state*. It computes a mental embedding. Finally, the *prediction net* uses the outputs of the *character* and the *mental nets* to update the observer's beliefs about the agent's behavior. While ToMNet offers significant opportunities to account for the complex dynamics of mental states to infer future behaviors, it does not capture the uncertainty and dynamics inherent to beliefs and other BDIs. As we will discuss in the next section, we propose that this uncertainty can be formally represented using quantum states evolving in a Hilbert space, inspired by quantum mechanics.

Table 1. ToMNet and selected ToMNet Extensions

Year	Name	Data	Description	Benefits/Extension
2018	ToMNet [13]	2-D grid	Learns a novel agent's current mental states and future actions in a new scene based on past trajectories	Learning the mental state of different agent types (species) based on commonalities
2021	ToMNet-G [16]	3-D scene	To learn and predict trajectories in 3-D environments	Extending ToMNet with GNN
2020	ToMNet+ [17]	2-D grid	Inferring human social interaction	Added preference inference to infer implicit human preferences
2022	Trait Based ToMNet [18]	2-D 2-agent grid	Inferring other actors' mental states and goals by observing their past and current behavior	Including stable personality traits (achieved through quick weights)
2025	ToMCAT [19]	2-D grid	Combines a ToM-predictive network with a multi-agent denoising-diffusion policy model	Extends ToMNet to make predictions about multiple agents, conditioned based on the observers' characteristics, and supports online replanning

3 Quantum Representation of Mental States

Prior research in mathematical psychology and related fields has provided evidence for the quantum-like nature of cognition [14,20–22]. Just like quantum systems, we as humans can occupy two or more mental states concurrently, with varying degrees of emphasis, and our mental states dynamically evolve as we act and learn.

To illustrate the quantum-like nature of the mind, we rely on *bra–ket notation* (Dirac notation), where a quantum state vector is written as a *ket* $|\psi\rangle$ and its conjugate transpose is the *bra* $\langle\psi|$. The product of two states, $\langle\phi|\psi\rangle$, is an inner product that measures overlap (similarity) between the two states. Specifically, we will use the *qubit*, a quantum analogue of a classical bit, to represent quantum belief states $|\psi\rangle$. While a classical bit is either 0 or 1, the computational basis states of a qubit are $|0\rangle$ and $|1\rangle$, and a general qubit can be in a superposition $c_1|0\rangle + c_2|1\rangle$.

Following this notation, a belief state can be mathematically formalized as a quantum state $|\psi\rangle$ in a complex Hilbert space $\mathcal{H}$. We assume that this space is finite-dimensional. For a single-qubit Hilbert space, $\mathcal{H} \cong \mathbb{C}^2$.

Human belief states can be represented using an orthonormal basis $\Omega = \{\, |\psi_j\rangle : j = 1, 2, \ldots, m \,\}$. These $|\psi_j\rangle$ are abstract elements of an m-dimensional Hilbert space

$$\Omega = \{\, |\psi_j\rangle : j = 1, 2, \ldots, m \,\}. \tag{1}$$

If we assume that human decision making follows principles of quantum theory, any general belief state can be represented as the *superposition* of these orthonormal basis states at time t. Mathematically,

$$|\phi(t)\rangle = \sum_{j=1}^{m} c_j(t)\, |\psi_j\rangle, \tag{2}$$

where $c_j(t) \in \mathbb{C}$ are the amplitudes of the state in this basis. The amplitudes can be written as $c_j(t) = |c_j(t)|\, e^{i\theta_j}$, where $|c_j(t)|$ is the magnitude and θ_j is the phase, and they satisfy the normalization $\sum_{j=1}^{m} |c_j(t)|^2 = 1$. Multiple ($n$) qubits form a joint system whose state lives in the tensor-product Hilbert space $\mathcal{H} \cong (\mathbb{C}^2)^{\otimes n}$.

The principle of *superposition* is essential for capturing the latent and hidden nature of belief states: Figuratively speaking, until an agent's belief becomes observable (is measured) or until an action is taken, multiple belief states may superimpose. Following quantum mechanics, we represent such *observables* using a Hermitian operator satisfying $M = M^\dagger$. The expectation value on state $|\psi\rangle$ is $\langle M\rangle = \langle\psi|M|\psi\rangle$. In the particular case of a qubit representation, we can use the Pauli-Z operator

$$Z = \begin{pmatrix} 1 & 0 \\ 0 & -1 \end{pmatrix}, \tag{3}$$

whose measurement outcomes are ± 1 and whose expectation value lies in $[-1, 1]$.

A quantum representation captures *uncertainties* underlying human decision making in a unique way, since amplitudes c_j are distinct from *probabilities* P_j, typically used to express uncertainty in classical theories of human decision making and belief formation [20,23]. The *probability* of outcome j is the squared magnitude

$$P_j(t) = |c_j(t)|^2. \tag{4}$$

We use the widely studied Sally–Anne task [24] to illustrate this further. The test examines an agent's ability to reason about another person's beliefs, including false beliefs. In this task, Sally puts her toy in the basket and leaves. While Sally is outside, Anne takes the toy out of Sally's basket and puts it in Anne's box. Then Sally returns. A machine ToM agent would have to reason about Sally's belief about the toy's location (*"Where does Sally find her toy?"*). Essentially, there are two basic belief states, $|\psi_1\rangle$ (*"The toy is in the basket."*) and $|\psi_2\rangle$ (*"The toy is in Anne's box."*). If we assume the amplitudes $c_1 = \sqrt{3/4}$ and $c_2 = \sqrt{1/4}$, then Sally's belief state can be represented as a superposition:

$$|\phi\rangle = \sqrt{\tfrac{3}{4}}\,|\psi_1\rangle + \sqrt{\tfrac{1}{4}}\,|\psi_2\rangle. \tag{5}$$

The amplitudes and probabilities are:

$$c_1 = \sqrt{\tfrac{3}{4}}, \quad c_2 = \sqrt{\tfrac{1}{4}}, \qquad P_1 = \tfrac{3}{4},\ P_2 = \tfrac{1}{4}. \tag{6}$$

In this context, amplitudes c_j can be interpreted as hidden, complex *confidence weights*, while probabilities are what can be observed (e.g. by asking a direct question). Then the belief collapses, is measurable with the expectation value $\langle\ \mathrm{M}\rangle = \langle\psi|M|\psi\rangle$. For example, if Sally reveals her belief as *"the toy is in the basket"*, the state collapses onto $|\psi_1\rangle$; if we encode *"in the basket"* as $|0\rangle$, a Pauli-Z measurement yields $+1$.

Importantly, Sally's belief state may evolve even before it becomes observable/is measurable, and thus, the amplitudes change dynamically. For example, Sally may use environmental cues and contextual factors (e.g. *the basket's location*) to adjust her confidence weights. Mathematically, such changes in amplitudes can be modeled as rotations by an angle θ using rotation operators (rotation gates) $R_a(\theta) = e^{-i\theta G}$, where G is a Hermitian generator. In a 2-dimensional qubit setting, $R_a(\theta) = e^{-i\theta\,\sigma_a/2}$ are single-qubit gates that rotate the state around the axis $a \in \{x, y, z\}$ using Pauli matrices:

$$\sigma_x = \begin{pmatrix} 0 & 1 \\ 1 & 0 \end{pmatrix}, \quad \sigma_y = \begin{pmatrix} 0 & -i \\ i & 0 \end{pmatrix}, \quad \sigma_z = \begin{pmatrix} 1 & 0 \\ 0 & -1 \end{pmatrix}. \tag{7}$$

For a rotation around the x-axis by angle θ we have

$$R_x(\theta) = e^{-i\theta\sigma_x/2} = \begin{pmatrix} \cos(\frac{\theta}{2}) & -i\sin(\frac{\theta}{2}) \\ -i\sin(\frac{\theta}{2}) & \cos(\frac{\theta}{2}) \end{pmatrix}, \tag{8}$$

so that the coefficient vector transforms as

$$\begin{pmatrix} c_1' \\ c_2' \end{pmatrix} = R_x(\theta) \begin{pmatrix} c_1 \\ c_2 \end{pmatrix}. \tag{9}$$

For $\theta = 180°$ this gives $c_1' = -i\,c_2$ and $c_2' = -i\,c_1$. On the Bloch sphere, this corresponds (up to a global phase factor) to flipping the state across the x-axis, e.g., an original polar angle of 60° becoming 120° (with the x-axis held fixed).

4 QToMNet: A Quantum-Enhanced ToMNet

4.1 Overview

To capture the benefits of a quantum representation of belief states, we next introduce a new Quantum-enhanced Theory of Mind (QToMNet) framework that extends the originally ToMNet architecture [13] by integrating a quantum perspective towards human belief formation into the architecture. In accordance with the original model, we consider a single-agent Partially Observable Markov Decision Process (POMDP) for evaluating Theory of Mind (ToM) capabilities for belief inference. We assume a distinct type of agents $A = U_i\, A_i$ with different policies, including random (guided by pre-specified probabilities), and learned ones (e.g. deep reinforcement Q-learning). Similar to the original paper, we assume a gridworld POMDP, described in more detail in Sect. 4.2 with a comparable action space (*up, down, left, right, stay*), in which the agents take actions to find objects and seek to meet multiple subgoals, while only seeing part of the world at any time. In analogy to the ToMNet meta-learning framework, our proposed Quantum-enhanced Theory of Mind (QToMNet) introduces an observer who infers beliefs, just like Sally in the Sally-Anne setting: The observer makes observations of the agent acting in the POMPD environment, including its past behavioral patterns over multiple episodes as well its recent actions, that are encoded in states that *characterize* the agent at a high level while also approximating the agent's recent *mental* state (For example, if agent a_i follows its policy π_i resulting in trajectory τ_{ij}, the observer would see certain information about τ_{ij}, such as past locations and following actions). The current context as it is perceived by the agent (its assumed state in the environment) further enhances the observer's ability to form a belief about the agent's next action. Just like Sally in the Sally-Anne experiment, we assume that the observer has unrestricted access to the environment (essentially turning it into an MPD) but does *not* have access to the agent's reward functions, policies or other identifiers (e.g. its perception of its state/context). To produce the agent's past behavior and current context, we train our agents with different policies and then produce *rollouts*, which are trajectories of observations, actions, and states used as input to fit the model. We use a supervised learning method with the input and output pairs from the rollout data (see Sect. 4.4 for more details).

Figure 1 describes how our QToMNet framework aligns with and extends the original architecture [13] using quantum embeddings.

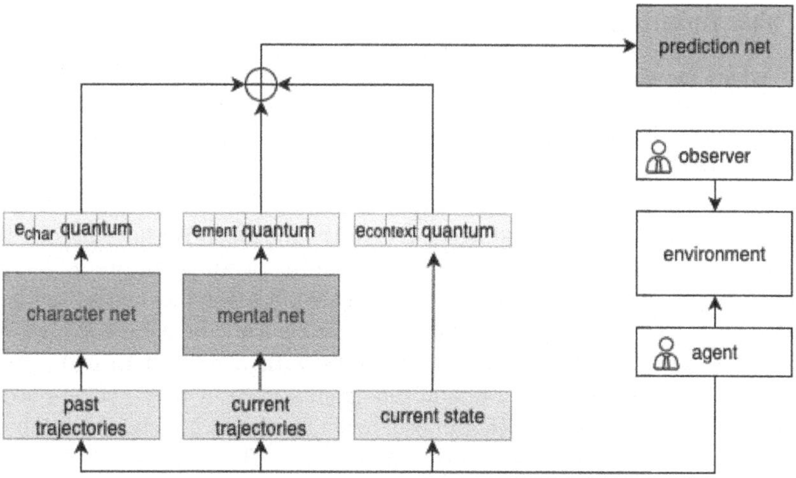

Fig. 1. QToMNet Overview.

Similar to ToMNet, QToMnet is composed of three modules: a *character net*, a *mental state net*, and a *prediction net*. The character net *characterises* the presented agent, by parsing observed past episode trajectories $\{\tau_{ij}\}_{j=1}^{past}$ into a character embedding $e_{char,i}$. The *mental state net* mentalises about the presented agent during the current episode to infer its current mental state by parsing the current episode trajectory τ_{ik} into a mental state embedding, $e_{mental,i}$. Additionally, we also have the current state of the environment, as it is perceived by the agent who has limited observability, as an input to the ToM; we denote this as the *current context*. The *prediction net* uses the output from the character net and the mental net, as well as the agent's current context, to predict the subsequent action of the agent.

A central contribution of our position paper is the representation of the output of the *character net*, *mental net* as well as the *current context* as quantum states before it is passed on to the observer's *prediction net*. As shown in Fig. 1, we plan to map $e_{car,i}$, $e_{mental,i}$, as well as $e_{context,i}$ into quantum embeddings. Specifically, we propose two ways to represent information that forms the agent's beliefs: (i) a *quantum-inspired embedding scheme* that applies circuit-style feature transformations before a small post-processing network, and (ii) a *hybrid embedding scheme* that fuses features from both the classical and quantum-inspired pathways. These alternatives provide different inductive biases for representing uncertainty and dynamics. In the implementation of the ToMNet framework experiments, we replace the original classical embedding scheme with these quantum (or hybrid) versions while keeping the rest of the setup fixed, to assess how quantum embeddings affect inference quality and action selection. Further, we also assess the impact of a quantum representation on the computational complexity, and study how performance scales when parameter-matched models compare.

4.2 The Environment

Our PMDP, in which the agent generates their trajectories, is an 9×9 grid with solid boundary walls and random interior walls; only walls block movement (see Fig. 2). One subgoal and four objects (types $0 - 3$) sit on traversable cells. Each step, the agent chooses up, down, left, or right to move but only if the target cell is in-bounds (meaning there is no wall or obstacle); otherwise, it stays. As mentioned above, the agent has only an egocentric field of view (fov) window of the environment showing nearby walls, the subgoal, and objects (no line–of–sight occlusion) while the external ToM observer has access to the full environment. After the agent touches the subgoal, the environment may (once per episode) swap the four objects' positions with a certain probability, e.g. $p_{swap} = 0.25$). If this happens outside the agent's view, it may act on outdated information when forming beliefs (a false–belief case). Each step returns the new observation, updated position, and termination flag. Rewards are not part of the environment. The transition dynamics are deterministic for agent movement but probabilistic for object swaps, creating a rich space of possible belief states that the agent must navigate.

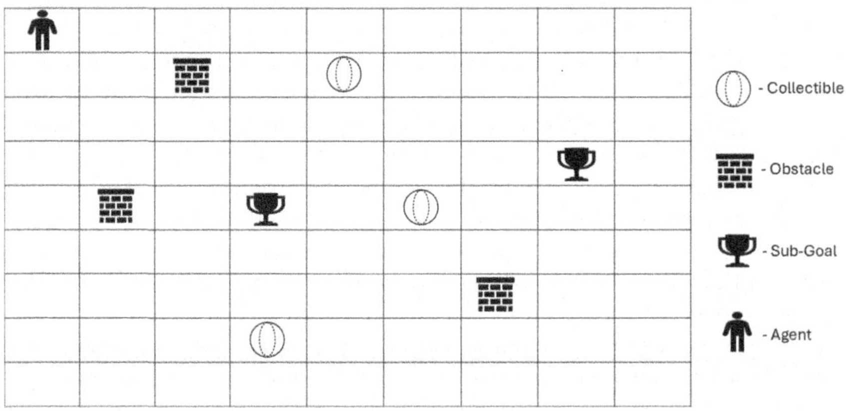

Fig. 2. The 9×9 grid world environment, in which *agents* move using 5 possible moves *left,right,up,down,stay*. If the agent reaches a *subgoal (e.g. 2)* it is positively rewarded, and the trajectory is terminated. Consuming a *collectible (e.g. 3)* object also implies a positive reward but without no termination. *Obstacle (e.g. 3)* result in a negative reward.

Our family of agents is modeled in alignment with ToMNet, and includes *random agents* who take actions with a Dirichlet distribution over the action space, as well as *deep learning agents* that use epsilon-greedy action selection and temporal difference (TD) updates with learning rate, discount, and exploration parameters, to learn a policy over a discretized, partially observable state space.

4.3 Theory of Mind Observer

As discussed in Fig. 1 the ToM observer predicts agent actions based on: $\pi(a_t|\mathbf{c}, \mathbf{m}, \mathbf{c_o}) = \text{softmax}(f_{head}(\mathbf{s} \oplus \mathbf{m} \oplus \mathbf{c_o}))$ where $\mathbf{c}$ is output (embedding $e_{car,i}$) of character net, $\mathbf{m}$ is the output (embedding $e_{mental,i}$) of the mental net, and $\mathbf{c_o}$ is the (perceived) current context of the environment. Some variants of our architecture will first pass the output through a quantum, or hybrid belief/state encoder to better capture the uncertainties inherent to belief inference.

The ToM observer serves as a meta-learning framework for belief state representations. Essentially, it seeks to learn to imitate the agent's choice. The policy head concatenates all three representations as input into a multi-layer perceptron to produce action probabilities over the 5-dimensional action space. Steps after an unseen swap are flagged to report accuracy under misinformation. At inference, it outputs probabilities over the five actions.

4.4 Training the ToM Observer

The observer is trained in a supervised fashion to predict the next action of an agent in a grid-world from three inputs: a character embedding c, a mental-window embedding m, and a current context vector (c_o). Each training example consists of

$$(c_i, m_i, (c_o)_i, y_i, s_i), \qquad y_i \in \{0, 1, 2, 3, 4\},\ s_i \in \{0, 1\} \tag{10}$$

where y_i is the discrete action label (*up/down/left/right/stay*) and s_i is the dataset-provided indicator of a "false-belief" scenario (recent swap outside field-of-view).

Intuitively, the model learns a mapping from multi-source context $(c, m, (c_o))$ to the agent's imminent action. The character vector summarizes longer-horizon tendencies; the mental window contains short-horizon contextual cues; the current context vector encodes the immediate world configuration as it is perceived by the agent. Supervision uses action choices generated by agents (random or learned) in simulated rollouts, allowing us to compare architectures under identical data.

Data and Batching Pipeline: Training data are produced by simulated rollouts and wrapped in a dataset class that yields tensors $(c, m, (c_o), y, s)$ per step. We shuffle training samples and form mini-batches of size B for stochastic optimization. This achieves two goals: (i) efficient vectorized computation on accelerators and (ii) noisy gradient estimates that often generalize better than full-batch gradients. At a high level, one epoch consists of iterating over all mini-batches once (order is reshuffled each epoch), computing the loss and its gradient on each batch, and stepping the optimizer. Validation follows the same batching pattern but disables gradient computation for speed and correctness.

Model and Outputs: Let f_θ denote the observer (classical, quantum, or hybrid), with parameters $\boldsymbol{\theta}$ collecting all classical weights and any quantum variational parameters. The model outputs a vector of logits $\boldsymbol{z}_i \in \mathbb{R}^5$ and induced class probabilities via the softmax

$$z_i = f_\theta(c_i, m_i, (c_o)_i), \qquad p_i(k) = \text{softmax}(z_i)_k = \frac{e^{z_{i,k}}}{\sum_{j=1}^{5} e^{z_{i,j}}}. \tag{11}$$

Predicted labels are $\hat{y}_i = \arg\max_k p_i(k)$ during evaluation. The 5 output logits $z_i \in \mathbb{R}^5$ correspond to the discrete action set. Using logits rather than probabilities is numerically stable and integrates cleanly with the cross-entropy loss. Internally, the observer fuses (c, m) with either a classical, quantum, or hybrid representation derived from (c_o); however, the training objective and optimizer are identical across these variants.

Training Objective: The primary loss is the cross-entropy between the predicted distribution and the true action label. For a batch $\mathcal{B}$ of size B,

$$\mathcal{L}(\theta; \mathcal{B}) = -\frac{1}{B} \sum_{(c,m,(c_o),s,y)\in\mathcal{B}} \log p_\theta(y \mid c, m, (c_o), s), \tag{12}$$

$$= \frac{1}{B} \sum_{(c,m,(c_o),s,y)\in\mathcal{B}} \ell(f_\theta(c, m, (c_o), s), y). \tag{13}$$

where $\ell(z, y)$ is the standard softmax cross-entropy. Dropout in some classical layers provides implicit regularization; no explicit weight decay is required in default runs.

Cross-entropy encourages the logit of the correct action to be larger than the others, shaping the decision boundaries in feature space. Because the labels are one of five classes, this objective naturally handles multi-class settings. If desired, we may incorporate class weights (for imbalanced data) or an auxiliary regularizer.

4.5 Quantum Embedding Framework

4.5.1 Terminology and Notation The central idea of this paper is the use of a quantum embedding applied to both the *character net*, the *mental net*, and the *current context* in the Enhanced ToM observer. Classical inputs are mapped to quantum states via angle encoding, processed by shallow variational entangling layers, and read out via bounded Pauli-Z expectations that are then post-processed by small MLPs. Next, we provide a concise overview of the terminology and notation needed to understand the structure of the embeddings. Some of those concepts have already been defined in Sect. 3. However, we are adding an overview for completion. The subsequent section details the technical structure of the quantum embedding structure using this terminology.

- **Qubit**: The quantum analogue of a classical bit. While a bit is either 0 or 1, a qubit can be in a *superposition* $\alpha|0\rangle + \beta|1\rangle$ where $\alpha, \beta \in \mathbb{C}$ and $|\alpha|^2 + |\beta|^2 = 1$. Multiple qubits form a joint system whose state lives in the tensor-product space $\mathcal{H} \cong (\mathbb{C}^2)^{\otimes n}$. This exponential state space is what gives quantum models their representational power.

- **Unitary** U: A linear transformation on states that preserves lengths and angles ($U^{\dagger}U = I$, where $\dagger$ means "conjugate transpose"). Quantum gates and time evolution are modeled by unitary operators, which means they are reversible and do not lose information
- **Rotation Gates** $R_a(\theta)$: Single-qubit gates that rotate the state around a chosen axis $a \in \{x, y, z\}$ on the *Bloch sphere* (a geometric picture of qubit states). They are defined by $R_a(\theta) = e^{-i\,\theta\,\sigma_a/2}$ where σ_a are the Pauli matrices (see Eq. 3). Intuitively, the angle θ approximates how far you "twist" the qubit around axis a, changing the superposition coefficients α, β.
- **Entanglement**: A uniquely quantum correlation between qubits. An entangled multi-qubit state cannot be written as a simple product of single-qubit states. Entanglement lets the circuit represent complex, joint patterns across features, which is crucial for expressive embeddings and uncertainty quantification.
- **Ansatz/Variational Circuit** U_{θ}: A parameterized circuit with learnable parameters $\boldsymbol{\theta}$. In practice we stack layers of local rotations and entangling gates to mix information across qubits. During training, $\boldsymbol{\theta}$ is tuned to produce features that help the downstream task.
- **Angle (Parameter) Encoding**: A simple and hardware-friendly way to load real numbers into a circuit. Each input value is turned into a rotation angle applied to a specific qubit. This directly influences the quantum state amplitudes and, after the variational circuit, changes the measured features.
- **Observable / Measurement**: An *observable* is a Hermitian matrix $M = M^{\dagger}$ describing a measurable quantity. The ideal measurement outcome (in expectation) on state $|\psi\rangle$ is $\langle M\rangle = \langle\psi|M|\psi\rangle$. Here we use local Pauli-Z operators Z_j (one per qubit), producing a real number in $[-1, 1]$ from each qubit and thus a feature vector in $\mathbb{R}^{n_q}$.
- **Expectation Estimation (Shots)**: On real hardware you do not get the expectation directly; you repeat the circuit many times (*shots*) and average the outcomes. More shots mean less noise in the average. In simulation, expectations can be computed exactly without sampling noise.
- **Parameter-Shift Rule**: A technique to compute exact gradients of measured expectations with respect to gate parameters without resorting to numerical finite differences. For gates $e^{-i\theta H/2}$ with $H^2 = I$ (which includes the rotation gates we use),

$$\frac{\partial\langle M\rangle}{\partial\theta} = \tfrac{1}{2}\Big(\langle M\rangle_{\theta+\pi/2} - \langle M\rangle_{\theta-\pi/2}\Big). \tag{14}$$

  This enables seamless end-to-end training of quantum layers inside deep learning frameworks.

4.6 Quantum Embedding Implementation

The Enhanced Theory-of-Mind (ToM) observer maps classical embeddings/encodings into a quantum embedding to produce action logits. For all our embeddings discussed in Fig. 1 the structure is identical as shown in Fig. 3 and moves through the following five steps:

1. Linear projection from $\mathbb{R}^d$ to $\mathbb{R}^{n_q}$ (one scalar per qubit) where d is the input data dimension and n_q is the number of qubits,
2. Angle-based data encoding on n_q qubits,
3. A variational entangling circuit with trainable parameters,
4. Measurement of local Pauli-Z observables to obtain $\mathbb{R}^{n_q}$ features,
5. A classical MLP mapping $\mathbb{R}^{n_q} \to \mathbb{R}^m$.

The resulting vector in $\mathbb{R}^m$ integrates with other embeddings and is fed to a classical policy head that outputs 5 logits (*up/down/left/right/stay*).

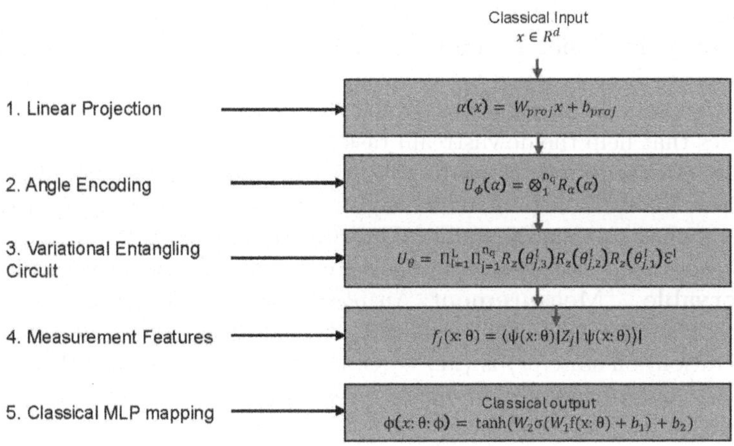

Fig. 3. Quantum Embedding Implementation.

Linear Projection to Qubit Angles: Given an input vector $x \in \mathbb{R}^d$, a trainable linear map with bias produces one angle per qubit (in radians):

$$\boldsymbol{\alpha}(x) \;=\; W_{\text{proj}} x + \boldsymbol{b}_{\text{proj}} \in \mathbb{R}^{n_q}, \tag{15}$$

with $W_{\text{proj}} \in \mathbb{R}^{n_q \times d}$ (a weight matrix) and $\boldsymbol{b}_{\text{proj}} \in \mathbb{R}^{n_q}$ (a bias vector). This stage is a standard dense layer trained end-to-end.

Angle Encoding: The angles in (15) are loaded into the circuit with single-qubit rotations. Writing $\mathcal{H} \cong (\mathbb{C}^2)^{\otimes n_q}$ for the n_q-qubit space and $|0\rangle^{\otimes n_q}$ for the all-zero starting state, the data-encoding unitary $U_\phi(\boldsymbol{\alpha})$ acts as

$$U_\phi(\boldsymbol{\alpha}) \;=\; \bigotimes_{j=1}^{n_q} R_a(\alpha_j)\,, \tag{16}$$

where $R_a(\cdot)$ denotes a single-qubit rotation about axis $a \in \{x, y, z\}$. The encoded state is

$$|\psi_{\text{enc}}(x)\rangle \;=\; U_\phi\big(\boldsymbol{\alpha}(x)\big)\,|0\rangle^{\otimes n_q} \tag{17}$$

Variational Entangling Circuit: To make the features expressive, we plan to add a trainable circuit $U_{\boldsymbol{\theta}}$ with L repeated layers. Let $\boldsymbol{\theta} \in \mathbb{R}^{L\times n_q\times 3}$ be its rotation parameters; schematically,

$$U_{\boldsymbol{\theta}} \;=\; \prod_{\ell=1}^{L}\Bigg[\,\Big(\prod_{j=1}^{n_q} R_z(\theta_{j,3}^{(\ell)})R_y(\theta_{j,2}^{(\ell)})R_x(\theta_{j,1}^{(\ell)})\Big)\;\cdot\;\mathcal{E}^{(\ell)}\,\Bigg], \tag{18}$$

where $\mathcal{E}^{(\ell)}$ entangles the qubits (two-qubit gates that create correlations). The full encoded state after this variational block is

$$|\psi(x;\boldsymbol{\theta})\rangle \;=\; U_{\boldsymbol{\theta}}\,U_\phi\big(\boldsymbol{\alpha}(x)\big)\,|0\rangle^{\otimes n_q} \tag{19}$$

Measurement Features: We convert the quantum state to ordinary numbers by measuring Pauli-Z on each qubit (one feature per qubit, always between -1 and $+1$):

$$f_j(x;\boldsymbol{\theta}) \;=\; \langle\psi(x;\boldsymbol{\theta})|\,Z_j\,|\psi(x;\boldsymbol{\theta})\rangle \in [-1,1], \qquad j = 1,\ldots,n_q. \tag{20}$$

Collecting all n_q values yields $\boldsymbol{f}(x;\boldsymbol{\theta}) \in \mathbb{R}^{n_q}$. In simulation, these are exact averages (expectations); on hardware, they are estimated by repeating the circuit and averaging the outcomes ("shots").

Classical Post-processing MLP: The measured vector is mapped to the model's embedding size m (assuming $m = 32$) by a small MLP:

$$\varphi(x;\boldsymbol{\theta},\Theta) \;=\; \tanh\Big(W_2\,\sigma\big(W_1\,\boldsymbol{f}(x;\boldsymbol{\theta}) + \boldsymbol{b}_1\big) + \boldsymbol{b}_2\Big) \;\in\; \mathbb{R}^m, \tag{21}$$

with $W_1 \in \mathbb{R}^{64\times n_q}$, $W_2 \in \mathbb{R}^{m\times 64}$, biases $\boldsymbol{b}_1, \boldsymbol{b}_2$, and activation $\sigma(\cdot) = \max\{\cdot, 0\}$ (ReLU). We write Θ for all classical post-processing parameters.

Quantum Gradients and Autodiff: When the model contains quantum layers, their outputs are differentiable w.r.t. circuit parameters through the *parameter-shift rule.* If an expectation feature has form $\langle M\rangle(\theta)$ for a gate $e^{-i\theta H/2}$ with $H^2 = I$, then

$$\frac{\partial\,\langle M\rangle}{\partial\theta} = \tfrac{1}{2}\Big(\langle M\rangle(\theta+\tfrac{\pi}{2}) - \langle M\rangle(\theta-\tfrac{\pi}{2})\Big). \tag{22}$$

Conceptually, the classical network sends a real-valued vector to the quantum layer (after a learned linear projection), the circuit produces expectation values (real features), and the classical post-processing maps these to the model's embedding size. Because the entire pipeline is differentiable, errors computed on the 5-way action prediction flow backward to update both classical weights and quantum rotation angles, aligning the quantum features with the task.

5 Proposed Experiments

This section proposes a set of experiments to evaluate the proposed QTomNet framework (see Fig. 1) that integrates quantum embeddings for Theory of Mind (ToM) observers in partially observable grid-worlds.

5.1 Experimental Overview

Experimental Goal: The goal of the evaluation is to rigorously examine whether quantum and hybrid embeddings can improve the performance of machine ToM, in comparison to pure classical embeddings in the original ToM-Net. At each setup, we contrast classical neural implementations with quantum and hybrid alternatives. Our quantum variants employ parameterized quantum circuits with angle encodings and trainable rotations interleaved with entangling gates, while hybrid designs situate a quantum feature map within classical pre- and post-processing layers. By systematically enabling a quantum or hybrid representation of the output of the *character net*, *mental net* or *current context*, we seek to examine the advantages of quantum embeddings.

Table 2. Hypotheses

Hypotheses	Description
H1 (Character net/mental net advantage/current context)	Applying quantum (or hybrid embeddings) to the character net or the mental net or the current context creates marginal gains compared to no quantum embedding, reflecting benefits when operating directly over aggregated current context.
H2 (Compositional synergy)	Joint application of quantum embeddings (mental net, character net, and current context) exhibit super-additive improvements for challenging ToM configurations.
H3 (False-belief sensitivity)	Quantum and hybrid current context modules improve false-belief accuracy due to richer interference-enabled feature maps over temporally aggregated observations.
H4 (Scaling)	Increasing qubits and entanglement depth initially improves accuracy with diminishing returns and rising compute costs.

Guiding Proposition and Hypotheses: The experiments are guided by an overarching proposition that QToMNet outperforms ToMNet in terms of accuracy due to the use of pure quantum or hybrid quantum embeddings. By learning quantum embeddings QToMNet formally represents the inherent uncertainty in inferring beliefs and actions from agents searching a POMPD. The reasoning behind this has been articulated in Sect. 3 and also 4. The quantum embeddings may also impact the computational complexity and scalability of the model, depending on the availability of quantum computing resources [25]. This leads to the following central proposition guiding the overall experimental plan: *Quantum embeddings in QToMNet's architecture improve prediction accuracy, while also having potential impact on computational complexity.* This proposition leads to a series of hypotheses that the experiments will investigate (see table 2).

5.2 Experimental Implementation

Our comparisons follow three setups. In each setup, one of the contributors out of the character net, mental net, and current context will be given either a classical, quantum or hybrid embedding scheme. This will allow us to isolate the effects of each embedding type for each component separately. Situations where different combinations are provided with quantum-classical embeddings may also be tested. We will also vary the sampling from our agents to increase the uncertainty associated with their actions.

Training Protocol and Early Stopping: All conditions are trained under a matched optimization protocol. We use the same optimizer, learning rate, batch size, and patience-based early stopping on validation accuracy so that training dynamics are comparable across representational choices. Each configuration is repeated over multiple random seeds to average over initialization and data ordering effects, and per-epoch logs of losses and sub-task accuracies are retained to identify the best validation epoch. This protocol mirrors our existing analyses and ensures that any differences we report can be attributed to the embeddings rather than to training idiosyncrasies.

Evaluation Metrics and Reporting: We report validation accuracy at the best epoch both in aggregate and decomposed into false-belief and visible subsets. To quantify computational overheads, we will record total training time and the average time per epoch, and we tally parameter counts to assess parameter efficiency. We further characterize sample efficiency by tracing accuracy as a function of the number of training episodes. For quantum and hybrid variants, we couple these metrics with scaling curves that track how accuracy and runtime evolve with qubit count and circuit depth. Results are communicated with concise six-panel visual summaries—covering accuracies, runtimes, and parameter counts—and with tables that present per-condition means and variability across seeds.

Reproducibility and Ablations: All experiments fix seeds, dataset generation parameters, and training hyperparameters across conditions. We ablate: encoder-only quantum vs. mental-net-only quantum; entanglement patterns; input modality; and patience/early-stopping settings. Where quantum toolchains are unavailable, we provide classical baselines and a hybrid fallback.

Expected Outcomes and Impact: If validated, these experiments will demonstrate that quantum or hybrid embeddings used to map the *current context* and outputs of the *mental net* as well as the *character net* provide measurable benefits on ToM tasks, particularly those requiring counterfactual perspective-taking (false beliefs), while quantifying the compute trade-offs and parameter efficiencies that make such methods practical in resource-constrained settings.

6 Concluding Remarks and Future Research

With this position paper, we have paved the way for a new direction for the emerging stream of machine and deep learning research focused on building a machine Theory of Mind (ToM). Our work responds to the grand challenge in AI research: to develop a performant machine ToM that enables artificial agents to infer the mental states and actions of other human or human-like agents, even when they are strangers. We draw our inspiration from scholars in the field of psychology, neuroscience and economics, who have demonstrated that our decision making follows principles of quantum dynamics [14]. Building on the influential ToMNet architecture proposed by Rabinowitz et al. [13], we extend the architecture with quantum and hybird embeddings to integrate the quantum-like nature of the mind. We invite others to extend our proposed theory and to explore the use of quantum embeddings in further adaptations of ToMNet, as discussed in Sect. 2. In addition, we encourage research into the generalizability of our QToMNet architecture, particularly in more complex task settings and with richer datasets (e.g., 3-D grids, real scenes, and images). We also see opportunities for integrating a quantum dynamics perspective into other streams of machine ToM. For example, those scholars crafting inverse deep reinforcement learning models for ToM (e.g. in alignment with research directions suggested by [5]) may seek alternative ways to integrate a quantum perspective in POMDPs [26]. Much research remains to be done before the grand challenge of a fully realized machine Theory of Mind can be solved. However, as computational resources for quantum computing continue to advance, we anticipate many additional opportunities for quantum-enhanced machine ToM.

References

1. Apperly, I.A.: What is "theory of mind"? Concepts, cognitive processes and individual differences. Q. J. Exp. Psychol. **65**(5), 825–839 (2012). https://doi.org/10.1080/17470218.2012.676055. Accessed 28 Sep 2024
2. Leslie, A.M.: Pretense and representation: the origins of "theory of mind.". Psychol. Rev. **94**(4), 412–426 (1987). https://doi.org/10.1037/0033-295X.94.4.412
3. Gopnik, A., Wellman, H.M.: Why the child's theory of mind really is a theory (1992)
4. Premack, D., Woodruff, G.: Does the chimpanzee have a theory of mind? Behav. Brain Sci. **1**(4), 515–526 (1978). https://doi.org/10.1017/S0140525X00076512. Accessed 30 Aug 2025
5. Jara-Ettinger, J.: Theory of mind as inverse reinforcement learning. Curr. Opin. Behav. Sci. **29**, 105–110 (2019). https://doi.org/10.1016/j.cobeha.2019.04.010 . Accessed 11 May 2025
6. Gordon, R.M.: The prior question: do human primates have a theory of mind? Behav. Brain Sci. **21**(1), 120–121 (1998). https://doi.org/10.1017/S0140525X98290704. Accessed 30 Aug 2025

7. Mao, Y., Liu, S., Ni, Q., Lin, X., He, L.: A review on machine theory of mind. IEEE Trans. Comput. Soc. Syst. 1–19 (2024). https://doi.org/10.1109/TCSS.2024.3416707. Accessed 28 Sep 2024
8. Baker, C., Saxe, R., Tenenbaum, J.: Bayesian theory of mind: modeling joint belief-desire attribution. Proc. Annu. Meet. Cogn. Sci. Soc. **33**(33) (2011). Accessed 20 Apr 2025
9. Baker, C.L.C.L.: Bayesian theory of mind : modeling human reasoning about beliefs, desires, goals, and social relations. Thesis, Massachusetts Institute of Technology (2012). Accepted: 2012-10-10T15:43:21Z Journal Abbreviation: BToM : modeling human reasoning about beliefs, desires, goals, and social relations. https://dspace.mit.edu/handle/1721.1/73768. Accessed 05 July 2025
10. Stacy, S., et al.: A Bayesian theory of mind approach to modeling cooperation and communication. Wiley Interdisc. Rev. Comput. Stat. **16**(1), 1631 (2024)
11. Westby, S., Riedl, C.: Collective intelligence in human-ai teams: A bayesian theory of mind approach. In: Proceedings of the AAAI Conference on Artificial Intelligence, vol. 37, pp. 6119–6127 (2023)
12. Dissing, L., Bolander, T.: Implementing theory of mind on a robot using dynamic epistemic logic. In: Proceedings of the Twenty-Ninth International Joint Conference on Artificial Intelligence. IJCAI'20, pp. 1615–1621. Yokohama, Yokohama, Japan (2021)
13. Rabinowitz, N., et al.: Machine theory of mind. In: Proceedings of the 35th International Conference on Machine Learning, pp. 4218–4227. PMLR, ??? (2018). ISSN: 2640-3498. https://proceedings.mlr.press/v80/rabinowitz18a.html. Accessed 24 Aug 2025
14. Busemeyer, J.R., Wang, Z., Townsend, J.T.: Quantum dynamics of human decision-making. J. Math. Psychol. **50**(3), 220–241 (2006). https://doi.org/10.1016/j.jmp.2006.01.003. Accessed 27 Feb 2025
15. Vanschoren, J.: Meta-Learning: A Survey. arXiv. arXiv:1810.03548 (2018). https://doi.org/10.48550/arXiv.1810.03548 . http://arxiv.org/abs/1810.03548. Accessed 04 Sep 2025
16. Swaby, L., Stewart, M., Harrold, D., Willis, C., Palmer, G.: Machine Theory of Mind for Autonomous Cyber-Defence (2024). https://arxiv.org/abs/2412.04367
17. Chuang, Y.S., et al.: Using machine theory of mind to learn agent social network structures from observed interactive behaviors with targets. In: 2020 29th IEEE International Conference on Robot and Human Interactive Communication (RO-MAN), pp. 1013–1019 (2020). https://doi.org/10.1109/RO-MAN47096.2020.9223453
18. Nguyen, D., et al.: Learning theory of mind via dynamic traits attribution. arXiv preprint arXiv:2204.09047 (2022)
19. Sequeira, P., Sadhu, V., Gervasio, M.: ToMCAT: Theory-of-Mind for Cooperative Agents in Teams via Multiagent Diffusion Policies. arXiv. arXiv:2502.18438 (2025). https://doi.org/10.48550/arXiv.2502.18438 . http://arxiv.org/abs/2502.18438 Accessed 04 Sep 2025
20. Ashtiani, M., Azgomi, M.A.: A survey of quantum-like approaches to decision making and cognition. Math. Soc. Sci. **75**, 49–80 (2015). https://doi.org/10.1016/j.mathsocsci.2015.02.004. Accessed 27 Feb 2025
21. Li, J.A., et al.: Quantum reinforcement learning during human decision-making. Nat. Hum. Behav. **4**(3), 294–307 (2020). https://doi.org/10.1038/s41562-019-0804-2. Accessed 02 May 2025

22. Chen, Y., Pan, Y., Dong, D.: Quantum language model with entanglement embedding for question answering. IEEE Trans. Cybern. **53**(6), 3467–3478 (2023). https://doi.org/10.1109/TCYB.2021.3131252. Accessed 02 May 2025
23. Meghdadi, A., Akbarzadeh-T, M.R., Javidan, K.: A quantum-like cognitive approach to modeling human biased selection behavior. Sci. Rep. **12**(1), 22545 (2022). https://doi.org/10.1038/s41598-022-13757-2. Accessed 06 Sep 2025
24. Wimmer, H., Perner, J.: Beliefs about beliefs: Representation and constraining function of wrong beliefs in young children's understanding of deception. Cognition **13**(1), 103–128 (1983). https://doi.org/10.1016/0010-0277(83)90004-5. Accessed 08 Sep 2025
25. Xu, Y., Aggarwal, V.: Accelerating Quantum Reinforcement Learning with a Quantum Natural Policy Gradient Based Approach (2025). https://arxiv.org/abs/2501.16243
26. Barry, J., Barry, D.T., Aaronson, S.: Quantum POMDPs. Phys. Rev. A **90**(3) (2014) https://doi.org/10.1103/PhysRevA.90.032311 . arXiv:1406.2858. Accessed 01 May 2025

Financial Fraud Detection with Entropy Computing

Babak Emami(✉), Wesley Dyk, David Haycraft, Carrie Spear, Lac Nguyen, and Nicholas Chancellor

Quantum Computing Inc (QCi), 5 Marine View Plaza, Hoboken, NJ 07030, USA
bemami@quantumcomputinginc.com

Abstract. We introduce CVQBoost, a novel classification algorithm that leverages early hardware implementing Quantum Computing Inc's Entropy Quantum Computing (EQC) paradigm, Dirac-3 [Nguyen et al. arXiv:2407.04512]. We apply CVQBoost to a fraud detection test case and benchmark its performance against XGBoost, a widely utilized ML method. Running on Dirac-3, CVQBoost demonstrates a significant runtime advantage over XGBoost, which we evaluate on high-performance hardware comprising up to 48 CPUs and four NVIDIA L4 GPUs using the RAPIDS AI framework. Our results show that CVQBoost maintains competitive accuracy (measured by AUC) while significantly reducing training time, particularly as dataset size and feature complexity increase. To assess scalability, we extend our study to large synthetic datasets ranging from 1M to 70M samples, demonstrating that CVQBoost on Dirac-3 is well-suited for large-scale classification tasks. These findings position CVQBoost as a promising alternative to gradient boosting methods, offering superior scalability and efficiency for high-dimensional ML applications such as fraud detection.

1 Introduction

Boosting is a powerful machine learning technique built on combining a set of "weak" classifiers which would be of little use individually to create a much stronger classifier which can perform well [5]. Boosting has been applied successfully to a variety of tasks, for example predicting chronic kidney disease [7], object detection [18], and email filtering [2] The fundamental task in boosting is deciding the weight to assign each weak classifier. A score is then determined form a linear weighted combination, and the score is used to determine the classification. Determining the weighting can be a hard task, since deciding which weak classifiers to include and how strongly to weight them should be done in such a way that it minimizes redundant information, leading to a naturally quadratic structure.

One advantage of boosting and in particular methods like QBoost and CVQBoost is that examining the highly weighted classifiers can allow researchers to understand how the algorithm is working. This is in contrast to methods such

V. Aggarwal et al. (Eds.): PQAI 2025, CCIS 2724, pp. 229–244, 2026.
https://doi.org/10.1007/978-981-95-7829-0_18

as deep neural networks where such insight is not as readily accessible[1] [17,20]. A concrete example of finding explanations from boosting can be found in [18], where the authors examined a few of the highest weighted classifiers and found that they could be intuitively understood. In this example for face detection one of the most highly weighted weak classifiers was the presence of two dark rectangles surrounding a bright rectangle, corresponding to two eyes on either side of the bridge of someone's nose.

More explainable AI techniques such as ours (which broadly falls on the axis of "simplifying algorithms" in the strategies discussed in [20]) are likely to be particularly important in highly regulated sectors such as finance, where compliance is highly important. A concrete example here is European Union regulations which form part of the General Data Protection Regulation (GDPR) and give consumers a "right to explanation" about algorithmic decisions made related to them [9], in situations where this legislation is relevant, AI which gives a decision but no explanation of why the decision was made, could cause legal difficulties. Likewise, in scientific settings [17] (for example chemistry and computational biology), the goal is often to uncover natural laws rather than just make a single predictions. In these settings explainable AI is much more valuable than a "black box" which simply returns an answer, even if the answer it returns is reliably correct.

The QBoost algorithm [14], converts the problem of weighting the weak classifiers into a quadratic unconstrained binary optimization (QUBO) problem, where diagonal elements correspond to the quality of classification, and off-diagonals correspond to measures of correlation between the classifiers. An advantage of QBoost is that it is constructed to naturally match the form of problems which could be solved by quantum annealers. In this work we further extend this concept to CVQBoost, an algorithm which can naturally be applied to the Dirac optical computing hardware developed by Quantum Computing Inc. [16] which encodes into continuous variables. We further demonstrate the performance of this algorithm on a fraud detection use case.

We find that when applied using QCi's commercially-available Dirac-3 devices (hereafter referred to as Dirac-3) CVQBoost to achieve superior speed and power efficiency compared to a state-of-the-art conventional approach, XGBoost [4] while maintaining comparable accuracy. We particularly find highly favorable scaling of the runtime versus both number of features and training data samples.

In traditional boosting algorithms like AdaBoost [6] the weights of weak classifiers are iteratively adjusted based on their performance, aiming to minimize the overall classification error. CVQBoost innovates by using a novel optical solver to solve this optimization problem. It encodes the boosting task as a quadratic optimization problem, leveraging the Dirac-3's capability to effectively search rough optimization landscapes.

[1] There is work on ways to make DNNs explainable (see citations in text), and this is an area of active research, but it would involve significant modifications to the methods. These explanations are furthermore often "fragile" in the sense that a small change to the data which shouldn't change anything, can completely change the explanation [8].

2 Implementation

2.1 Dirac-3 Optimization Machine

Entropy quantum computing (EQC) has emerged as a promising optimization technique, using artificial dissipation to find the optimal solutions of a Hamiltonian [16]. In a photonic implementation, qudits are represented as superpositions of photon number states within time bins, which evolve in an optical fiber loop. This loop incorporates a target Hamiltonian as a dissipative operator, effectively simulating imaginary time evolution. During this process, high-energy eigenstates experience dissipation and decoherence, while lower-energy eigenstates are favored in the evolution. The latest iteration of EQC hardware at Quantum Computing Inc (QCi), the Dirac-3 system, is a hybrid architecture that combines the high-speed parallel processing capabilities of photonic systems with the precise control and programmability of electronic circuits. By exploiting the natural ability of photonics to manipulate complex optical fields for entropy-driven state evolution, alongside the use of electronics for state initialization, feedback, and fine-tuning, hybrid entropy computing enables efficient optimization. With analog computing approach, we explore if Dirac-3 provides advantages in accelerating training time while maintain low power consumption.

The relevant computational mode for this work is when it is used as a continuous quadratic solver, where it solves a problem of the form

$$\min_{w} \sum_{i} \sum_{j} J_{ij} w_i w_j + \sum_{i} C_i w_i \tag{1}$$

where J and C are a user specified problem definition and $w_i \in \mathbb{R}^+$ are continuous variables subject to $w_i > 0$ and an overall sum constraint

$$\sum_{i=1}^{N} w_i = 1. \tag{2}$$

Dirac-3 has a number of other capabilities such as higher order coupling, which are not used in this study.

2.2 Formulation

Our methods build on the conventional formulation of boosting [5], most similar to the quadratic formulation used in QBoost [14]. Let us assume that we have a collection of N "weak" classifiers h_i where $i = 1, 2, ..., N$. Depending on the method used to build weak classifiers, the values of h_i can be discrete or continuous. The goal is to construct a "strong" classifier as a linear superposition of these weak classifiers, that is,

$$y = \sum_{i=1}^{N} w_i h_i(\mathbf{x}) \tag{3}$$

where $\mathbf{x}$ is a vector of input features and $y \in \{-1, 1\}$. The goal is to find w_i, continuous positive weights associated with the weak classifiers.

We use a training set $\{(\mathbf{x_s}, y_s)|s = 1, 2, ..., S\}$ of size S. We can determine optimal weights $w_i \geq 0$ by minimizing,

$$\min_{\mathbf{w}} \sum_{s=1}^{S} \left| \sum_{i=1}^{N} w_i h_i(\mathbf{x_s}) - y_s \right|^2 + \lambda \sum_{i=1}^{N} (w_i)^2 \tag{4}$$

where the regularization term $\lambda \sum_{i=1}^{N} (w_i)^2$ penalizes non-zero weights; λ is the regularization coefficient which is used to control the relative importance of regularisation. Note that this is similar to the formulation in the QBoost [14] algorithm. The key difference is that we do not need to encode the weights into binary variables since the Dirac machine we use is based on a native continuum encoding. We now observe that the objective function in Eq. 4 can be cast (up to an irrelevant constant offset) in the form of Eq. 1 required by Dirac-3 by setting

$$J_{ij} = \sum_{s=1}^{S} h_i(\mathbf{x_s}) h_j(\mathbf{x_s}) + \delta_{ij}\lambda \tag{5}$$

$$C_i = -2 \sum_{s=1}^{S} y_s h_i(\mathbf{x_s}) \tag{6}$$

where δ_{ij} is a Kronecker delta, defined such that $\delta_{ii} = 1$ and $\delta_{i \neq j} = 0$.

2.3 Choices of Weak Classifiers

There are many ways to design a subset of weak classifiers. We have tested CVQBoost using logistic regression, decision tree, naive Bayesian, and Gaussian process classifiers. Each weak classifier is constructed using one or two of the features chosen from all features. This yields a set of weak classifiers that can be used to construct a strong classifier. In the present study, we have used logistic regression weak classifiers. In principle the highly weighted features within the most highly weighted regression classifiers could be used to explain the automated decisions made by our method, but we have elected not to examine this further as the main focus of the present work is to judge performance, not verify explainability.

2.4 Dataset

The dataset used in this study is the Kaggle Credit Card Fraud Detection dataset [13], a widely recognized resource for machine learning research, particularly in imbalanced datasets where the minority class constitutes a small fraction of the data.

This dataset comprises transactions made by European credit cardholders over a two-day period in September 2013. It contains approximately 200,000 transactions, of which approximately 0.1% are labeled as fraudulent. The significant class imbalance makes it ideal for exploring techniques tailored to imbalanced classification problems.

A total of 38 features are included in the dataset.

3 Results

In this section we present two sets of results. First in Sect. 3.1 we analyze the accuracy of CVQBoost and find that at least when balancing has been applied the performance can be comparable to XGBoost. This is followed by analysis of the scaling of CVQBoost compared with XGBoost. We start with a simpler comparison with single core instances of XGBoost in Sect. 3.2. Here we see substantial improvement, we also discuss the relative time CVQBoost spends on pre- and post- processing versus Dirac-3 runtime. We find that the time spent in the former scales modestly with both training data count and number of features, while the Dirac-3 time remains almost constant. Next in Sect. 3.3, we present some scaling results compared with GPU and multicore processors on the Kaggle data. Finally, in Sect. 3.4 we test scalability against training data count with a synthetic data set and compare with XGBoost implementations on muticore devices and GPUs. In this analysis we find favorable performance even when compared to state-of-the-art multi-core and GPU implementations of XGBoost. Unless stated otherwise, data are averaged over 10 runs and error bars are two standard deviations, so 95% of data should lie within the bars assuming normally distributed results. The 95% confidence intervals are therefore $1/\sqrt{10} \approx 1/3$ the size of the depicted bars.

3.1 Accuracy

Handling class imbalance is a central challenge in machine learning, particularly in applications such as fraud detection, where the minority class represents rare events. In the original dataset used for this study, fraud cases (the minority class) constitute less than 0.1% of the data. To address this extreme imbalance, four widely-used data balancing strategies-ADASYN [10], SMOTE [3], SMOTE-SVM [15], and majority class downsampling-were applied to the training data [12]. The test data remained unchanged to ensure an unbiased evaluation of model performance. The accuracy of CVQBoost and XGBoost was assessed across six levels of class imbalance, characterized by minority-to-majority class ratios of 0.01, 0.02, 0.05, 0.1, 0.5, and 1.0. The Area Under the Curve (AUC) [1], calculated on the test data, was used to measure model accuracy. Figure 1 illustrates the trends for each balancing strategy.

The results reveal interesting patterns that provide insight into the behavior of the two algorithms under varying levels of class imbalance. When the balancing is minimal-that is, when the minority-to-majority class ratio remains small (0.01 or 0.02)-XGBoost achieves higher AUC scores compared to CVQBoost. This suggests that XGBoost is better able to leverage the heavily skewed class distribution in the early stages of balancing. However, as the class ratio increases and the training data becomes more balanced, CVQBoost's performance improves steadily and catches up to that of XGBoost, eventually achieving comparable AUC scores.

The effect of balancing strategies is particularly noticeable when using ADASYN. ADASYN generates synthetic samples by focusing on regions of the feature space that are prone to misclassification, thereby improving the representation of the minority class [10]. As the minority-to-majority class ratio increases, CVQBoost not only matches but eventually surpasses XGBoost in terms of AUC. This can be seen clearly in Fig. 1(a), where CVQBoost achieves a slight but consistent improvement over XGBoost at higher class ratios. This suggests that CVQBoost is particularly effective at learning from the diversity introduced by ADASYN-generated synthetic samples.

A similar trend is observed with SMOTE and SMOTE-SVM. While XGBoost maintains a slight edge at smaller ratios, CVQBoost consistently narrows the gap as the balancing becomes more pronounced. The superior performance of CVQBoost at higher class ratios reflects its ability to generalize better when sufficient representation of the minority class is provided.

In the case of majority class downsampling, where no synthetic data is introduced and the majority class is simply reduced in size, CVQBoost algorithms demonstrate gradual improvement in AUC as the class ratio increases. However, the rate of improvement is less pronounced compared to the oversampling methods.

In summary CVQBoost becomes increasingly advantageous as class imbalance is mitigated, offering competitive performance for moderately to highly balanced training data. While the accuracy is comparable, in the findings we report later in this section we show that CVQBoost offers a major advantage in runtime even when compared to XGBoost run on state-of-the-art hardware.

3.2 Runtime Versus Single Core XGBoost

Impact of Training Data Count on Runtimes. Figure 2 presents the runtimes of CVQBoost and XGBoost across varying training dataset sizes, ranging from $1,000$ to $150,000$ samples. All runtimes are reported in seconds, with each experiment repeated at least 5 times and averaged to ensure reliability. The table provides both the mean and standard deviation of the runtimes, as well as a breakdown of CVQBoost's runtime components. Notably, the Dirac-3 component accounts for a substantial portion of the total runtime for CVQBoost.

A comparison between CVQBoost and XGBoost reveals that XGBoost achieves shorter runtimes on smaller datasets. However, as the training dataset size increases, XGBoost's runtime grows significantly faster than CVQBoost's. This trend is further illustrated in Fig. 2a, that plots training runtimes against dataset size. Figure 2b shows a detailed barplot which visualizes the fraction of time spent by Dirac-3.

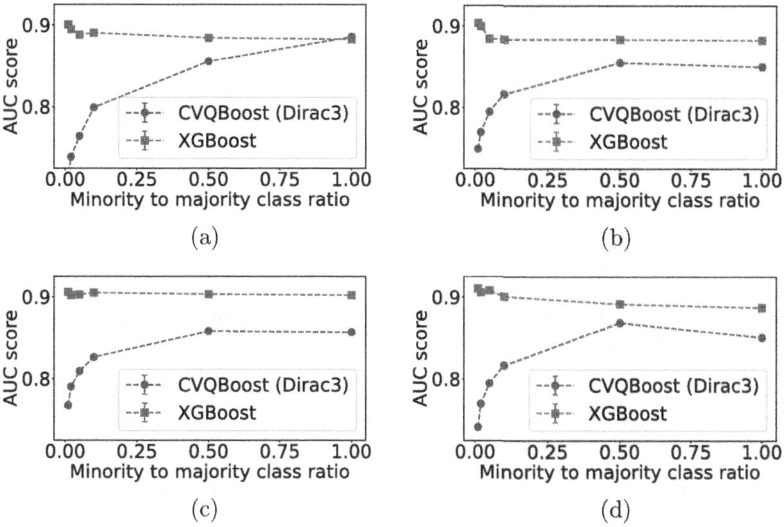

Fig. 1. AUC score of CVQBoost vs. minority-to-majority class ratio in training datasets using different balancing strategies: (a) ADASYN, (b) SMOTE, (c) SMOTE-SVM, and (d) majority class downsampling. The bars (obscured by the symbol at most points) are intervals in which 95% of instances should lie. These data can be found in tabular form in Table 1 of the appendix.

These results suggest that while XGBoost excels in speed for smaller datasets, CVQBoost offers a clear performance advantage as dataset sizes increase. Specifically, CVQBoost begins to outperform XGBoost when the training sample size approaches $20,000$ samples, highlighting its scalability for larger datasets.

Number of Features. Figures 2c and 2d visualize the runtimes of CVQBoost and XGBoost for varying numbers of features, ranging from 5 to 38. All runtimes are reported in seconds, with each experiment repeated multiple times to ensure reliability.

Figure 2c illustrates the relationship between runtime and feature count for both methods. As the number of features increases, the runtimes of both CVQBoost and XGBoost grow. However, CVQBoost consistently demonstrates significantly shorter runtimes compared to XGBoost. Figure 2d breaks down the time spent by Dirac-3 versus other processes in CVQBoost. It is again notable that the Dirac-3 component accounts for a substantial portion of the total runtime for CVQBoost.

3.3 Comparison with GPU and Multicore Implementations

XGBoost on Multiple Cores. The performance of XGBoost was evaluated a high performance machine to assess its scalability in a multi-core environment.

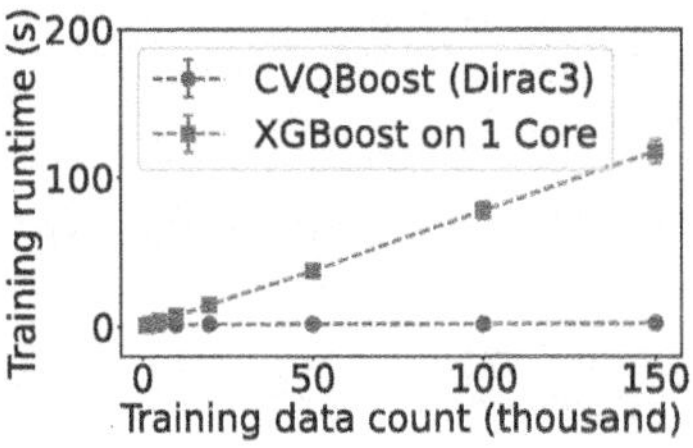

(a) runtime versus training counts

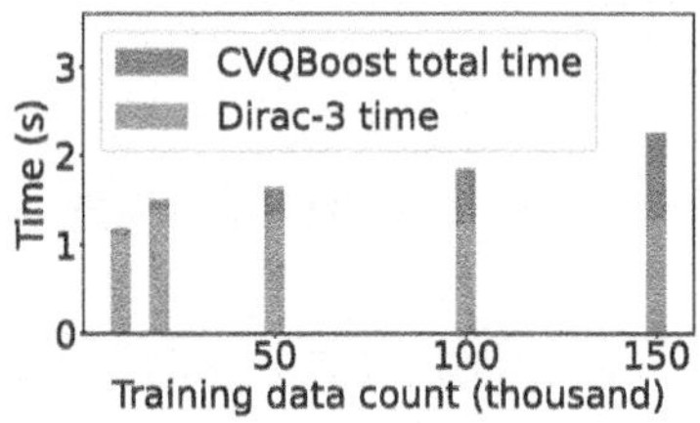

(b) runtime versus training counts

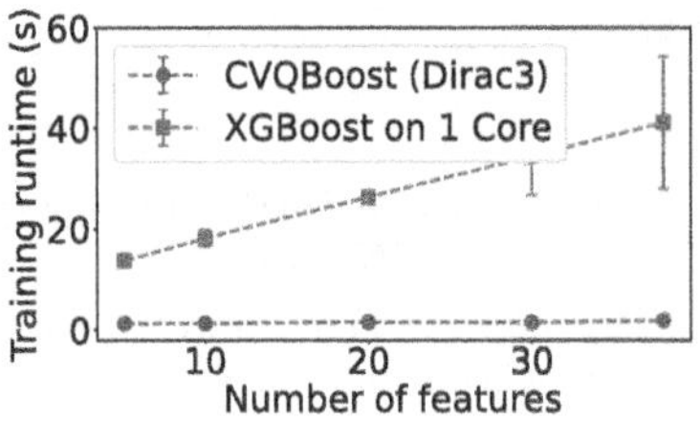

(c) runtime versus number of features

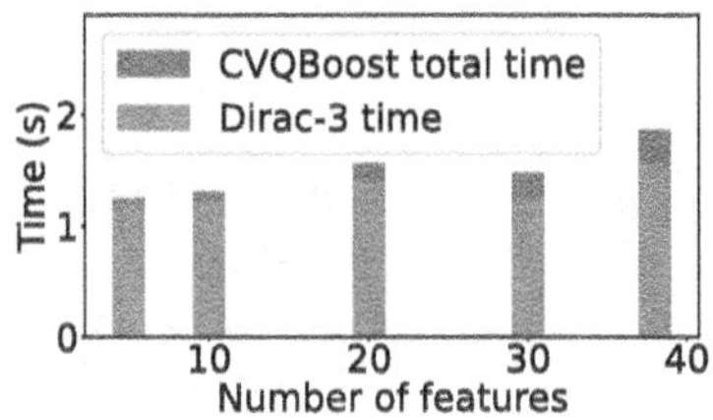

(d) runtime versus number of features

Fig. 2. (a) training runtime of CVQBoost and XGBoost vs. count of training data samples. The bars are intervals in which 95% of instances should lie. (b) fraction of CVQBoost runtime which was comprised by running on Dirac-3. (c) Training runtime of CVQBoost and XGBoost vs. number of features. The bars are intervals in which 95% of instances should lie. (d) fraction of CVQBoost runtime which was comprised by running on Dirac-3. These data can be found in tabular form in tables 2 (training counts) and 3 (feature counts) of the appendix.

Figure 3 illustrates the training runtimes of XGBoost as a function of the number of cores utilized on a high-performance machine with 16 cores.

In Fig. 3, diminishing returns are seen as more cores are added; while the runtime reduces with the addition of cores, the rate of improvement diminishes significantly beyond 8 cores, and further scaling provides only marginal benefits. There is also a degradation in runtime as the number of cores is increased beyond 8, this is likely due to inefficiencies caused by communication.

This saturation in scalability highlights a fundamental limitation of XGBoost in leveraging large numbers of cores for training speedup. In contrast, CVQBoost on the Dirac-3 platform exhibits remarkable scalability. Unlike XGBoost, which faces hardware limitations in multi-core environments, CVQBoost's design allows it to harness the power of optical hardware effectively, avoiding the saturation observed with multi-core scaling.

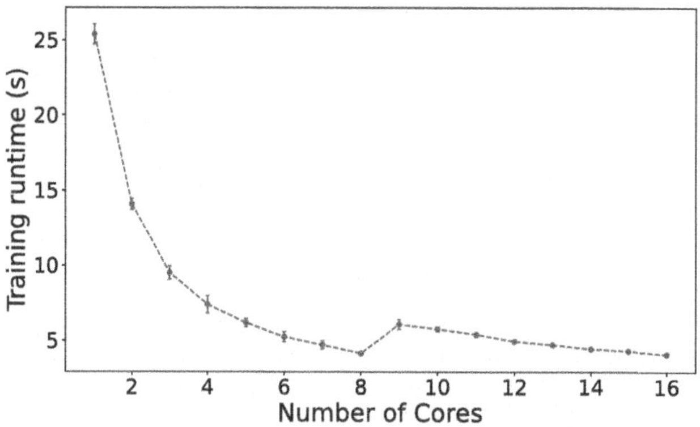

Fig. 3. Training runtime of XGBoost vs. number of cores on a high-performance machine with 16 cores. Intervals within which 95% of the data should lie are shown; training data count: 150, 000, number of features: 38.

As shown in Fig. 3, using 8 cores for XGBoost appears to be optimal for minimizing training runtime.

Building on the scalability experiments, we conducted further tests to compare the performance of CVQBoost and XGBoost running in parallel on 8 cores, focusing on varying dataset sizes. As shown in Fig. 4a, XGBoost achieves shorter training runtimes for smaller datasets due to its efficient parallelization, however, as the data size increases CVQBoost runtimes become more favorable due to better scaling. It is also worth noting that CVQBoost runtimes are highly variable. Note that because each data point represents the average of 10 samples the 95% confidence intervals will be $1/\sqrt{10} \approx 1/3$ times the depicted bars. While the training runtimes of CVQBoost and parallelized XGBoost are similar for these relatively small datasets (ranging from 1,000 to 150,000 samples), experiments with much larger datasets reveal that CVQBoost scales significantly better with dataset size. This will be discussed further in Sect. 3.4, where a clearer scaling advantage can be seen at larger sizes.

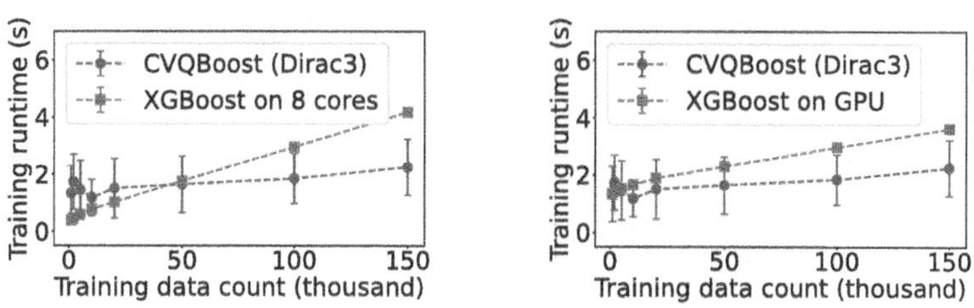

(a) Eight CPU cores

(b) GPU rapids

Fig. 4. comparison of training runtimes for CVQBoost and XGBoost on (a) run on eight cores in parallel (b) GPU using the NVIDIA RAPIDS AI package. Bars represent intervals where 95% of data should lie. These data can be found in tabular form in tables (CPU) and 5 (GPU) of the appendix.

XGBoost on GPU. To evaluate the performance of XGBoost in a GPU-accelerated environment, we ran the algorithm on an NVIDIA L4 GPU for various training dataset sizes, as shown in Fig. 4b. The NVIDIA RAPIDS AI package, version 24.12.00, was used. The GPU implementation demonstrates reduced runtimes compared to CVQBoost for smaller datasets. This efficiency arises from the GPU's ability to parallelize computations, which aligns well with the nature of XGBoost's gradient boosting framework. As the data size grows, CVQBoost shows slightly better scalability than XGBoost trained on GPU. While CVQBoost and XGBoost on GPU exhibit similar runtimes for the relatively small dataset sizes used here (ranging from $1,000$ to $150,000$ samples), further investigation reveals that CVQBoost scales significantly better on much larger datasets. This is discussed in Sect. 3.4.

3.4 Scalability on Synthetic Data

To evaluate the scalability of CVQBoost, we generated large-scale synthetic datasets using the *make_classification* functionality in the Scikit-Learn package. The dataset sizes range from 1M to 70M samples, with 80% allocated for training. Each dataset consists of 100 to 900 features.

Experiments were conducted on a high-performance computing system equipped with 48 CPU cores (2.64 GHz) and four NVIDIA L4 GPUs. Each GPU unit has 24 GB of memory, totaling 96 GB across all four units. XGBoost was tested under two configurations: (1) using 48 CPU cores and (2) running on four NVIDIA L4 GPUs via the NVIDIA RAPIDS AI framework. The results of these experiments can be seen in Fig. 5 with a comparison of scaling with training data count for 100 features in Fig. 5a and the analogous plot for 500 features in Fig. 5b. Both show a substantial advantage from CVQBoost. This is even true in the 500 feature case, where the XGBoost appears to be better able to take advantage of the massive parallelization provided by the GPU.

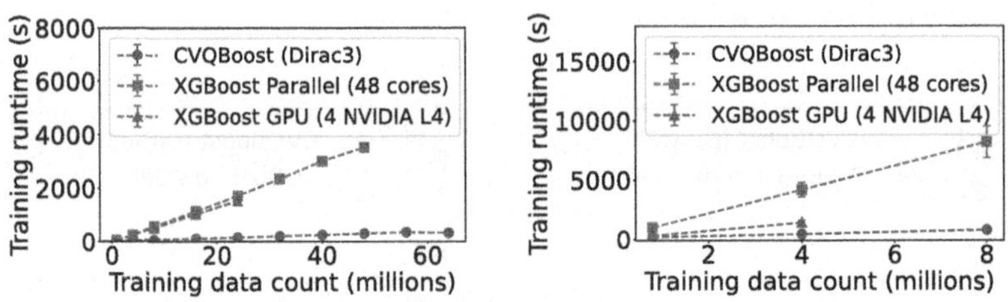

(a) 100 features (b) 500 features

Fig. 5. Training runtime of CVQBoost vs. XGBoost on 48 CPUs and four NVIDIA L4 GPUs. The bars are intervals in which 95% of instances should lie. CVQBoost exhibits superior scalability for large datasets.

Figure 6 further demonstrates that CVQBoost maintains linear scalability with increasing feature dimensions, whereas XGBoost exhibits quadratic runtime growth.

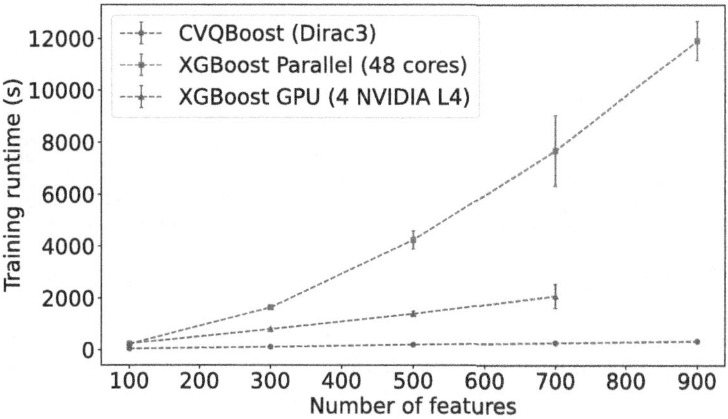

Fig. 6. Training runtime of CVQBoost and XGBoost vs. number of features. The bars are intervals in which 95% of instances should lie.

These results highlight the scalability advantage of CVQBoost for large-scale classification tasks, where computational efficiency is critical.

3.5 Training CVQBoost with Other Solvers

While the CVQBoost algorithm is designed to run on QCi's Dirac-3 device, the algorithm can be run using other solvers as well. We tested CVQBoost using two additional solvers: Hexaly [11], a commercial solver, and the Sequential Least Squares Programming (SLSQP) algorithm implemented in Scipy [19]. The objective of this comparison is to assess the scalability and efficiency of Dirac-3 relative to traditional optimization methods when solving Hamiltonians derived building a CVQBoost classification model.

The Hamiltonians used in this study were built from the synthetic datasets used in Sect. 3.4 and range from 100 to 900 dimensions, with dynamic ranges spanning 40 to 70 dB. The dynamic range, defined as the ratio between the largest and smallest coefficients in the Hamiltonian (expressed in decibels), significantly influences numerical stability and optimization difficulty. Notably, Dirac-3 has a known performance limitation at 23 dB, making it particularly relevant to evaluate how it performs when handling Hamiltonians that exceed this range.

For Hexaly, where we cannot directly control the optimality gap, we allow the solver to terminate based on either achieving the default optimality gap or reaching a predefined maximum iteration/runtime threshold. For a fair comparison, we use the default tolerances in Scipy's SLSQP solver and cap the maximum number of iterations for both solvers. This approach ensures that comparisons focus on fundamental scalability rather than fine-tuning solver-specific parameters.

Figure 7 presents a detailed comparison of Dirac-3 and the other solvers in terms of runtime and solution quality. The experiments were conducted across five trials per problem size, with the results averaged.

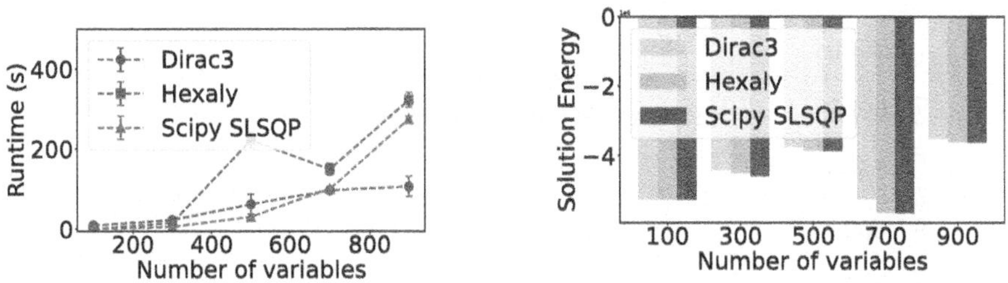

(a) Runtime comparison

(b) Solution energy comparison

Fig. 7. Comparison of Dirac-3, Hexaly, and Scipy (SLSQP) solvers in terms of (a) runtime scaling and (b) solution energy performance across different problem sizes.

From Fig. 7a, we observe that while the other solvers achieve shorter runtimes for small Hamiltonian sizes, their scalability deteriorates significantly for larger problem instances. They exhibit quadratic or worse runtime growth with the number of variables, whereas Dirac-3 maintains a more favorable linear scaling. At a Hamiltonian size of 900, Dirac-3 outperforms the other solvers in runtime efficiency.

Figure 7b compares the final solution energies, as measured against the quadratic objective functions. While all solvers achieve similar energy levels for small problem sizes (e.g., 100 variables), for larger problem instances, a slight energy gap emerges in favor of the other methods. Recall from our previous results that the solutions found by Dirac-3 are good enough quality to provide competitive performance with more traditional boosting methods. This discrepancy is likely due to the large dynamic ranges of these Hamiltonians, which exceed the 23 dB operational limit of Dirac-3, potentially impacting its optimization performance.

4 Conclusions and Outlook

In this work we have proposed and demonstrated an new boosting algorithm called CVQBoost which can leverage the unique features of QCi's Dirac-3 computer. The results show that comparable accuracy with traditional boosting methods (XGBoost) when balancing techniques have been applied. In particular, we found that CVQBoost can even exceed the performance of XGBoost when the ADASYN balancing strategy is used. This strategy involves generating artificial examples to improve regions of the feature space prone to misclassification suggesting that CVQBoost is particularly well adapted to learning from this method. We also find that CVQBoost can be trained much faster, even when compared to state-of-the-art implementations on GPU's and multicore systems (Tables 4, 6).

While the specific demonstration used here was related to fraud detection, there will be many other potential applications. A particularly appealing future direction is in science, for example computational bioscience or chemistry. This is due to the potential for explainability of boosting techniques by examining the weak classifiers which were used in cases where the classifiers have more explanatory potential.

Acknowledgments. All authors were completely supported by Quantum Computing Inc. in writing this paper. In particular NC did not receive any UKRI support in the production of this manuscript. The authors thank Mohammad-Ali Miri for helpful comments and suggestions. For transparency and reproducibility, we have created a code repository with code associated with these studies at https://github.com/qci-github/eqc-studies/tree/main/CVQBoost.

A Data in Tabular Form

Table 1. AUC scores for CVQBoost and XGBoost using different balancing strategies and minority to majority class ratios.

sampling strategy	minority to majority class ratio	cvqboost auc	xgboost auc
ADASYN	0.01	0.7105 +/− 0.0	0.9003 +/− 0.0
ADASYN	0.02	0.7388 +/− 0.0	0.8946 +/− 0.0
ADASYN	0.05	0.7642 +/− 0.0	0.8881 +/− 0.0
ADASYN	0.1	0.799 +/− 0.0	0.8904 +/− 0.0
ADASYN	0.5	0.8555 +/− 0.0002	0.8843 +/− 0.0
ADASYN	1	0.8855 +/− 0.0019	0.8826 +/− 0.0
SMOTE	0.01	0.7493 +/− 0.0	0.9037 +/− 0.0
SMOTE	0.02	0.7697 +/− 0.0	0.8998 +/− 0.0
SMOTE	0.05	0.7948 +/− 0.0	0.8842 +/− 0.0
SMOTE	0.1	0.8164 +/− 0.0	0.8836 +/− 0.0
SMOTE	0.5	0.855 +/− 0.0004	0.8836 +/− 0.0
SMOTE	1	0.8498 +/− 0.0021	0.8824 +/− 0.0
SVMSMOTE	0.01	0.7674 +/− 0.0	0.9058 +/− 0.0
SVMSMOTE	0.02	0.7902 +/− 0.0	0.9019 +/− 0.0
SVMSMOTE	0.05	0.8092 +/− 0.0	0.9027 +/− 0.0
SVMSMOTE	0.1	0.8263 +/− 0.0	0.9051 +/− 0.0
SVMSMOTE	0.5	0.8582 +/− 0.0014	0.9033 +/− 0.0
SVMSMOTE	1	0.8569 +/− 0.001	0.902 +/− 0.0
Downsampling	0.01	0.7415 +/− 0.0	0.9105 +/− 0.0
Downsampling	0.02	0.7698 +/− 0.0	0.9056 +/− 0.0
Downsampling	0.05	0.7948 +/− 0.0	0.9079 +/− 0.0
Downsampling	0.1	0.8167 +/− 0.0	0.9007 +/− 0.0
Downsampling	0.5	0.8689 +/− 0.0004	0.8915 +/− 0.0
Downsampling	1	0.8508 +/− 0.0016	0.8872 +/− 0.0

Table 2. Breakdown of CVQBoost and XGBoost runtimes for different training data counts; number of features: 38.

train data count	cvqboost train time	xgboost train time	dirac3 time
1000	1.3 +/− 0.5	0.9 +/− 0.1	1.3 +/− 0.5
2000	1.7 +/− 0.5	1.8 +/− 0.1	1.7 +/− 0.5
5000	1.4 +/− 0.5	3.7 +/− 0.1	1.4 +/− 0.5
10000	1.2 +/− 0.3	7.2 +/− 0.2	1.1 +/− 0.3
20000	1.8 +/− 0.9	14.6 +/− 0.3	1.7 +/− 0.9
50000	2.6 +/− 3.1	37.4 +/− 1.2	2.3 +/− 3.1
100000	1.8 +/− 0.4	78.0 +/− 2.9	1.2 +/− 0.4
150000	2.2 +/− 0.5	117.4 +/− 4.0	1.3 +/− 0.5

Table 3. Breakdown of CVQBoost and XGBoost runtimes for different counts of features; training data count: 50, 000.

num features	cvqboost train time	xgboost train time	dirac3 time
5	1.2 +/− 0.4	13.7 +/− 0.3	1.2 +/− 0.4
10	1.3 +/− 0.4	18.2 +/− 0.8	1.2 +/− 0.4
20	1.6 +/− 0.5	26.3 +/− 0.5	1.4 +/− 0.5
30	1.5 +/− 0.4	34.4 +/− 3.8	1.2 +/− 0.4
38	1.9 +/− 0.5	41.1 +/− 6.6	1.6 +/− 0.5

Table 4. Parallelized XGBoost runtimes on 8 cores for different training data counts; number of features: 38.

train data count	xgboost total train time
1000	0.4 +/− 0.1
2000	0.5 +/− 0.0
5000	0.6 +/− 0.0
10000	0.8 +/− 0.0
20000	1.0 +/− 0.0
50000	1.8 +/− 0.0
100000	3.0 +/− 0.0
150000	4.2 +/− 0.1

Table 5. XGBoost runtimes on GPU for different training data counts; number of features: 38.

train data count	xgboost total train time
1000	1.3 +/− 0.1
2000	1.4 +/− 0.0
5000	1.5 +/− 0.0
10000	1.7 +/− 0.0
20000	1.9 +/− 0.0
50000	2.3 +/− 0.0
100000	3.0 +/− 0.0
150000	3.6 +/− 0.0

Table 6. Comparison of runtimes for Dirac 3, Hexaly, and Scipy (SLSQP) solvers across different Hamiltonian sizes.

num vars	convex	dynamic range	Dirac3 runtime	Hexaly runtime	Scipy runtime
100	FALSE	45.9	9.4 +/− 1.5	0.2 +/− 0.0	0.2 +/− 0.0
300	FALSE	41.7	22.6 +/− 2.1	15.7 +/− 1.7	5.7 +/− 1.1
500	FALSE	64.3	62.0 +/− 12.4	221.7 +/− 10.0	30.6 +/− 0.6
700	FALSE	65.2	96.8 +/− 3.5	149.6 +/− 7.1	100.3 +/− 2.4
900	FALSE	57.6	106.2 +/− 12.5	322.1 +/− 9.1	274.8 +/− 3.2

References

1. Bradley, A.P.: The use of the area under the ROC curve in the evaluation of machine learning algorithms. Pattern Recogn. **30**(7), 1145–1159 (1997)
2. Carreras, X., Marquez, L.: Boosting trees for anti-spam email filtering. In: Proceedings of RANLP 2001, pp. 58–64 (2001)
3. Chawla, N.V., Bowyer, K.W., Hall, L.O., Kegelmeyer, W.P.: Smote: synthetic minority over-sampling technique. J. Artif. Intell. Res. **16**, 321–357 (2002). https://doi.org/10.1613/jair.953
4. Chen, T., Guestrin, C.: XGBoost: a scalable tree boosting system. In: Proceedings of the 22nd ACM SIGKDD International Conference on Knowledge Discovery and Data Mining, pp. 785–794. KDD 2016, Association for Computing Machinery, New York (2016). https://doi.org/10.1145/2939672.2939785
5. Freund, Y., Schapire, R., Abe, N.: A short introduction to boosting. J. Jpn. Soc. Artif. Intell. 14(5), 771–780 (1999). https://doi.org/10.11517/jjsai.14.5_771
6. Freund, Y., Schapire, R.E.: A desicion-theoretic generalization of on-line learning and an application to boosting. In: Vitányi, P. (ed.) EuroCOLT 1995. LNCS, vol. 904, pp. 23–37. Springer, Heidelberg (1995). https://doi.org/10.1007/3-540-59119-2_166
7. Ganie, S.M., Dutta Pramanik, P.K., Mallik, S., Zhao, Z.: Chronic kidney disease prediction using boosting techniques based on clinical parameters. PLoS ONE **18**(12), 1–21 (2023)

8. Ghorbani, A., Abid, A., Zou, J.: Interpretation of neural networks is fragile. In: Proceedings of the AAAI Conference on Artificial Intelligence, vol. 33, no. 01, pp. 3681–3688 (2019). https://doi.org/10.1609/aaai.v33i01.33013681, https://ojs.aaai.org/index.php/AAAI/article/view/4252
9. Goodman, B., Flaxman, S.: European union regulations on algorithmic decision making and a "right to explanation". AI Mag. **38**(3), 50–57 (2017). https://doi.org/10.1609/aimag.v38i3.2741, https://onlinelibrary.wiley.com/doi/abs/10.1609/aimag.v38i3.2741
10. He, H., Bai, Y., Garcia, E.A., Li, S.: Adasyn: adaptive synthetic sampling approach for imbalanced learning. In: 2008 IEEE International Joint Conference on Neural Networks (IEEE World Congress on Computational Intelligence), pp. 1322–1328 (2008). https://doi.org/10.1109/IJCNN.2008.4633969
11. Hexaly: Hexaly optimizer (2025). https://www.hexaly.com/hexaly-optimizer
12. Lee, W., Seo, K.: Downsampling for binary classification with a highly imbalanced dataset using active learning. Big Data Res. **28**, 100314 (2022)
13. MACHINE LEARNING GROUP - ULB: Credit card fraud detection (2013). https://www.kaggle.com/datasets/mlg-ulb/creditcardfraud
14. Neven, H., Denchev, V.S., Rose, G., Macready, W.G.: Qboost: large scale classifier training withadiabatic quantum optimization. In: Hoi, S.C.H., Buntine, W. (eds.) Proceedings of the Asian Conference on Machine Learning. Proceedings of Machine Learning Research, vol. 25, pp. 333–348. PMLR, Singapore Management University, Singapore (2012). https://proceedings.mlr.press/v25/neven12.html
15. Nguyen, H.M., Cooper, E.W., Kamei, K.: Borderline over-sampling for imbalanced data classification. Int. J. Knowl. Eng. Soft Data Paradigm. **3**(1), 4–21 (2011). https://doi.org/10.1504/IJKESDP.2011.039875
16. Nguyen, L., et al.: Entropy computing: a paradigm for optimization in an open quantum system (2024). https://arxiv.org/abs/2407.04512, arXiv:2407.04512
17. Samek, W., Wiegand, T., Müller, K.: Explainable artificial intelligence: understanding, visualizing and interpreting deep learning models. CoRR arXiv:1708.08296 (2017). http://arxiv.org/abs/1708.08296
18. Viola, P., Jones, M.: Rapid object detection using a boosted cascade of simple features. In: Proceedings of the 2001 IEEE Computer Society Conference on Computer Vision and Pattern Recognition. CVPR 2001, vol. 1, pp. I–I (2001)
19. Virtanen, P., et. al., SciPy 1.0 Contributors: SciPy 1.0: fundamental algorithms for scientific computing in python. Nature Methods **17**, 261–272 (2020). https://doi.org/10.1038/s41592-019-0686-2
20. Yu, S.: Towards trustworthy and understandable AI: unraveling explainability strategies on simplifying algorithms, appropriate information disclosure, and high-level collaboration. In: Proceedings of the 26th International Academic Mindtrek Conference, pp. 133–143. Mindtrek 2023. Association for Computing Machinery, New York (2023). https://doi.org/10.1145/3616961.3616965, https://doi.org/10.1145/3616961.3616965

Author Index

V. Aggarwal et al. (Eds.): PQAI 2025, CCIS 2724, pp. 245–246, 2026.
https://doi.org/10.1007/978-981-95-7829-0

The manufacturer's authorised representative in the EU is Springer
Nature Customer Service Centre GmbH, Europaplatz 3, 69115 Heidelberg,
Germany. If you have any concerns regarding our products, please
contact ProductSafety@springernature.com

Printed and bound by CPI Group (UK) Ltd, Croydon, CR0 4YY
30/07/2026
02181608-0002